University Texts in the Mathematical Sciences

Editors-in-Chief

Raju K. George, Department of Mathematics, Indian Institute of Space Science and Technology, Valiamala, Kerala, India

S. Kesavan, Department of Mathematics, Institute of Mathematical Sciences, Chennai, Tamil Nadu, India

Sujatha Ramdorai, Department of Mathematics, University of British Columbia, Vancouver, BC, Canada

Shalabh, Department of Mathematics and Statistics, Indian Institute of Technology Kanpur, Kanpur, Uttar Pradesh, India

Associate Editors

Kapil Hari Paranjape, Department of Mathematics, Indian Institute of Science Education and Research Mohali, Mohali, Chandigarh, India

K. N. Raghavan, Department of Mathematics, Institute of Mathematical Sciences, Chennai, Tamil Nadu, India

V. Ravichandran, Department of Mathematics, National Institute of Technology Tiruchirappalli, Tiruchirappalli, India

Riddhi Shah, School of Physical Sciences, Jawaharlal Nehru University, New Delhi, Delhi, India

Kaneenika Sinha, Department of Mathematics, Indian Institute of Science Education and Research, Pune, Maharashtra, India

Kaushal Verma, Department of Mathematics, Indian Institute of Science Bangalore, Bengaluru, Karnataka, India

Enrique Zuazua, Department of Mathematics, Friedrich-Alexander-Universität Erlangen-Nürnberg (FAU), Erlangen, Germany

Textbooks in this series cover a wide variety of courses in mathematics, statistics and computational methods. Ranging across undergraduate and graduate levels, books may focus on theoretical or applied aspects. All texts include frequent examples and exercises of varying complexity. Illustrations, projects, historical remarks, program code and real-world examples may offer additional opportunities for engagement. Texts may be used as a primary or supplemental resource for coursework and are often suitable for independent study.

Satya N. Mukhopadhyay · Subhasis Ray

Measure and Integration

An Introduction

 Springer

Satya N. Mukhopadhyay
Department of Mathematics
University of Burdwan
Burdwan, West Bengal, India

Subhasis Ray
Department of Mathematics
Visva-Bharati University
Santiniketan, West Bengal, India

ISSN 2731-9318 ISSN 2731-9326 (electronic)
University Texts in the Mathematical Sciences
ISBN 978-981-97-2510-6 ISBN 978-981-97-2511-3 (eBook)
https://doi.org/10.1007/978-981-97-2511-3

Mathematics Subject Classification: 26-01, 26A24, 26A42, 26A46, 28A25, 28A33, 28A99

© The Editor(s) (if applicable) and The Author(s), under exclusive license to Springer Nature Singapore Pte Ltd. 2025

This work is subject to copyright. All rights are solely and exclusively licensed by the Publisher, whether the whole or part of the material is concerned, specifically the rights of translation, reprinting, reuse of illustrations, recitation, broadcasting, reproduction on microfilms or in any other physical way, and transmission or information storage and retrieval, electronic adaptation, computer software, or by similar or dissimilar methodology now known or hereafter developed.
The use of general descriptive names, registered names, trademarks, service marks, etc. in this publication does not imply, even in the absence of a specific statement, that such names are exempt from the relevant protective laws and regulations and therefore free for general use.
The publisher, the authors and the editors are safe to assume that the advice and information in this book are believed to be true and accurate at the date of publication. Neither the publisher nor the authors or the editors give a warranty, expressed or implied, with respect to the material contained herein or for any errors or omissions that may have been made. The publisher remains neutral with regard to jurisdictional claims in published maps and institutional affiliations.

This Springer imprint is published by the registered company Springer Nature Singapore Pte Ltd.
The registered company address is: 152 Beach Road, #21-01/04 Gateway East, Singapore 189721, Singapore

If disposing of this product, please recycle the paper.

Introduction

Measure and integration is most essential for higher studies in Mathematical Analysis. This book presents a detailed discussion on measure and integration beginning from Riemann integral to reach Lebesgue integral so that the readers, can easily enter into the subject. Since differentiation is related to integration, a complete chapter is devoted for discussion on differentiation. General measure and function spaces which are related to integration theory are also discussed. Signed measure and complex measure are also included. The material is presented with details and helpful examples given so that it is suitable for introductory course as well as for self-study for those who begin to study the subject for the first time. Only knowledge of differential and integral calculus and of Riemann integration is necessary.

This book contains ten chapters:

Chapter 1 is devoted to discuss preliminaries which are needed for study of the book.

In Chap. 2, Lebesgue measure on the real line is introduced. Lebesgue measurable sets are defined and various properties of Lebesgue measurable sets are proved. Nonmeasurable sets are constructed and Vitali's covering theorem is proved.

In Chap. 3, measurable functions are defined and various properties of measurable functions are proved. Convergence theorems for measurable functions are also considered.

Chapter 4 deals with further properties of sets and functions. Cardinality of the classes of all Borel sets, of all Lebesgue measurable sets, and of all nonmeasurable sets are studied. Cantor sets and functions and Lebesgue functions are discussed.

Chapter 5 is devoted to the study of the Lebesgue integral of functions from $\mathbb{R}$ to $\mathbb{R}$. Considering first a bounded function f (measurability of f not assumed) on a measurable set E of finite measure, the upper and lower Lebesgue integral of f are defined on E as in the case of Riemann integral on a bounded closed interval $[a, b]$. This method will help to understand how Lebesgue integral is a generalization of Riemann integral, measurability of Lebesgue integrable functions will follow from this definition just as almost everywhere continuity of Riemann integral functions follows from definitions. The upper and lower Lebesgue integrals are also defined using measurable simple functions and it is shown that those two definitions are equivalent (Theorem 5.6). Lebesgue integral of unbounded functions and unbounded measurable sets are also considered. Various properties of Lebesgue integral including convergence theorems are proved. Comparison of Lebesgue integral, improper Riemann integral, and Newton integral are also discussed and it is shown that these integrals are independent of each other.

In Chap. 6, various properties of derivatives are studied which is helpful to get further information regarding Lebesgue integral. Indefinite Lebesgue integral is defined. Absolutely continuous functions are discussed which helped to get a descriptive definition of Lebesgue integral (Theorem 6.41). Integration by parts and change of variable for Lebesgue integral are studied. Points of density of sets and approximately continuous functions are defined and their influence on sets and Lebesgue integral are studied. It is shown that a function f is measurable if and only if f is approximately continuous almost everywhere (Theorem 6.65).

In Chap. 7, Lebesgue outer measure and measure in $\mathbb{R}^N$, $N \geq 2$, are considered. Structure of open sets in $\mathbb{R}^N$ is studied. It is shown that outer Lebesgue measure and measure in $\mathbb{R}^N$ are, respectively, regular and complete. Various properties of measurable sets in $\mathbb{R}^N$ and measurable functions from $\mathbb{R}^N$ to $\mathbb{R}$ are proved. Integration for functions from $\mathbb{R}^N$ to $\mathbb{R}$ are also considered.

In Chap. 8, general measure and general outer measure are defined independently. While outer measure generates a measure (as in the case of Lebesgue outer measure), it is shown that a measure can also induce an outer measure. Extension of measure and outer measure are studied. Existence of incomplete measure space and irregular outer measure are shown in Theorem 8.5 and in Theorem 8.9, respectively. Extension of measure and of outer measure can remove these deficiencies (see Note 1 and Note 2, respectively, under Theorem 8.13). More precisely, an incomplete measure space can be extended to a complete measure space and an irregular outer measure can give by extension a regular outer measure. Lebesgue-Stieltje's outer measure and Hausdorff measure on the real line are studied. Measurable functions and integration are considered. Measure and integration in a product space are studied. Product measure and Lebesgue measure in $\mathbb{R}^N$ are discussed. Integration of complex-valued functions on a measure space are also considered.

Chapter 9 contains function spaces. Various properties of L^P—spaces, space of measurable functions, space of continuous functions, and space of Riemann integrable functions are studied. Completeness property of L^P—spaces and the space of measurable functions are proved while incompleteness of the spaces of continuous functions and of Riemann integrable functions with metric similar to L'—metric are discussed. Convergence theorems in these spaces are proved.

Chapter 10 contains signed measure and complex measure. Radon-Nikodym theorems relating to measure and signed measure are proved. Radon-Nikodym derivatives and their properties are studied. Integration with respect to signed measure and complex measure is considered. Pointwise differentiation of measures is also studied.

Contents

1 Preliminaries 1
    1.1   Sets 1
    1.2   Rings and Algebras of Sets 2
    1.3 F_σ and G_δ-sets, Borel Sets 5
    1.4   Sequence of Sets 7
    1.5   Cartesian Product 9
    1.6   Completely Additive Set Function 10
    1.7   Construction of σ-Algebra 13
    1.8   Exercises 14

2 Lebesgue Measure on Real Line 17
    2.1   Outer Measure 17
    2.2   Measurable Sets and Their Properties 20
    2.3   Nonmeasurable Sets 29
    2.4   Further Properties of Measurable Sets 30
    2.5   Vitali's Covering 33
    2.6   Exercises 36

3 Measurable Functions 37
    3.1   Measurable Functions 38
    3.2   Properties of Measurable Function 38
    3.3   Class of Measurable Functions 43
    3.4   Further Properties of Measurable Functions. Simple
           Functions 46
    3.5   Convergence Theorems for Measurable Functions 51
    3.6   Exercises 59

4	**More About Sets and Functions**	61
	4.1 Cardinal Members	61
	4.2 The Cardinal Numbers $\underline{a}$ and $\underline{c}$	64
	4.3 Properties of Set	69
	4.4 Cantor Sets	72
	4.5 Cantor Ternary Set	73
	4.6 Cardinality of the σ–Algebra of All Borel Sets	74
	4.7 Cardinality of the σ–Algebra of All Lebesgue Measurable Sets	76
	4.8 Cardinality of the Class of All Nonmeasurable Sets	76
	4.9 Cantor Function	77
	4.10 Lebesgue Function	78
	4.11 Properties of Cantor Function	79
	4.12 Associated Cantor Function and Its Inverse Function and Some Consequences	79
	4.13 Functions Similar to Cantor Function	81
	4.14 Exercises	81
5	**The Lebesgue Integral**	83
	5.1 Lebesgue Integral of Bounded Functions	83
	5.2 Properties of the Lebesgue Integral of Bounded Functions	91
	5.3 Lebesgue's Criterion for Riemann Integrability	97
	5.4 Riemann Integrability of Bounded Derivatives. Volterra Function	99
	5.5 Lebesgue Integral of Unbounded Function	103
	5.6 Properties of Lebesgue Integral	105
	5.7 Convergence Theorems for Lebesgue Integral	114
	5.8 Lebesgue Integral of Functions on Sets of Infinite Measure	123
	5.9 Improper Riemann Integral and Lebesgue Integral	124
	5.10 Newton Integral and Lebesgue Integral	130
	5.11 Conclusion	133
	5.12 Exercises	134
6	**Differentiation of Functions**	137
	6.1 Limits of a Function and Their Properties	137
	6.2 Derivates of a Function and Their Properties	140
	6.3 Measurablity of Dini Derivates	142
	6.4 Differentiability of Monotone Functions	148
	6.5 Functions of Bounded Variation and Their Properties	154
	6.6 Absolutely Continuous Functions	161
	6.7 Monotonicity Theorems and Their Consequences	167

6.8	The Indefinite Lebesgue Integral	171
6.9	Characterization of Indefinite Lebesgue Integral and Indefinite Riemann Integral	174
6.10	Integration by Parts for Lebesgue Integral	182
6.11	Change of Variable for Lebesgue Integral	183
6.12	Lebesgue Set	187
6.13	Singular Function	190
6.14	Points of Density and Approximate Continuity	191
6.15	Properties of Approximately Continuous Function	193
6.16	Exercises	203

7 Lebesgue Measure and Integration in $\mathbb{R}^N$ ... 207

7.1	Structure of Open Sets in $\mathbb{R}^2$	207
7.2	Lebesgue Outer Measure and Measure in $\mathbb{R}^N$	208
7.3	Lebesgue Measurable Function in $\mathbb{R}^N$	218
7.4	Lebesgue Integral in $\mathbb{R}^N$	219
7.5	Exercises	224

8 General Measure and Outer Measure ... 225

8.1	Measure	225
8.2	Outer Measure	228
8.3	Outer Measure Induced by a Measure	232
8.4	Extension of Measure. Interplay Between Measure and Outer Measure	234
8.5	Construction of Outer Measure in $\mathbb{R}$	236
8.6	Lebesgue Stieltje's Outer Measure and Measure	237
8.7	Hausdorff Measure on the Real Line	242
8.8	Measurable Function and Integration	251
8.9	Measure and Integration in a Product Space	251
8.10	Product Measure and Lebesgue Measure in $\mathbb{R}^N$	259
8.11	Integration of Complex Valued Functions	260
8.12	Exercises	261

9 Function Spaces ... 265

9.1	Metric Space and Linear Space	265
9.2	The L^p-spaces	266
9.3	Counting Measure and Application	271
9.4	Completeness of the L^p-spaces	272
9.5	Other Properties of L^p-spaces	275
9.6	Space of Measurable Functions	279
9.7	Spaces of Continuous Functions and Riemann Integrable Functions	281
9.8	Convergence Theorems (2)	283
9.9	Exercises	285

10 Signed Measure and Complex Measure 287
    10.1   Signed Measure 287
    10.2   Hahn and Jordan Decomposition 288
    10.3   Integration with Respect to Signed Measure 292
    10.4   Absolute Continuity of Measures 297
    10.5   Radon-Nikodym Theorems 300
    10.6   Application of Radon-Nikodym Theorem 305
    10.7   Radon Nikodym Derivative 307
    10.8   Complex Measure and Integration 311
    10.9   Point Wise Differentiation of Measures 316
    10.10   Exercises 320

References 321

Chapter 1
Preliminaries

1.1 Sets

The definition of sets and their operation, namely, union, intersection complement, etc. are assumed to be known. For any set A the complement of A will be denoted by $\sim A$ or $\tilde{A}$. For any two sets A and B the set of points in A which are not in B will be denoted by $A \backslash B$ or $A \sim B$ or $A \cap \tilde{B}$. Clearly $\tilde{\tilde{A}} = A$, $\widetilde{A \cup B} = \tilde{A} \cap \tilde{B}$, $\widetilde{A \cap B} = \tilde{A} \cup \tilde{B}$. The symmetric difference of two sets A and B are defined by $A \triangle B = (A \sim B) \cup (B \sim A)$. Clearly $A \triangle B = (A \cap \tilde{B}) \cup (B \cap \tilde{A})$. The following relations are true:

(i) $A \triangle B = B \triangle A$,
(ii) $A \triangle B = (A \cup B) \sim (A \cap B) = (A \cup B) \cap (\tilde{A} \cup \tilde{B})$;
(iii) $(A \triangle B) \triangle C = A \triangle (B \triangle C)$
(iv) $(A \triangle B) \triangle (C \triangle D) = (A \triangle C) \triangle (B \triangle D)$
(v) $A \triangle B \subset (A \triangle C) \cup (B \triangle C)$
(vi) $E \cap (F \triangle C) = (E \cap F) \triangle (E \cap C)$

The relations (i) and (ii) are clear. To obtain (iii) note that

$$\widetilde{A \triangle B} = \widetilde{(A \cap \tilde{B}) \cup (B \cap \tilde{A})} = \widetilde{(A \cap \tilde{B})} \cap \widetilde{(B \cap \tilde{A})}$$
$$= (\tilde{A} \cup B) \cap (\tilde{B} \cup A) = (\tilde{A} \cap (\tilde{B} \cup A)) \cup (B \cap (\tilde{B} \cup A))$$
$$= (\tilde{A} \cap \tilde{B}) \cup (A \cap B).$$

Using this we get

$$(A \triangle B) \triangle C = (A \cap \tilde{B} \cap \tilde{C}) \cup (\tilde{A} \cap B \cap \tilde{C}) \cup (\tilde{A} \cap \tilde{B} \cap C) \cup (A \cap B \cap C)$$

By symmetry, the right-hand side equals to $(B \triangle C) \triangle A$ which by (i) is $A \triangle (B \triangle C)$.

To obtain (iv) we get by applying (i) and (iii)

$$(A \Delta B)\Delta(C \Delta D) = ((B \Delta A)\Delta C)\Delta D = (B \Delta (A \Delta C))\Delta D$$
$$= ((A \Delta C)\Delta B)\Delta D = (A \Delta C)\Delta(B \Delta D)$$

To get (v) note that $(A \sim B) \subset (A \sim C) \cup (C \sim B)$ and $(B \sim A) \subset (B \sim C) \cup (C \sim A)$ and taking the union gives the result. The proof of (vi) is left as an exercise.

1.2 Rings and Algebras of Sets

Definition 1.1 Let X be a non-void set and let $\mathcal{A}$ be a non-void family of subsets of X. Then $\mathcal{A}$ is called a ring if

(i) $A, B \in \mathcal{A}$ implies $A \cup B \in \mathcal{A}$
(ii) $A, B \in \mathcal{A}$, implies $A \cap \tilde{B} \in \mathcal{A}$.

A ring $\mathcal{A}$ is called a σ-ring, if it is closed under countable union, i.e. if $\{A_n\}$ is any countable collection of members of $\mathcal{A}$ then $\bigcup_n A_n \in \mathcal{A}$.

Definition 1.2 Let X be a non-void set and let $\mathcal{A}$ be a non-void family of subsets of X. Then $\mathcal{A}$ is called an algebra if

(i) $A, B \in \mathcal{A}$ implies $A \cup B \in \mathcal{A}$
(ii) $A \in \mathcal{A}$ implies $\tilde{A} \in \mathcal{A}$.

An algebra $\mathcal{A}$ is called a σ-algebra if $\{A_n\}$ is any countable collection of members of $\mathcal{A}$ then $\bigcup_n A_n \in \mathcal{A}$. A σ-algebra is also called a σ-field. The condition $\bigcup_n A_n \in \mathcal{A}$ in the definition of σ-algebra can be replaced by condition $\bigcap_n A_n \in \mathcal{A}$. For, if $\bigcup_n A_n \in \mathcal{A}$ then since $A_n \in \mathcal{A}$, $\tilde{A}_n \in \mathcal{A}$ and so $\bigcup_n \tilde{A}_n \in \mathcal{A}$ and hence $\widetilde{\bigcap_n A_n} = \bigcup_n \tilde{A}_n \in \mathcal{A}$ and therefore by (ii) $\bigcap_n A_n \in \mathcal{A}$. Conversely if $\bigcap_n A_n \in \mathcal{A}$ for every $\{A_n\}$ then taking $\{\tilde{A}_n\}$, $\bigcap_n \tilde{A}_n \in \mathcal{A}$ and by (ii) $\widetilde{\bigcap_n \tilde{A}_n} \in \mathcal{A}$ i.e $\bigcup_n A_n \in \mathcal{A}$.

Theorem 1.1 *If $\mathcal{A}$ is an algebra of subsets of X then, $\phi, X \in \mathcal{A}$, when ϕ is the null set.*

Proof Let $A \in \mathcal{A}$. Then $\tilde{A} \in \mathcal{A}$ and so $A \cup \tilde{A} = X \in \mathcal{A}$. Again since $X \in \mathcal{A}$, $\tilde{X} = \phi \in \mathcal{A}$. $\square$

Theorem 1.2 *If $\mathcal{A}$ is an algebra then*

(i) $\bigcup_{n=1}^{N} A_n \in \mathcal{A}$; (ii) $\bigcap_{n=1}^{N} A_n \in \mathcal{A}$.
whenever $A_1, A_2, ... A_N \in \mathcal{A}$.

1.2 Rings and Algebras of Sets

Proof Since $A_1, A_2 \in \mathcal{A}$, $A_1 \cup A_2 \in \mathcal{A}$ and $\tilde{A}_1, \tilde{A}_2 \in \mathcal{A}$ and so $\tilde{A}_1 \cup \tilde{A}_2 \in \mathcal{A}$ and hence $\widetilde{\tilde{A}_1 \cup \tilde{A}_2} = A_1 \cap A_2 \in \mathcal{A}$. So the result is true if N = 2. Suppose the result is true for N = r. Then $\bigcup_{n=1}^{r} A_n \in \mathcal{A}$ and $\bigcap_{n=1}^{r} A_n \in \mathcal{A}$. If $A_{r+1} \in \mathcal{A}$, then since the result is true for N = 2, $(\bigcup_{n=1}^{r} A_n) \cup A_{r+1} \in \mathcal{A}$ and $(\bigcap_{n=1}^{r} A_n) \cap A_{r+1} \in \mathcal{A}$ and so $\bigcup_{n=1}^{r+1} A_n \in \mathcal{A}$ and $\bigcap_{n=1}^{r+1} A_n \in \mathcal{A}$ showing that the result is true for $N = r + 1$ and the proof follows by finite induction. $\square$

Theorem 1.3 *Every algebra is a ring. If a ring contains the whole set X then it is an algebra.*

Proof Let $\mathcal{A}$ be an algebra of subsets of X. Then the condition (i) of ring is satisfied. Let $A, B \in \mathcal{A}$. Then $\tilde{A} \in \mathcal{A}$ and so $\tilde{A} \cup B \in \mathcal{A}$ and so $\widetilde{\tilde{A} \cup B} = A \cap \tilde{B} \in \mathcal{A}$. Hence $\mathcal{A}$ is a ring.

Now suppose that $\mathcal{A}$ is a ring containing the whole set X. The condition (i) of algebra is satisfied. Let $A \in \mathcal{A}$. Then since $X \in \mathcal{A}$ by condition (ii) of ring $X \cap \tilde{A} = \tilde{A} \in \mathcal{A}$ and so $\mathcal{A}$ is an algebra. $\square$

Remark From the above theorem it is clear that every σ-algebra is a σ-ring. The converse is not true. To see this let $\mathcal{A}$ be the family of all subsets of [0, 1] which are countable. Then $\mathcal{A}$ is a σ-ring. For, if $A, B \in \mathcal{A}$ then $A \cup B$ and $A \cap \tilde{B}$ being countable, $A \cup B$, $A \cap \tilde{B} \in \mathcal{A}$ and so $\mathcal{A}$ is a ring. Also if $A_i \in \mathcal{A}$ for each $i, i = 1, 2, \ldots$ then $\bigcup_{i=1}^{\infty} A_i$ being the countable union of countable sets belongs to $\mathcal{A}$. So $\mathcal{A}$ is a σ-ring. But $\mathcal{A}$ is not an algebra. For if $A \in \mathcal{A}$ then since $A \cup \tilde{A} = [0, 1]$, $\tilde{A}$ cannot be countable and so $\tilde{A} \notin \mathcal{A}$. (See also Example 1.5 below).

Definition 1.3 Let X be any set. Then the family of all subsets of X is called the power set of X and is denoted by $\mathcal{P}^X$.
That is $\mathcal{P}^X = \{A : A \subset X\}$.

Example 1.1 Let X be any set. The largest σ-algebra of subsets of X is the power set $\mathcal{P}^X$ and the smallest σ-algebra of subsets of X is $\{\phi, X\}$, ϕ being the null set.

The proof is easy.

Theorem 1.4 *Let $\mathcal{A}$ be an algebra and let $\{A_i\}$ be any countable collection of members of $\mathcal{A}$. Then there is a collection $\{B_i\}$ of members of $\mathcal{A}$ such that $B_m \cap B_n = \phi$ for $m \neq n$, $B_n \subset A_n$ for all n and $\cup A_i = \cup B_i$.*

Proof Set

$$B_1 = A_1, \quad B_n = A_n \sim \bigcup_{i=1}^{n-1} A_i = A_n \cap \bigcap_{i=1}^{n-1} \tilde{A}_i \quad \text{for } n \geq 2$$

Since $\tilde{A}_i \in \mathcal{A}$, by Theorem 1.2 $B_n \in \mathcal{A}$. Clearly $B_m \cap B_n = \phi$ for $m \neq n$. For, if $m < n$ and if $\xi \in B_n$ then $\xi \notin A_i$ for $i = 1, 2....n - 1$ and hence $\xi \notin B_m$ and if $\xi \in B_m$ then $\xi \in A_m$ and hence $\xi \notin B_n$. Also $B_n \subset A_n$ for all n. Finally, let $\xi \in \cup A_i$. Let n be the smallest of i for which $\xi \in A_i$. Then $\xi \in A_n$ but $\xi \notin A_i$ for $i = 1, 2....n - 1$ and so $\xi \in B_n$. Hence $\cup A_i \subset \cup B_i$. Since $B_n \subset A_n$ for all n, the reverse inclusion is obvious, completing the proof. $\square$

Example 1.2 Let X be any non-void set and let $A \subset X$. Then the smallest σ-algebra containing A is $\mathcal{A} = \{\phi, X, A, \tilde{A}\}$.

Example 1.3 Let X be any non-void set and let $\mathcal{A}$ be the class of all countable subsets of X and those subsets of X whose complements are countable. Then $\mathcal{A}$ is a σ-algebra.

If $A \in \mathcal{A}$ then A is either countable or is the complement of a countable set and hence $\tilde{A} \in \mathcal{A}$. Let $\{A_i\} \in \mathcal{A}$. If all A_i are countable then $\cup A_i$ is countable and so $\cup A_i \in \mathcal{A}$. If some of A_i say A_n is not countable then A_n is the complement of a countable subset of X and then $\widetilde{\cup A_i} = \cap \tilde{A}_i \subset \tilde{A}_n$. Since $\tilde{A}_n$ is countable, $\widetilde{\cup A_i}$ is countable and so $\cup A_i \in \mathcal{A}$. Hence $\mathcal{A}$ is a σ-algebra.

Example 1.4 Let X be any set containing infinite number of elements and let $\mathcal{A}$ be the family of all finite subsets of X and those subsets of X whose complements are finite. Then $\mathcal{A}$ is an algebra but not a σ-algebra.

Clearly if $A \in \mathcal{A}$ then $\tilde{A} \in \mathcal{A}$. Let $A, B \in \mathcal{A}$. If A and B are finite then $A \cup B$ is finite. If one of A and B is not finite, say B is not finite then the complement of B is finite. Since $\widetilde{A \cup B} = \tilde{A} \cap \tilde{B} \subset \tilde{B}$, $\widetilde{A \cup B}$ is finite and so $A \cup B \in \mathcal{A}$. So, $\mathcal{A}$ is an algebra. To show that $\mathcal{A}$ is not a σ-algebra, consider any countable infinite subset $\{x_1, x_2, ..., x_n, ...\}$ of X. The singleton sets $\{x_{2i-1}\}, i = 1, 2, 3, ...$ are members of $\mathcal{A}$ but their union $\bigcup_{i=1}^{\infty}\{x_{2i-1}\} = \{x_1, x_3, x_5, ...\}$ does not belong to $\mathcal{A}$, since $\{x_1, x_3, x_5, ...\}$ is neither finite nor its complement $\{x_2, x_4, x_6, ...\}$ is finite.

Example 1.5 Let X be an uncountable set and let $\mathcal{A}$ be the family of all countable subsets of X. Then $\mathcal{A}$ is a σ-ring but not an algebra. Let $A, B \in \mathcal{A}$. Since A and B are countable $A \cup B$ is also countable and hence $A \cup B \in \mathcal{A}$. Since $A \cap \tilde{B} \subset A$, $A \cap \tilde{B}$ is countable and so $A \cap \tilde{B} \in \mathcal{A}$. So $\mathcal{A}$ is a ring. If $A_n \in \mathcal{A}$ for each n then since the countable union of countable sets is countable, $\cup A_n \in \mathcal{A}$. Hence $\mathcal{A}$ is a σ-ring. Finally if $A \in \mathcal{A}$ then A is countable and since X is uncountable $\tilde{A}$ is uncountable and so $\tilde{A} \notin \mathcal{A}$ and hence $\mathcal{A}$ is not an algebra.

1.3 F_σ and G_δ-sets, Borel Sets

Let $\mathcal{E}$ be any collection of subsets of a given set X. Define

$$\mathcal{E}_\sigma = \{\bigcup_{n=1}^{\infty} E_n : E_n \in \mathcal{E}; n = 1, 2, ..., \}, \quad \mathcal{E}_\delta = \{\bigcap_{n=1}^{\infty} E_n : E_n \in \mathcal{E}; n = 1, 2, ...\}$$

That is $\mathcal{E}_\sigma$ and $\mathcal{E}_\delta$ are respectively the class of all subsets of X which are countable union and countable intersection of members of $\mathcal{E}$. Clearly $\mathcal{E} \subset \mathcal{E}_\sigma$ and $\mathcal{E} \subset \mathcal{E}_\delta$ since if $E \in \mathcal{E}$ then taking $E_n = E$ for each n, $E = \cup E_n = \cap E_n$ but the reverse inclusion need not be true. Since the countable union of countable union is a countable union and countable intersection of countable intersection is a countable intersection, $(\mathcal{E}_\sigma)_\sigma = \mathcal{E}_\sigma$ and $(\mathcal{E}_\delta)_\delta = \mathcal{E}_\delta$ although as above $\mathcal{E}_\sigma \subset (\mathcal{E}_\sigma)_\delta$ and $\mathcal{E}_\delta \subset (\mathcal{E}_\delta)_\sigma$. Thus consecutive operation of the same type σ or δ produce no new class but alternate application of σ and δ may give larger classes.

Example 1.6 If $\mathcal{A}$ is a $\sigma-$ algebra then $\mathcal{A}_\sigma = \mathcal{A}_\delta = \mathcal{A}$. Let $A \in \mathcal{A}_\sigma$. Then $A = \bigcup_{n=1}^{\infty} A_n$, $A_n \in \mathcal{A}$. Since $\mathcal{A}$ is a σ-algebra $A \in \mathcal{A}$ and hence $\mathcal{A}_\sigma \subset \mathcal{A}$. Similarly if $A \in \mathcal{A}_\delta$ then $A = \bigcap_{n=1}^{\infty} A_n$, $A_n \in \mathcal{A}$ and so $\tilde{A} = \bigcup_{n=1}^{\infty} \tilde{A}_n$. Since $\tilde{A}_n \in \mathcal{A}$, $\tilde{A}_n \in \mathcal{A}$ and so $\bigcup_{n=1}^{\infty} \tilde{A}_n \in \mathcal{A}$ which gives $\tilde{A} \in \mathcal{A}$ and hence $A \in \mathcal{A}$. Thus $\mathcal{A}_\delta \subset \mathcal{A}$. The other inclusion being obvious, the result follows.

Example 1.7 If $\mathcal{A}$ and $\mathcal{B}$ are two classes and if $\mathcal{A} \subset \mathcal{B}$ then $\mathcal{A}_\sigma \subset \mathcal{B}_\sigma$ and $\mathcal{A}_\delta \subset \mathcal{B}_\delta$. The proof is clear.

Let $\mathcal{F}$ and $\mathcal{G}$ denote respectively the families of all closed and all open sets in a topological space (in particular, the space of real numbers $\mathbb{R}$). Then $\mathcal{F}_\delta = \mathcal{F}$ and $\mathcal{G}_\sigma = \mathcal{G}$, since any intersection of closed sets is a closed set and any union of open sets is an open set. Clearly $\mathcal{F}_\sigma$ is larger than $\mathcal{F}$ and $\mathcal{G}_\delta$ is larger than $\mathcal{G}$ since countable union of closed sets need not be a closed set (consider $\bigcup_{n=1}^{\infty}[\frac{1}{n}, 1] = (0, 1]$) and countable intersection of open sets need not be an open set (Consider $\bigcap_{n=1}^{\infty}(-\frac{1}{n}, 1+\frac{1}{n}) = [0, 1]$). Members of $\mathcal{F}_\sigma$ and $\mathcal{G}_\delta$ are respectively called F_σ - and G_δ - sets.

Definition 1.4 The smallest σ-algebra containing all open sets is called Borel σ-algebra or Borel σ-field. The members of the Bolel σ-algebra are called Borel sets.

Since the open sets and closed sets are complements of each other, Borel σ-algebra is also the smallest σ-algebra containing all closed sets. It is also the smallest σ-algebra containing (i) all open intervals; (ii) all closed intervals; (iii) all half-open and half-closed intervals.

Example 1.8 Every open interval is in $\mathcal{F}_\sigma$ and every closed interval is in $\mathcal{G}_\delta$. Every half-open interval is in $\mathcal{F}_\sigma$ and in $\mathcal{G}_\delta$.

For
$$(a,b) = \bigcup_{n=N}^{\infty} [a + \tfrac{1}{n}, b - \tfrac{1}{n}], \ N > \tfrac{2}{b-a} \text{ and } [a,b] = \bigcap_{n=1}^{\infty}(a - \tfrac{1}{n}, b + \tfrac{1}{n})$$

also
$$[a,b) = \bigcup_{n=N}^{\infty} [a, b - \tfrac{1}{n}] = \bigcap_{n=1}^{\infty}(a - \tfrac{1}{n}, b), \ N > \tfrac{1}{b-a}$$

Example 1.9 If $E \in \mathcal{G}_\delta$ then its complement $\tilde{E} \in \mathcal{F}_\sigma$ and if $E \in \mathcal{F}_\sigma$ then $\tilde{E} \in \mathcal{G}_\delta$. For, if $E \in \mathcal{G}_\delta$ then $E = \bigcap_{n=1}^{\infty} E_n$ when $E_n \in \mathcal{G}$. So, $\tilde{E}_n \in \mathcal{F}$. Since $\tilde{E} = \bigcup_{n=1}^{\infty} \tilde{E}_n$, $\tilde{E} \in \mathcal{F}_\delta$. The other part is similar.

Example 1.10 Every open set is in $\mathcal{F}_\sigma$ and every closed set is in $\mathcal{G}_\delta$.

Since every open set is a countable union of open intervals the first part follows from Example 1.8. The second part now follows from Example 1.9.

Example 1.11 The set of rational numbers in [0, 1] is an $\mathcal{F}_\sigma$-set but not a $\mathcal{G}_\delta$-set.

Let Q be the set of rational numbers in [0, 1]. Since Q is countable, let $Q = \{r_1, r_2, \ldots, r_n, \ldots\}$ where each r_n is a rational number in [0, 1]. Since each singleton set is closed and $Q = \bigcup_{n=1}^{\infty} \{r_n\}$, Q is an $\mathcal{F}_\sigma-$ set. If possible, suppose $Q \in \mathcal{G}_\delta$. Then $Q = \bigcap_{n=1}^{\infty} G_n$ where $G_n \in \mathcal{G}$. Since every open set is a countable union of disjoint open intervals, $G_n = \bigcup_{i=1}^{\infty}(a_{ni}, b_{ni})$. Since $Q \subset G_n$ for all n, the endpoints a_{ni}, and b_{ni} are irrational numbers for all i and all n. Let C be the collection of all these endpoints i.e $C = (\bigcup_{n=1}^{\infty}\bigcup_{i=1}^{\infty}\{a_{ni}\}) \cup (\bigcup_{n=1}^{\infty}\bigcup_{i=1}^{\infty} b_{ni})$. If T is the set of irrational numbers in [0, 1] then since Q is everywhere dense in [0, 1] and $Q \subset G_n$ for all n, $T \sim C \subset G_n$ for all n. So, $(T \sim C) \subset \bigcap_{n=1}^{\infty} G_n = Q$. But this is a contradiction since T being uncountable, $T \sim C$ is uncountable.

Example 1.12 The set of irrational numbers in [0, 1] is a $\mathcal{G}_\delta-$ set but not an $\mathcal{F}_\sigma-$ set.

This follows from Examples 1.11 and 1.9.

1.4 Sequence of Sets

Example 1.13 The following relations hold :

$$\mathcal{F}_\sigma \not\subset \mathcal{G}_\delta, \mathcal{G}_\delta \not\subset \mathcal{F}_\sigma, \mathcal{F}_\sigma \neq \mathcal{F}_{\sigma\delta}, \mathcal{G}_\delta \neq \mathcal{G}_{\delta\sigma}.$$

The first two cases follow from Examples 1.11 and 1.12. For the third case let $E = Q \cup T$ where Q is the set of rationals in $[0, 1]$ and T is the set of irrationals in $[-1, 0]$. Then $Q \in \mathcal{F}_\sigma \subset \mathcal{F}_{\sigma\delta}$ and $T \in \mathcal{G}_\delta \subset \mathcal{F}_{\sigma\delta}$. So, $Q = \bigcap_{i=1}^\infty \bigcup_{j=1}^\infty Q_{ij}$ and $T = \bigcap_{i=1}^\infty \bigcup_{j=1}^\infty T_{ij}$ where Q_{ij} and T_{ij} are closed sets. Hence if $E = Q \cup T = \bigcap_{i=1}^\infty \bigcup_{j=1}^\infty (Q_{ij} \cup T_{ij})$ then $E \in \mathcal{F}_{\sigma\delta}$. But $E \notin \mathcal{F}_\sigma$. For if $E \in \mathcal{F}_\sigma$ then $E = \bigcup_{i=1}^\infty E_i$, where E_i is closed for each i and so $E \cap [-1, 0] = \bigcup_{i=1}^\infty (E_i \cap [-1, 0]) \in \mathcal{F}_\sigma$ and hence $T \in \mathcal{F}_\sigma$. But as in Example 1.12, $T \notin \mathcal{F}_\sigma$ which is a contradiction.

1.4 Sequence of Sets

For any collection of sets C the upper and lower bounds of C, denoted by $\sup C$ and $\inf C$ respectively are defined by

$$\sup C = \cup\{E : E \in C\}, \text{ and } \inf C = \cap\{E : E \in C\}.$$

If a sequence of sets $\{E_n\}$ is given then its upper and lower bounds are respectively

$$\sup\{E_n\} = \bigcup_{n=1}^\infty E_n, \text{ and } \inf\{E_n\} = \bigcap_{n=1}^\infty E_n.$$

If there is no confusion we shall write $\sup_n E_n$ and $\inf_n E_n$ for the upper and lower bounds of $\{E_n\}$. The upper and lower limits of $\{E_n\}$ are defined by

$$\limsup_n E_n = \bigcap_{n=1}^\infty \bigcup_{k=n}^\infty E_k, \text{ and } \liminf_n E_n = \bigcup_{n=1}^\infty \bigcap_{k=n}^\infty E_k.$$

So, an element $x \in \sup_n E_n$ if and only if $x \in E_n$ for some n and $x \in \inf_n E_n$ if and only if $x \in E_n$ for all n while an element $x \in \limsup_n E_n$ if and only if for each n there is $k \geq n$ such that $x \in E_k$ i.e $x \in E_n$ for an infinite number of values on n and $x \in \liminf_n E_n$ if and only if there is n such that $x \in E_k$ for all $k \geq n$, i.e $x \in E_n$ for all but a finite number of n.

If $\mathcal{A}$ is a σ-algebra and $\{E_n\} \subset \mathcal{A}$ then it follows that $\inf_n E_n$, $\sup_n E_n$, $\limsup_n E_n$ and $\liminf_n E_n$ all belong to $\mathcal{A}$.

Definition 1.5 If $\limsup_n E_n = \liminf_n E_n$ then the sequence $\{E_n\}$ is said to be convergent and the common value is called the limit of the sequence $\{E_n\}$ and is denoted by $\lim_n E_n$.

Definition 1.6 A sequence $\{E_n\}$ is said to be increasing if $E_n \subset E_{n+1}$ for all n and $\{E_n\}$ is said to be decreasing if $E_{n+1} \subset E_n$ for all n. A sequence is said to be monotone if it is either increasing or decreasing.

The following theorem shows that monotone sequences of sets behave just like monotone sequences of real numbers.

Theorem 1.5 *Every monotone sequence of sets is convergent and it converges to its upper or lower bound according as it is increasing or decreasing.*

Proof Let $\{E_n\}$ be a monotone increasing sequence of sets. Let $\xi \in \bigcup_{n=1}^{\infty} E_n$. Then $\xi \in E_n$ for some n, say for $n = n_0$. Since $\{E_n\}$ is increasing $\xi \in E_n$ for all $n \geq n_0$ and so $\xi \in \bigcap_{n=n_0}^{\infty} E_n$ and hence $\xi \in \bigcup_{k=1}^{\infty} \bigcap_{n=k}^{\infty} E_k$. So, $\bigcup_{n=1}^{\infty} E_n \subset \bigcup_{k=1}^{\infty} \bigcap_{n=k}^{\infty} E_k$, i.e $\sup_n E_n \subset \liminf_n E_n$. Since $\liminf_n E_n \subset \limsup_n E_n \subset \sup_n E_n$ for every sequence of sets, we have,

$$\liminf_n E_n = \limsup_n E_n = \sup_n E_n$$

and so $\{E_n\}$ converges to its upper bound. The proof for decreasing sequence is similar. □

Example 1.14 If $\{G_n\}$ and $\{F_n\}$ are sequences of members of $\mathcal{G}$ and $\mathcal{F}$ respectively then $\limsup_n G_n \in \mathcal{G}_\delta$ and $\liminf_n F_n \in \mathcal{F}_\sigma$, while $\liminf_n G_n \in \mathcal{G}_{\delta\sigma}$ and $\limsup_n F_n \in \mathcal{F}_{\sigma\delta}$.

The proof follows from the definition and so is left as exercises.

Definition 1.7 Let E be any subset of $\mathbb{R}$. The function $\chi_E : \mathbb{R} \to \mathbb{R}$ defined by $\chi_E(x) = 1$ if $x \in E$ and $\chi_E(x) = 0$ if $x \in \mathbb{R} \sim E$, is called characteristic function or indicator function of E.

Theorem 1.6 *If $\{E_n\}$ is any sequence of subsets of $\mathbb{R}$ and if $A = \sup_n E_n$, $B = \inf_n E_n$, $P = \limsup_n E_n$ and $Q = \liminf_n E_n$, then for all $x \in \mathbb{R}$.*
(i) $\chi_A(x) = \sup_n \chi_{E_n}(x)$; (ii) $\chi_B(x) = \inf_n \chi_{E_n}(x)$; (iii) $\chi_P(x) = \limsup_n \chi_{E_n}(x)$; (iv) $\chi_Q = \liminf_n \chi_{E_n}(x)$.

1.5 Cartesian Product

Proof We prove (iii), the rest being left as exercises.

If $x \in P$ then $x \in \bigcap_{n=1}^{\infty} \bigcup_{k=n}^{\infty} E_k$ and so for each n there is $k \geq n$ such that $x \in E_k$ and hence $x \in E_k$ for infinite number of values of k and hence $\chi_{E_k}(x) = 1$ for infinite number of values of k. Since $\chi_{E_n}(x) = 0$ or 1, $\limsup_{n} \chi_{E_n}(x) = 1$. Also since $x \in P$, $\chi_P(x) = 1$. If $x \in \mathbb{R} \sim P$ then $x \notin \bigcap_{n=1}^{\infty} \bigcup_{k=n}^{\infty} E_k$ and hence $x \notin \bigcup_{k=n}^{\infty} E_k$ for at least one values of n say for $n = n_0$ and therefore $x \notin E_k$ for all $k \geq n_0$. Hence $\chi_{E_k}(x) = 0$ for all $k \geq n_0$ and so $\limsup_{n} \chi_{E_n}(x) = 0$. Also since $x \notin P$, $\chi_P(x) = 0$, completing the proof of (iii). □

Example 1.15 For any two sets A and B

(i) $\chi_{A \cup B}(x) = \chi_A(x) + \chi_B(x) - \chi_{A \cap B}(x)$
(ii) $\chi_{A \cap B}(x) = \chi_A(x) + \chi_B(x) - \chi_{A \cup B}(x)$
(iii) $\chi_{A \sim B}(x) = \chi_A(x) + \chi_{\widetilde{B}}(x) - \chi_{A \cup \widetilde{B}}(x)$.

If $x \in A \cup B$ then $\chi_{A \cup B}(x) = 1$. Also if $x \in A$ but $x \notin B$ then $\chi_A(x) = 1$ and $\chi_B(x) = 0$ and $\chi_{A \cap B}(x) = 0$. If $x \in A$ and $x \in B$ then $\chi_A(x) = \chi_B(x) = 1 = \chi_{A \cap B}(x)$. Finally if $x \notin A$ but $x \in B$ then $\chi_A(x) = 0$ and $\chi_B(x) = 1$ and $\chi_{A \cap B}(x) = 0$. So (i) is proved ; (ii) and (iii) follow from (i).

Theorem 1.7 *A sequence of sets $\{E_n\}$ is monotone if and only if the sequence of its characteristic function $\{\chi_{E_n}\}$ is monotone of the same type.*

Proof Let $\{E_n\}$ be increasing then $E_n \subset E_{n+1}$ for all n. So, if $x \in E_n$ then $x \in E_{n+1}$ and hence $\chi_{E_n}(x) = 1 = \chi_{E_{n+1}}(x)$ and if $x \notin E_n$ then when $x \notin E_{n+1}$, $\chi_{E_n}(x) = 0 = \chi_{E_{n+1}}(x)$ and when $x \in E_{n+1}$, $\chi_{E_n}(x) = 0 < 1 = \chi_{E_{n+1}}(x)$. Then $\chi_{E_n}(x) \leq \chi_{E_{n+1}}(x)$. Conversely suppose that $\{\chi_{E_n}\}$ is increasing. If possible suppose that $\{E_n\}$ is not increasing. So there is n such that $E_n \not\subset E_{n+1}$. Then there is $x \in E_n$ but $x \notin E_{n+1}$ and so $\chi_{E_n}(x) = 1$ and $\chi_{E_{n+1}}(x) = 0$ which shows that $\{\chi_{E_n}\}$ is not increasing which is a contradiction. The proof for decreasing $\{E_n\}$ is similar. □

1.5 Cartesian Product

Given a non-empty collection of non-empty sets $C = \{A_i : i \in I\}$ the problem whether it is possible to choose one element from each set A_i cannot be solved without further assumption on the collection C; this is because which element of the set A_i is to be chosen. This is assumed as axiom known as the axiom of choice. This is precisely stated as.

Axiom of choice Given a non-empty collection of non-empty sets $C = \{A_i : i \in I\}$ there is a function $c : C \to \bigcup_{i \in I} A_i$ such that $c(A_i) \in A_i$ for each $i \in I$. The function c is called a choice function.

The axiom of choice is useful for many purposes. This will be used to define Cartesian products and also will be used elsewhere.

If A_1 and A_2 are non-void sets, their Cartesian product $A_1 \times A_2$ is defined to be the set of all ordered pairs (x_1, x_2) such that $x_1 \in A_1$ and $x_2 \in A_2$. That is,

$$A_1 \times A_2 = \{(x_1, x_2) : x_1 \in A_1, x_2 \in A_2\}.$$

For any finite number of sets $A_1, A_2, \ldots, A_n$ their Cartesian product is defined as

$$\prod_{i=1}^{n} A_i = A_1 \times A_2 \times \ldots A_n = \{(x_1, x_2, \ldots, x_n) : x_i \in A_i, i = 1, 2, \ldots n\}.$$

Clearly each member $(x_1, x_2, \ldots x_n)$ of the set $\prod_{i=1}^{n} A_i$ may be considered as a functions f defined on the set $\{1, 2, \ldots n\}$ such that $f(i) = x_i \in A_i$ and so $\prod_{i=1}^{n} A_i$ is the family of all functions f on $\{1, 2, \ldots n\}$ with $f(i) \in A_i$. Thus we can define the Cartesian product of non-void family of non-void sets as follows.

Definition 1.8 Let $C = \{A_i : i \in I\}$ be a non-void family of non-void sets. Then the Cartesian product of the family C, denoted by $\prod_{i \in I} A_i$ is the set of all functions f defined on I such that $f(i) = x_i \in A_i$.

By the axiom of choice, there is a function $c : C \to \bigcup_{i \in I} A_i$ such that $c(A_i) \in A_i$ for all $i \in I$. So letting $f(i) = c(A_i)$ for $i \in I$ we get a function f defined on I such that $f(i) \in A_i$. Therefore $\prod_{i \in I} A_i$ is non-void.

If $A_i = A$ for all $i \in I$ where A and I are fixed sets then from the definition $\prod_{i \in I} A_i$ is the set of all functions defined on I with values in A and is denoted by A^I.

1.6 Completely Additive Set Function

Definition 1.9 Let $\mathcal{A}$ be an algebra and let $\psi : \mathcal{A} \to \overline{\mathbb{R}}$ where $\overline{\mathbb{R}}$ is the set of extended real numbers. Then ψ is called completely additive or countably additive set function if
(i) $\psi(\phi) = 0$, ϕ being the null set.
(ii) for any countable collection of disjoint sets $\{E_i\} \subset \mathcal{A}$, $\sum_i \psi(E_i)$ is defined and $\psi(\cup E_i) = \sum \psi(E_i)$ whenever $\cup E_i \in \mathcal{A}$, the function ψ is called finitely additive if (ii) holds for finite collection $\{E_i\}$.

Note: If $\mathcal{A}$ is a σ-algebra then $\cup E_i \in \mathcal{A}$ is always true and so it is not necessary to write 'whenever $\cup E_i \in \mathcal{A}$' in (ii). Also in (ii) '$\sum \psi(E_i)$ is defined' means that

1.6 Completely Additive Set Function

it is not of the form $\infty - \infty$ which may occur only when ψ assumes both of ∞ and $-\infty$. Finally, since $\cup E_i$ does not depend on the order of the union, the series $\sum \psi(E_i)$ should either be absolutely convergent or it should be properly divergent. Also if ψ assumes only finite values then (i) is redundant, since $\psi(A) = \psi(A \cup \phi) = \psi(A) + \psi(\phi)$.

Theorem 1.8 *If ψ is a completely additive set function defined on an algebra $\mathcal{A}$ and if $A, B \in \mathcal{A}$, $B \subset A$ and if $\psi(A)$ is finite then $\psi(B)$ is also finite and $\psi(A \sim B) = \psi(A) - \psi(B)$.*

Proof Since $A = (A \sim B) \cup B$ and $A \sim B \in \mathcal{A}$ we have $\psi(A) = \psi(A \sim B) + \psi(B)$. Since $\psi(A)$ is finite, $\psi(A \sim B)$ and $\psi(B)$ are both finite. For, otherwise both are infinite and of opposite sign which contradicts the definition of ψ. Hence $\psi(A \sim B) = \psi(A) - \psi(B)$. $\square$

Theorem 1.9 *If ψ is a completely additive set function defined on an algebra $\mathcal{A}$ and if $\{E_n\} \subset \mathcal{A}$ is an increasing sequence of sets such that $\cup E_n \in \mathcal{A}$, then*

$$\lim_{n \to \infty} \psi(E_n) = \psi(\lim_{n \to \infty} E_n).$$

Proof Since $\{E_n\}$ is increasing $\lim_{n \to \infty} E_n = \cup E_n$. Also $\cup E_n = E_1 \cup (E_2 \sim E_1) \cup (E_3 \sim E_2)\dots$. Since the right-hand side is a union of disjoint sets and ψ is completely additive

$$\psi(\lim E_n) = \psi(\cup E_n) = \psi(E_1) + \sum_{n=2}^{\infty} \psi(E_n \sim E_{n-1})$$

$$= \psi(E_1) + \lim_{k \to \infty} \sum_{n=2}^{k} \psi(E_n \sim E_{n-1})$$

$$= \lim_{k \to \infty} [\psi(E_1) + \sum_{n=2}^{k} \psi(E_n \sim E_{n-1})]$$

$$= \lim_{k \to \infty} [\psi(E_1 \cup (E_2 \sim E_1) \cup (E_3 \sim E_2) \cup \dots \cup (E_k \sim E_{k-1}))]$$

$$= \lim_{k \to \infty} \psi(E_k).$$

$\square$

Theorem 1.10 *If ψ is a completely additive set function defined on an algebra $\mathcal{A}$ and if $\{E_n\} \subset \mathcal{A}$ is a decreasing sequence of sets such that $\cap E_n \in \mathcal{A}$ and $\psi(E_1)$ is finite then*

$$\lim_{n \to \infty} \psi(E_n) = \psi(\lim_{n \to \infty} E_n).$$

Proof Since $\lim_{n \to \infty} E_n = \cap E_n \subset E_1$ and $\psi(E_1)$ is finite, by Theorem 1.8 $\psi(\cap E_n)$ is finite and

$$\psi(E_1 \sim \cap E_n) = \psi(E_1) - \psi(\cap E_n).$$

Since $\{E_n\}$ is decreasing, the sequence $\{E_1 \sim E_n\}$ is increasing and hence by Theorem 1.9

$$\lim \psi(E_1 \sim E_n) = \psi(\lim(E_1 \sim E_n)) = \psi(\cup(E_1 \sim E_n)) = \psi(E_1 \sim \cap E_n).$$

Hence

$$\psi(E_1) - \lim \psi(E_n) = \lim[\psi(E_1) - \psi(E_n)] = \lim[\psi(E_1 \sim E_n)] = \psi(E_1 \sim \cap E_n).$$

Since $\psi(E_1)$ is finite the result follows. $\square$

Remark If instead of $\psi(E_1)$ being finite, $\psi(E_n)$ is finite for some n, say $n = n_0$ then also the theorem is true. For, consider the sequence $\{F_k\}$ where $F_k = E_{n_0+k-1}$ and since $\lim_{k \to \infty} F_k = \lim_{n \to \infty} E_n$ the result follows.

Definition 1.10 Let C be a collection of sets and $\psi : C \to \overline{\mathbb{R}}$, where $\overline{\mathbb{R}}$ is the set of extended real numbers. Then the set function ψ is called increasing (or non-decreasing) if $\psi(A) \leq \psi(B)$ whenever $A \subset B$, $A, B \in C$ and it is called decreasing (or non-increasing) if $\psi(A) \geq \psi(B)$ whenever $A \subset B$, $A, B \in C$.

Theorem 1.11 *A completely additive set function $\psi : \mathcal{A} \to \overline{\mathbb{R}}$, where $\mathcal{A}$ is an algebra is increasing if and only if ψ is non-negative.*

Proof Let ψ be increasing and let $A \in \mathcal{A}$. Since the null set ϕ is a subset of every set and since $\phi \in \mathcal{A}$, $\psi(A) \geq \psi(\phi) = 0$. Conversely, let ψ be non-negative and let $A \subset B$, $A, B \in \mathcal{A}$. Then $\psi(A) \leq \psi(A) + \psi(B \sim A) = \psi(A \cup (B \sim A)) = \psi(B)$. So, ψ is increasing. $\square$

Definition 1.11 Let $\mathcal{A}$ be a σ-algebra and let $\psi : \mathcal{A} \to \mathbb{R}$ be a set function. Then ψ is said to be continuous at $E \in \mathcal{A}$ if for every sequence $\{E_n\} \subset \mathcal{A}$ which converges to E, $\lim_{n \to \infty} \psi(E_n) = \psi(E)$.

Theorem 1.12 *If $\mathcal{A}$ is a σ-algebra and $\psi : \mathcal{A} \to \mathbb{R}$ is a completely additive finite and non-negative set function then ψ is continuous at E for each $E \in \mathcal{A}$.*

Proof Let $E \in \mathcal{A}$, and $\{E_n\}$ be any sequence in $\mathcal{A}$ such that $\lim_{n \to \infty} E_n = E$. Then by definition $E = \bigcup_{n=1}^{\infty} \bigcap_{k=n}^{\infty} E_k = \bigcap_{n=1}^{\infty} \bigcup_{k=n}^{\infty} E_k$, (See Definition 1.5). Let $U_n = \bigcap_{k=n}^{\infty} E_k$, $V_n = \bigcup_{k=n}^{\infty} E_k$. Since $\{U_n\}$ is increasing and $\lim_{n \to \infty} U_n = E$ by Theorem 1.9, $\lim_{n \to \infty} \psi(U_n) = \psi(E)$. Again since $\{V_n\}$ is decreasing and $\lim_{n \to \infty} V_n = E$ by Theorem 1.10. $\lim_{n \to \infty} \psi(V_n) = \psi(E)$. Since $U_n \subset E_n \subset V_n$ for all n, and since ψ is non-negative, by Theorem 1.11, $\psi(U_n) \leq \psi(E_n) \leq \psi(V_n)$. Hence

$$\psi(E) = \lim_{n \to \infty} \psi(U_n) \leq \lim_{n \to \infty} \psi(E_n) \leq \lim_{n \to \infty} \psi(V_n) = \psi(E).$$

So, $\lim_{n \to \infty} \psi(E_n) = \psi(E)$. So ψ is continuous at E. $\square$

1.7 Construction of σ-Algebra

Theorem 1.13 *Let X be a non-void set and let C be any non-void collection of subset of X. Then there exists a smallest σ-algebra containing C. That is, there is a σ-algebra $\mathcal{G}$ containing C such that if $\mathcal{J}$ is any σ-algebra containing C then $\mathcal{G} \subset \mathcal{J}$.*

Proof Let $\mathcal{F}$ be the family of all σ-algebras of subsets of X containing C. Then $\mathcal{F}$ is non-void since the power set $\mathcal{P}(X)$ of X is a member of $\mathcal{F}$. Let $\mathcal{G} = \cap\{\mathcal{B} : \mathcal{B} \in \mathcal{F}\}$. Since $C \subset \mathcal{B}$ for all $\mathcal{B} \in \mathcal{F}$, $C \subset \mathcal{G}$. If $A, B \in \mathcal{G}$ then $A, B \in \mathcal{B}$ for all $\mathcal{B} \in \mathcal{F}$ and so $A \cup B \in \mathcal{B}$ for all $\mathcal{B} \in \mathcal{F}$ and hence $A \cup B \in \mathcal{G}$. Also, if $A \in \mathcal{G}$ then $A \in \mathcal{B}$ for all $\mathcal{B} \in \mathcal{F}$ and hence $\widetilde{A} \in \mathcal{B}$ for all $\mathcal{B} \in \mathcal{F}$ and so $\widetilde{A} \in \mathcal{G}$. So $\mathcal{G}$ is an algebra. Let $\{A_i\}$ be any countable collection of members of $\mathcal{G}$. Then for each $\mathcal{B} \in \mathcal{F}$, $\{A_i\} \subset \mathcal{B}$. Since $\mathcal{B}$ is a σ-algebra, $\cup A_i \in \mathcal{B}$ for each $\mathcal{B} \in \mathcal{F}$ and hence $\cup A_i \in \mathcal{G}$. So, $\mathcal{G}$ is a σ-algebra containing C. From the definition of $\mathcal{G}$, it is the smallest σ-algebra containing C.

The above theorem proves the existence of the smallest σ-algebra containing C. The following theorem presents the actual smallest σ-algebra containing C that can be derived from C. $\square$

Theorem 1.14 *Let X be a non-void set and let C be any non-void collection of subset of X. Then there exists smallest σ-algebra containing C.*

Proof Let $C_1 = \{E : \widetilde{E} \in C\}$. Let $\mathcal{A}_0 = C \cup C_1$. Then it is clear that $E \in \mathcal{A}_0$ if and only if $\widetilde{E} \in \mathcal{A}_0$. Let $\mathcal{A}_1$ be the collection of all subsets of X that can be expressed as a countable union or a countable intersection of members of $\mathcal{A}_0$. Clearly, if $E \in \mathcal{A}_1$ then $\widetilde{E} \in \mathcal{A}_1$. For, let $E \in \mathcal{A}_1$ and suppose $E = \bigcup_{i=1}^{\infty} E_i$ where $E_i \in \mathcal{A}_0$ for all i. Then $\widetilde{E}_i \in \mathcal{A}_0$ for all i and so $\bigcap_{i=1}^{\infty} \widetilde{E}_i \in \mathcal{A}_1$ and hence $\widetilde{E} = \widetilde{\bigcup_{i=1}^{\infty} E_i} = \bigcap_{i=1}^{\infty} \widetilde{E}_i \in \mathcal{A}_1$. Let $\mathcal{A}_2$ be the collection of all subsets of X that can be expressed as a countable union or a countable intersection of members of $\mathcal{A}_1$. Clearly if $E \subset \mathcal{A}_2$ then $\widetilde{E} \in \mathcal{A}_2$. Continuing this process we get a sequence $\{\mathcal{A}_n\}$ of collection of subsets of X such that if $E \in \mathcal{A}_n$ then E is either a countable union or a countable intersection of members of $\mathcal{A}_{n-1}$ and if $E \in \mathcal{A}_n$ then $\widetilde{E} \in \mathcal{A}_n$. The process being continued indefinitely we get $\mathcal{A}_0 \subset \mathcal{A}_1 \subset \mathcal{A}_2 \subset ... \subset \mathcal{A}_n \subset ....$ Let $\mathcal{G}(C) = \bigcup_{n=0}^{\infty} \mathcal{A}_n$. Clearly $C \subset \mathcal{G}(C)$. Also $\mathcal{G}(C)$ is a σ-algebra. For let $E \subset \mathcal{G}(C)$. Then $E \in \mathcal{A}_n$ for some n. So, $\widetilde{E} \in \mathcal{A}_n$ and hence $\widetilde{E} \in \mathcal{G}(C)$. If $E_1, E_2 \in \mathcal{G}(C)$ then $E_1, E_2 \in \mathcal{A}_n$ for some n and hence $E_1 \cup E_2 \in \mathcal{A}_{n+1}$ and so $E_1 \cup E_2 \in \mathcal{G}(C)$. So $\mathcal{G}(C)$ is an algebra. Let $\{E_i\}$ be any countable collection of members of $\mathcal{G}(C)$. Then for any fixed k there is n such that $\bigcup_{i=1}^{k} E_i \subset \mathcal{A}_n$ and so $\bigcup_{k=1}^{\infty} E_i \subset \bigcup_{n=0}^{\infty} \mathcal{A}_n = \mathcal{G}(C)$. Hence $\mathcal{G}(C)$ is a σ-algebra. Clearly $\mathcal{G}(C)$ is the smallest σ-algebra containing C. For, suppose that $\Omega(C)$ is the smallest σ-algebra containing C. Then by the definition of $\mathcal{A}_n$, $\mathcal{A}_n \subset \Omega(C)$ for all n and so $\bigcup_{n=0}^{\infty} \mathcal{A}_n \subset \Omega(C)$ and hence $\mathcal{G}(C) \subset \Omega(C)$. Since $\mathcal{G}(C)$ is a σ-algebra containing C

and $\Omega(C)$ is the smallest σ-algebra containing C, $\Omega(C) \subset \mathcal{G}(C)$. So, $\mathcal{G}(C) = \Omega(C)$, completing the proof. $\square$

Definition 1.12 Let X be a non-void set and let $\mathcal{J}$ be a non-void collection of subsets of X. Then $\mathcal{J}$ is called a monotone class if for every monotone sequence (increasing or decreasing) $\{E_n\}$ in $\mathcal{J}$, $\lim_n E_n \in \mathcal{J}$.

Theorem 1.15 *If $\mathcal{J}$ is a monotone class then every countable union and every countable intersection of members of $\mathcal{J}$ is in $\mathcal{J}$.*

Proof Let $\{u_i\}$ be any countable collection of members of $\mathcal{J}$. For each n, let $E_n = \bigcup_{i=1}^{n} u_i$ and $F_n = \bigcap_{i=1}^{n} u_i$. Then $\{E_n\}$ is increasing and $\{F_n\}$ is decreasing and so $\bigcup_{i=1}^{\infty} u_i = \lim_n E_n \in \mathcal{J}$ and $\bigcap_{i=1}^{\infty} u_i = \lim_n F_n \in \mathcal{J}$. $\square$

Corollary 1.16 *Every σ-algebra is a monotone class.*

Theorem 1.17 *Let X be a non-void set and let C be any collection of subset of X. Then there is a smallest monotone class containing C.*

Proof Clearly the set of all subsets of X is a monotone class containing C. Also intersection of all monotone classes containing C is a monotone class containing C, which is the smallest monotone class containing C. $\square$

Theorem 1.18 *Let C be any collection of subsets of a non-void set X such that if $E \in C$ then its complement $\widetilde{E} \in C$. Then the smallest σ-algebra containing C is the smallest monotone class containing C.*

Proof Let $\mathcal{G}(C)$ be the smallest σ-algebra containing C and let $\mathcal{J}(C)$ be the smallest monotone class containing C. Since every σ algebra is a monotone class, $\mathcal{G}(C)$ is a monotone class containing C and hence $\mathcal{J}(C) \subset \mathcal{G}(C)$. To prove the opposite inclusion we consider the sets $\mathcal{A}_n$ of Theorem 1.14.
Note that to construct $\mathcal{G}(C)$ in Theorem 1.14 we considered $\mathcal{A}_0 = C \cup C_1$ where $C_1 = \{E, \widetilde{E} \in C\}$. Here $C = C_1$ and so we take $\mathcal{A}_0 = C$. Then $\mathcal{A}_0 \subset \mathcal{J}(C)$. Since every member of $\mathcal{A}_n$, $n = 1, 2, \ldots$ is the countable union or countable intersection of members of $\mathcal{A}_{n-1}$, every $\mathcal{A}_n \subset \mathcal{J}(C)$ and so $\mathcal{G}(C) = \bigcup_{n=0}^{\infty} \mathcal{A}_n \subset \mathcal{J}(C)$, completing the proof. $\square$

1.8 Exercises

1. If $\mathcal{G}$ and $\mathcal{F}$ are the classes of open sets and closed sets respectively and if $\{G_n\}$ and $\{F_n\}$ are sequences of members of $\mathcal{G}$ and $\mathcal{F}$ respectively then prove that the following relations may not hold:
   (i) $\limsup_n G_n \in \mathcal{G}$; (ii) $\liminf_n G_n \in \mathcal{G}_\delta$; (iii) $\liminf_n F_n \in \mathcal{F}$; (iv) $\limsup_n F_n \in \mathcal{F}_\sigma$.

1.8 Exercises

2. If $\{E_n\}$ is any sequence of sets prove that $\inf_n E_n \subset \liminf_n E_n \subset \limsup_n E_n \subset \sup_n E_n$.

3. Let $\mathcal{A}$ be an algebra of sets. For $E \in \mathcal{A}$ define $\psi(E) =$ the number of elements of E if E is finite and $\psi(E) = \infty$ otherwise. Show that ψ is a completely additive set function on $\mathcal{A}$.

4. Show that if ψ is a completely additive set functions defined on the class of all subsets of a countable set C then ψ is identically 0 if and only if $\psi(\{x\}) = 0$ for every singleton sets $\{x\} \in C$.

5. Let X be any set and let $\mathcal{A}$ be the collection of all subsets of X which are either countable or their complements are countable. Show that $\mathcal{A}$ is σ-algebra (The null set ϕ is assumed to be countable).

6. Let X be any infinite set and let $\mathcal{A}$ be the collection of all subsets of X which are either finite or their complements are finite. Show that $\mathcal{A}$ is an algebra but not a σ-algebra.

7. Show that the set of all rational numbers is in $\mathcal{F}_\sigma$ and the set of all irrational numbers is in $\mathcal{G}_\delta$.

8. If $\mathcal{A}$ is an algebra of sets then examine whether $\mathcal{A}_\sigma$ is a σ-algebra.

9. If ψ is an additive set function on a σ-algebra $\mathcal{A}$ and if for every increasing sequence $\{E_n\}$ of sets in $\mathcal{A}$ $\lim_n \psi(E_n) = \psi(\lim_n E_n)$ then prove that ψ is completely additive.

Chapter 2
Lebesgue Measure on Real Line

Every finite interval I on the real line $\mathbb{R}$ has its length which is defined to be the difference of the endpoints of I and the length of I is its measure. Since an interval is a subset of $\mathbb{R}$ it is desirable that every subset of $\mathbb{R}$ should have some measure. Being motivated by this idea E. Borel and H. Lebesgue attempted to give definition of measure of a set. But it was found that not all sets can have a measure in that sense and example of nonmeasurable sets were found from which the concepts of σ-algebra σ-ring etc. were evolved and measure theory developed. We start with the Lebesgue measure on the real line. Throughout this chapter 'measure' and 'outer measure' mean Lebesgue measure and Lebesgue outer measure.

2.1 Outer Measure

Definition 2.1 Let E be any subset of $\mathbb{R}$. Let $\{I_n\}$ be a countable collection of open intervals such that $E \subset \bigcup_{n=1}^{\infty} I_n$. Denoting by $|I_n|$ the length of I_n, consider for each such collection $\{I_n\}$, the sum $\sum_{n=1}^{\infty} |I_n|$. The outer measure of E is the infimum of all such sums and is denoted by $\mu^*(E)$. That is

$$\mu^*(E) = \inf \left\{ \sum_{n=1}^{\infty} |I_n| : E \subset \bigcup_{n=1}^{\infty} I_n \right\},$$

the infimum being taken over all such collections.

The outer measure has the following properties the first three of which are used as definition of general outer measure which we shall consider later.

Theorem 2.1 (i) $\mu^*(\phi) = 0$ where ϕ is the null set.
(ii) If $E_1 \subset E_2$ then $\mu^*(E_1) \leq \mu^*(E_2)$
(iii) If $\{E_n\}$ is any countable collection of subsets of $\mathbb{R}$ then $\mu^*(\bigcup E_n) \leq \sum \mu^*(E_n)$

(iv) If E is a singleton set, i. e if $E = \{x\}, x \in \mathbb{R}$, then $\mu^*(E) = 0$ and if C is any countable subset of $\mathbb{R}$ then $\mu^*(C) = 0$.
(v) For every interval $I \subset \mathbb{R}$, $\mu^*(I) = |I|$ where $|I|$ is the length of I.
(vi) (Caratheodory property), If A and B are subsets of $\mathbb{R}$ such that $\inf\{|x - y| : x \in A, y \in B\} > 0$ than $\mu^*(A \cup B) = \mu^*(A) + \mu^*(B)$.

Proof

(i) Let $\epsilon > 0$ be arbitrary. Since the null set ϕ is a subset of every set, $\phi \subset (-\epsilon, \epsilon)$ and so by definition $\mu^*(\phi) \leq 2\epsilon$. Since ϵ is arbitrary, $\mu^*(\phi) = 0$.

(ii) If a collection $\{I_n\}$ covers E_2 then $\{I_n\}$ also covers E_1 and hence

$$\left\{\sum |I_n| : E_2 \subset \bigcup I_n\right\} \subset \left\{\sum |I_n| : E_1 \subset \bigcup I_n\right\}$$

and so by definition $\mu^*(E_1) \leq \mu^*(E_2)$.

(iii) If $\sum \mu^*(E_n) = \infty$ the proof is complete. So, we suppose that $\mu^*(E_n)$ is finite for all n. Let $\epsilon > 0$ be arbitrary. For each n there is a countable collection of open intervals $\{I_{n,k} : k = 1, 2, ...\}$ such that $E_n \subset \bigcup_k I_{n,k}$ and $\sum_k |I_{n,k}| < \mu^*(E_n) + \frac{\epsilon}{2^{n+1}}$. So, $\bigcup_n E_n \subset \bigcup_n \bigcup_k I_{n,k}$ and hence

$$\mu^*(\bigcup_n E_n) \leq \sum_n \sum_k |I_{n,k}| < \sum_n (\mu^*(E_n) + \frac{\epsilon}{2^{n+1}}) \leq \sum_n \mu^*(E_n) + \epsilon.$$

Since ϵ is arbitrary, the result follows.

(iv) Let $E = \{x\}$ be a singleton set. Let $\epsilon > 0$ be arbitrary. Since $E \subset (x - \epsilon, x + \epsilon)$, $\mu^*(E) \leq 2\epsilon$, and since ϵ is arbitrary, $\mu^*(E) = 0$. Let $C = \{x_1, x_2, ..., x_n..\}$. Then $C = \bigcup_n E_n$ where $E_n = \{x_n\}$. Since $\mu^*(E_n) = 0$ for all n, by (iii) $\mu^*(C) = \mu^*(\bigcup_n E_n) \leq \sum_n \mu^*(E_n) = 0$. Since $\mu^*(C) \geq 0$, $\mu^*(C) = 0$.

(v) Let $I = (a, b)$. Since $I \subset I$, by definition $\mu^*(I) \leq |I|$. Let $\epsilon > 0$ be arbitrary. Then by definition there is a countable collection $\{I_n\}$ of open intervals such that $I \subset \bigcup_n I_n$ and $\sum_n |I_n| < \mu^*(I) + \epsilon$. Since $|I| \leq \sum_n |I_n|$, $|I| < \mu^*(I) + \epsilon$. Since ϵ is arbitrary $|I| \leq \mu^*(I)$. So $\mu^*(I) = |I|$.

Let $I = [a, b]$. Then $I = (a, b) \cup \{a\} \cup \{b\}$ and so by (iii) and (iv) $\mu^*(I) \leq \mu^*((a, b)) + \mu^*(\{a\}) + \mu^*(\{b\}) = \mu^*((a, b))$. Also since $(a, b) \subset [a, b]$, by (ii) $\mu^*((a, b)) \leq \mu^*(I)$. Hence $\mu^*(I) = \mu^*((a, b)) = |I|$. If $I = (a, b]$ or $[a, b)$ then $(a, b) \subset I \subset [a, b]$ and so $\mu^*(I) = |I|$.

(vi) Let $\mathcal{C} = \{I_n\}$ be any countable collection of open intervals such that $A \cup B \subset \bigcup I_n$. We divide the collection $\{I_n\}$ in three sub collections $\mathcal{L}, \mathcal{M}$ and $\mathcal{N}$ such that if an interval I_n in $\mathcal{C}$ contains points of A only then we put I_n in $\mathcal{L}$, if I_n contains points of B only then we put I_n in $\mathcal{M}$ and if I_n contains points of both A and B then we put I_n in $\mathcal{N}$. Let $I_n \in \mathcal{N}$. Then since $\inf\{|x - y| : x \in A; y \in B\} > 0$ there are closed sub-intervals of I_n which contain no points of $A \cup B$. Removing all these sub-intervals of I_n we get two sub collections $\{J_{n,i} : i = 1, 2, ...\}$ and $\{K_{n,r} : r = 1, 2, ...\}$ of open sub-intervals of I_n such that each member of $\{J_{n,i} : i = 1, 2, ...\}$ contains points of A only and each

2.1 Outer Measure

member of $\{K_{n,r} : r = 1, 2, ...\}$ contains points of B only. So, the collection $\{I_n : I_n \in \mathcal{L}\} \bigcup \{J_{n,i} : i = 1, 2, ...\}$ is a countable collection of open intervals such that

$$A \subset (\bigcup_{I_n \in \mathcal{L}} I_n) \bigcup (\bigcup_{I_n \in \mathcal{N}} \bigcup_i J_{n,i}) \text{ and}$$

$$B \subset (\bigcup_{I_n \in \mathcal{M}} I_n) \bigcup (\bigcup_{I_n \in \mathcal{N}} \bigcup_r K_{n,r})$$

Hence

$$\mu^*(A) \leq \sum_{I_n \in \mathcal{L}} |I_n| + \sum_{I_n \in \mathcal{N}} \sum_i |J_{n,i}|$$

$$\mu^*(B) \leq \sum_{I_n \in \mathcal{M}} |I_n| + \sum_{I_n \in \mathcal{N}} \sum_r |K_{n,r}|$$

Since $\sum_i |J_{n,i}| + \sum_r |K_{n,r}| \leq |I_n|$ we have

$$\sum_{I_n \in \mathcal{N}} \sum_i |J_{n,i}| + \sum_{I_n \in \mathcal{N}} \sum_r |K_{n,r}| \leq \sum_{I_n \in \mathcal{N}} |I_n|.$$

So, from above

$$\mu^*(A) + \mu^*(B) \leq \sum_{I_n \in \mathcal{L}} |I_n| + \sum_{I_n \in \mathcal{M}} |I_n| + \sum_{I_n \in \mathcal{N}} |I_n| = \sum_{I_n \in \mathcal{C}} |I_n|.$$

Since the collection $\mathcal{C}$ is arbitrary, considering all countable collections of open intervals which covers $A \bigcup B$ and taking infimum we get

$$\mu^*(A) + \mu^*(B) \leq \mu^*(A \bigcup B).$$

Since the other inequality follows from (iii), the proof is complete. □

Example 2.1 If $E \subset \mathbb{R}$ and $\mu^*(E) = 0$ then for any set $A \subset \mathbb{R}$

$$\mu^*(A) \geq \mu^*(A \cap E) + \mu^*(A \cap \widetilde{E})$$

Let $A \subset \mathbb{R}$ be any set. Since $A \cap E \subset E$, by Theorem 2.1 (ii) $\mu^*(A \cap E) \leq \mu^*(E) = 0$ and so $\mu^*(A \cap E) = 0$. Also $A \cap \widetilde{E} \subset A$ and so $\mu^*(A \cap \widetilde{E}) \leq \mu^*(A)$. These two together gives the above result.

Example 2.2 If $\mu^*(A \triangle B) = 0$ then $\mu^*(A \cap \widetilde{B}) = 0$ and $\mu^*(\widetilde{A} \cap B) = 0$. Since $A \triangle B = (A \cap \widetilde{B}) \bigcup (\widetilde{A} \cap B)$, $A \cap \widetilde{B} \subset A \triangle B$ and so $\mu^*(A \cap \widetilde{B}) = 0$. The other part is similar.

2.2 Measurable Sets and Their Properties

Definition 2.2 A set $E \subset \mathbb{R}$ is said to be measurable if for every set $A \subset \mathbb{R}$ the relation $\mu^*(A) = \mu^*(E \cap A) + \mu^*(\widetilde{E} \cap A)$ holds, where $\widetilde{E}$ is the complement of E. If E is measurable then the outer measure $\mu^*(E)$ of E is called the measure of E and is denoted by $\mu(E)$.

Theorem 2.2 *A set $E \subset \mathbb{R}$ is measurable if and only if for every set $A \subset \mathbb{R}$, $\mu^*(A) \geq \mu^*(E \cap A) + \mu^*(\widetilde{E} \cap A)$.*

Proof Suppose that the given condition hold. Then by Theorem 2.1 (iii) the equality holds and so E is measurable. Conversely, suppose that E is measurable then by definition equality holds and so the given condition is satisfied, completing the proof. $\square$

Theorem 2.3

(i) *If E is measurable then its complement $\widetilde{E}$ is also measurable.*
(ii) *If $\mu^*(E) = 0$ then E is measurable.*
(iii) *If E_1 and E_2 are measurable then $E_1 \cup E_2$ and $E_1 \cap E_2$ are also measurable.*

Proof

(i) Since E is measurable, for any set A, $\mu^*(A) = \mu^*(E \cap A) + \mu^*(\widetilde{E} \cap A)$ and since $\widetilde{\widetilde{E}} = E$, this gives $\mu^*(A) = \mu^*(\widetilde{\widetilde{E}} \cap A) + \mu^*(\widetilde{E} \cap A)$ and so $\widetilde{E}$ is measurable.
(ii) Since for every set A, $A \cap E \subset E$, $\mu^*(A \cap E) \leq \mu^*(E) = 0$. Also since $A \cap \widetilde{E} \subset A$, $\mu^*(A \cap \widetilde{E}) \leq \mu^*(A)$. Hence $\mu^*(A) \geq \mu^*(E \cap A) + \mu^*(\widetilde{E} \cap A)$. So by Theorem 2.2, the set E is measurable.
(iii) Let A be any set, Since $A \cap (E_1 \cup E_2) = (A \cap E_1) \cup (A \cap E_2 \cap \widetilde{E}_1)$, we have $\mu^*(A \cap (E_1 \cup E_2)) \leq \mu^*(A \cap E_1) + \mu^*(A \cap E_2 \cap \widetilde{E}_1)$. So

$$\mu^*(A \cap (E_1 \cup E_2)) + \mu^*(A \cap (\widetilde{E_1 \cup E_2}))$$
$$\leq \mu^*(A \cap E_1) + \mu(A \cap E_2 \cap \widetilde{E}_1) + \mu^*(A \cap \widetilde{E}_1 \cap \widetilde{E}_2)$$
$$= \mu^*(A \cap E_1) + \mu^*(A \cap \widetilde{E}_1) \text{ since } E_2 \text{ is measurable}$$
$$= \mu^*(A) \text{ since } E_1 \text{ is measurable}.$$

Hence by Theorem 2.2, $E_1 \cup E_2$ is measurable.

For the second part, since E_1 and E_2 are measurable then by (i) $\widetilde{E}_1$ and $\widetilde{E}_2$ are measurable and so by the first part $\widetilde{E}_1 \cup \widetilde{E}_2$ is measurable and hence $\widetilde{\widetilde{E}_1 \cap \widetilde{E}_2}$ is measurable and so $E_1 \cap E_2$ is measurable by (i). $\square$

Theorem 2.4 *If $\{E_n\}$ is a countable collection of measurable sets then $\bigcup_n E_n$ and $\bigcap_n E_n$ are measurable.*

2.2 Measurable Sets and Their Properties

Proof We first suppose that the sets E_n are disjoint. Let $E = \bigcup_n E_n$ and let for fixed n, $S_n = \bigcup_{i=1}^{n} E_i$ Then by repeated application of Theorem 2.3(iii) S_n is measurable. Since $S_n \subset E$, $\widetilde{E} \subset \widetilde{S}_n$ and so for any set A, $A \cap \widetilde{E} \subset A \cap \widetilde{S}_n$. Hence

$$\mu^*(A) = \mu^*(A \cap S_n) + \mu^*(A \cap \widetilde{S}_n) \geq \mu^*(A \cap S_n) + \mu^*(A \cap \widetilde{E}) \quad (2.1)$$

Since $S_i \cap \widetilde{E}_i = S_{i-1}$ for $i = 2, 3, .., n$, by the measurability of E_i

$$\mu^*(A \cap S_i) = \mu^*(A \cap S_i \cap E_i) + \mu^*(A \cap S_i \cap \widetilde{E}_i)$$
$$= \mu^*(A \cap E_i) + \mu^*(A \cap S_{i-1})$$

Putting $i = 2, 3, ..., n$ and adding

$$\mu^*(A \cap S_n) = \sum_{i=1}^{n} \mu^*(A \cap E_i) \quad (2.2)$$

From (2.1) and (2.2)

$$\mu^*(A) \geq \sum_{i=1}^{n} \mu^*(A \cap E_i) + \mu^*(A \cap \widetilde{E})$$

Since this is true for all n, by Theorem 2.1(iii)

$$\mu^*(A) \geq \sum_{i=1}^{\infty} \mu^*(A \cap E_i) + \mu^*(A \cap \widetilde{E}) \geq \mu^*(A \cap E) + \mu^*(A \cap \widetilde{E})$$

Hence E is measurable.

For the general case, let $F_1 = E_1$, $F_n = E_n \sim \bigcup_{i=1}^{n-1} E_i = E_n \cap (\bigcap_{i=1}^{n-1} \widetilde{E}_i)$ for $n \geq 2$. Then $\{F_n\}$ is a countable collection of disjoint measurable sets. So, by the above special case $\bigcup F_n$ is measurable. Since $\bigcup F_n = \bigcup E_n$, $\bigcup E_n$ is a measurable.

For the second part, since E_n are measurable, $\widetilde{E}_n$ are measurable and so by the first part $\bigcup \widetilde{E}_n$ is measurable. Since $\bigcup \widetilde{E}_n = \widetilde{\bigcap E_n}$, $\widetilde{\bigcap E_n}$ is measurable and hence $\bigcap E_n$ is measurable. □

Theorem 2.5 *Let $\mathcal{M}$ be the collection of all measurable sets. Then $\mathcal{M}$ is a σ-algebra.*

The proof follows from the definition of σ-algebra (see Sect. 1.2) and from Theorems 2.3 and 2.4.

This σ-algebra is called Lebesgue σ-algebra, since we are considering Lebesgue measurable sets only. The triplet $(\mathbb{R}, \mathcal{M}, \mu)$ is called Lebesgue measure space where $\mathcal{M}$ is the collection of all Lebesgue measurable sets and μ is the Lebesgue measure on $\mathcal{M}$.

Theorem 2.6 *If $\{E_n\}$ is a countable collection of disjoint measurable sets then*

$$\mu(\bigcup_n E_n) = \sum_n \mu(E_n).$$

Proof Since E_1 and E_2 are measurable $E_1 \cup E_2$ is measurable and so

$$\begin{aligned}
\mu(E_1 \cup E_2) &= \mu^*(E_1 \cup E_2) \\
&= \mu^*((E_1 \cup E_2) \cap E_1) + \mu^*((E_1 \cup E_2) \cap \widetilde{E_1}) \\
&= \mu^*(E_1) + \mu^*(E_2) \\
&= \mu(E_1) + \mu(E_2).
\end{aligned}$$

Repeating this argument we conclude that for any n

$$\mu(\bigcup_{i=1}^n E_i) = \sum_{i=1}^n \mu(E_i).$$

Since $\bigcup_{i=1}^n E_i \subset \bigcup_{i=1}^\infty E_i$, this gives

$$\mu(\bigcup_{i=1}^\infty E_i) = \mu^*(\bigcup_{i=1}^\infty E_i) \geq \mu^*(\bigcup_{i=1}^n E_i) = \mu(\bigcup_{i=1}^n E_i) = \sum_{i=1}^n \mu(E_i)$$

This being true for all n

$$\mu(\bigcup_{i=1}^\infty E_i) \geq \sum_{i=1}^\infty \mu(E_i).$$

Since

$$\mu(\bigcup_{i=1}^\infty E_i) = \mu^*(\bigcup_{i=1}^\infty E_i) \leq \sum_{i=1}^\infty \mu^*(E_i) = \sum_{i=1}^\infty \mu(E_i)$$

the result follows. $\square$

Corollary 2.7 *If A and B are measurable sets with $\mu(A)$ finite and $A \subset B$, then $B \sim A$ is measurable and $\mu(B \sim A) = \mu(B) - \mu(A)$.*

Proof Since $B \sim A = B \cap \widetilde{A}$, the first part follows. For the second part, since $A \subset B$, $B = A \cup (B \sim A)$ and so $\mu(B) = \mu(A) + \mu(B \sim A)$, since A and $B \sim A$ are disjoint. This gives the result, since $\mu(A)$ is finite. $\square$

Theorem 2.8 *If $\{E_n\}$ is an increasing sequence of measurable sets then $\bigcup_{n=1}^\infty E_n$ is measurable and*

2.2 Measurable Sets and Their Properties

$$\lim_{n\to\infty} \mu(E_n) = \mu(\bigcup_n E_n) = \mu(\lim_{n\to\infty} E_n).$$

Proof If $\mu(E_n) = \infty$ for some n, say for $n = k$, then since $\{E_n\}$ is increasing $\mu(E_n) = \infty$ for all $n \geq k$ and $\mu(\bigcup_{n=1}^{\infty} E_n) = \infty$ and so the result follows. So, we suppose that $\mu(E_n) < \infty$ for all n. Since $\{E_n\}$ is increasing $\bigcup_{n=1}^{\infty} E_n = E_1 \cup (E_2 \sim E_1) \cup (E_3 \sim E_2) \cup \ldots$ and so by Theorem 2.6 and Corollary 2.7

$$\mu(\bigcup_{n=1}^{\infty} E_n) = \mu(E_1) + \sum_{n=2}^{\infty} \mu(E_n \sim E_{n-1})$$

$$= \mu(E_1) + \sum_{n=2}^{\infty} (\mu(E_n) - \mu(E_{n-1}))$$

$$= \lim_{N\to\infty} \left[\mu(E_1) + \sum_{n=2}^{N} (\mu(E_n) - \mu(E_{n-1})) \right]$$

$$= \lim_{N\to\infty} \mu(E_N)$$

proving the first equality. The second equality follows from the definition of $\lim_{n\to\infty} E_n$ (See Theorem 1.5). □

Theorem 2.9 *If $\{E_n\}$ is a decreasing sequence of measurable sets such that $\mu(E_n)$ is finite for at least one n then $\bigcap_{n=1}^{\infty} E_n$ is measurable and*

$$\lim_{n\to\infty} \mu(E_n) = \mu(\bigcap_{n=1}^{\infty} E_n) = \mu(\lim_{n\to\infty} E_n)$$

Proof Let $\mu(E_k)$ be finite. Since $\lim_{n\to\infty} E_n = \bigcap_{n=k}^{\infty} E_n \subset E_k$ by Corollary 2.7 $\mu(E_k \sim \bigcap_{n=k}^{\infty} E_n) = \mu(E_k) - \mu(\bigcap_{n=k}^{\infty} E_n)$. Since $\{E_n\}$ is decreasing, $\{E_k \sim E_n\}$ is increasing with n, $n \geq k$, and hence by Theorem 2.8

$$\lim_{n\to\infty} \mu(E_k \sim E_n) = \mu(\bigcup_{n=k}^{\infty} (E_k \sim E_n)) = \mu(E_k \sim \bigcap_{n=k}^{\infty} E_n) = \mu(E_k) - \mu(\bigcap_{n=k}^{\infty} E_n).$$

But

$$\lim_{n\to\infty} \mu(E_k \sim E_n) = \lim_{n\to\infty} [\mu(E_k) - \mu(E_n)] = \mu(E_k) - \lim_{n\to\infty} \mu(E_n)$$

and since $\{E_n\}$ is decreasing, $\bigcap_{n=k}^{\infty} E_n = \bigcap_{n=1}^{\infty} E_n$ and so

$$\lim_{n\to\infty} \mu(E_n) = \mu(\bigcap_{n=k}^{\infty} E_n) = \mu(\bigcap_{n=1}^{\infty} E_n)$$

Theorem 2.10 *Every interval is measurable and its measure equals its length.*

Proof We first prove that (a, ∞) is measurable for every $a \in \mathbb{R}$. Let A be any set and let $A_1 = A \cap (a, \infty)$, $A_2 = A \cap (-\infty, a]$ Then we are to prove that $\mu^*(A) \geq \mu^*(A_1) + \mu^*(A_2)$. If $\mu^*(A) = \infty$, this is obvious. So, we suppose that $\mu^*(A) < \infty$. Let $\epsilon > 0$ be arbitrary. Then there is a countable collection of open intervals $\{I_n\}$ such that $A \subset \bigcup_n I_n$ and $\sum_n |I_n| < \mu^*(A) + \epsilon$. Let $I'_n = I_n \cap (a, \infty)$, $I''_n = I_n \cap (-\infty, a]$. Then I'_n and I''_n are intervals or empty and $I_n = I'_n \cup I''_n$. Since $A_1 \subset \cup I'_n$ and $A_2 \subset \cup I''_n$ we have $\mu^*(A_1) \leq \mu^*(\cup I'_n) \leq \sum \mu^*(I'_n) = \sum |I'_n|$ by Theorem 2.1 (v). Similarly $\mu^*(A_2) \leq \sum |I''_n|$. Hence

$$\mu^*(A_1) + \mu^*(A_2) \leq \sum |I'_n| + \sum |I''_n| = \sum (|I'_n| + |I''_n|) = \sum |I_n| < \mu^*(A) + \epsilon.$$

Since ϵ is arbitrary, $\mu^*(A_1) + \mu^*(A_2) \leq \mu^*(A)$. This shows that the interval (a, ∞) is measurable. Since the singleton set $\{a\}$ is measurable, $\{a\} \cup (a, \infty) = [a, \infty)$ is measurable. Hence their complementary sets $(-\infty, a]$ and $(-\infty, a)$ are measurable. Since a is arbitrary, the measurability of the intervals (a, b), $[a, b)$, $(a, b]$ and $[a, b]$ follows from

$$(a, b) = (-\infty, b) \cap (a, \infty), \; [a, b) = (-\infty, b) \cap [a, \infty), \; (a, b] = (-\infty, b] \cap (a, \infty)$$
$$\text{and } [a, b] = (-\infty, b] \cap [a, \infty),$$

completing the proof. □

Corollary 2.11 *All open sets, all closed sets, all F_σ—sets and all G_δ—sets are measurable and in general all Borel sets are Lebesgue measurable.*

Proof Since an open set is a countable union of open intervals, every open set is measurable. A closed set being the complement of an open set, every closed set is measurable. An F_σ—set being a countable union of closed sets and a G_δ—set being a countable intersection of open sets, the measurability of F_σ—sets and G_δ—sets follows. Finally, since the collection of all Lebesgue measurable sets is a σ-algebra $\mathcal{M}$ by Theorem 2.5 and since by the above theorem all intervals are in $\mathcal{M}$, $\mathcal{M}$ is a σ-algebra containing all open intervals and hence all open sets. Since Borel σ-algebra $\mathcal{B}$ is the smallest σ-algebra containing all open sets (Sec Definition 1.4) it follows that $\mathcal{B} \subset \mathcal{M}$. So, all Borel sets are Lebesgue measurable. □

Theorem 2.12 *(Regularity of Lebesgue outer measure) For any set A there is a G_δ set (and hence a measurable set) E such that $A \subset E$ and $\mu^*(A) = \mu(E)$.*

Proof For each n there is a countable collection of open intervals $\{I_{n,k}; k = 1, 2, ..\}$ such that $A \subset \bigcup_{k=1}^{\infty} I_{nk}$ and $\sum_{k=1}^{\infty} |I_{n,k}| < \mu^*(A) + \frac{1}{n}$. Let $E = \bigcap_{n=1}^{\infty} \bigcup_{k=1}^{\infty} I_{n,k}$. Then E is a

2.2 Measurable Sets and Their Properties

G_δ set and hence by Corollary 2.11, E is measurable. Also $A \subset E$ and so $\mu^*(A) \le \mu(E)$. Again since $E \subset \bigcup_{k=1}^{\infty} I_{n,k}$ for each n, $\mu(E) \le \sum_{k=1}^{\infty} |I_{n,k}| < \mu^*(A) + \frac{1}{n}$ for each n and so letting $n \to \infty$, $\mu(E) \le \mu^*(A)$. Hence $\mu^*(A) = \mu(E)$.

The following theorems give necessary and sufficient condition for a set to be measurable. □

Theorem 2.13 *The following statements are equivalent :*

(i) E is measurable
(ii) For every $\epsilon > 0$ there is an open set G such that $E \subset G$ and $\mu^*(G \sim E) < \epsilon$
(iii) There exists a G_δ-set G such that $E \subset G$ and $\mu^*(G \sim E) = 0$.

Proof (i) $\Rightarrow$ (ii). Let E be measurable. First suppose that $\mu(E) < \infty$. Let $\epsilon > 0$ be arbitrary. Then there is a countable collection $\{I_n\}$ of open intervals such that $E \subset \bigcup_n I_n$ and $\sum_n |I_n| < \mu(E) + \epsilon$. Let $G = \bigcup_n I_n$. Then G is an open set and $\mu(G) \le \sum_n |I_n|$ and so $\mu(G) < \mu(E) + \epsilon$. Since $E \subset G$ this gives $\mu(G \sim E) = \mu(G) - \mu(E) < \epsilon$.

Let us now suppose that $\mu(E) = \infty$. Let $\{I_n\}$ be any countable collection of open intervals such that $E \subset \bigcup_n I_n$. Let $E_n = E \cap I_n$. Then $\mu(E_n) < \infty$. So, by the first part, for every $\epsilon > 0$ there is an open set G_n such that $E_n \subset G_n$ and $\mu(G_n \sim E_n) < \frac{\epsilon}{2^n}$. Let $G = \bigcup_{n=1}^{\infty} G_n$. Then G is open and $E = \bigcup_n E_n \subset \bigcup_n G_n = G$. Also, $G \sim E \subset \bigcup_n (G_n \sim E_n)$ and hence $\mu(G \sim E) \le \sum_n \mu(G_n \sim E_n) < \sum_n \frac{\epsilon}{2^n} = \epsilon$.

(ii) $\Rightarrow$ (iii). Let (ii) be true. Then for each n there is an open set G_n such that $E \subset G_n$ and $\mu^*(G_n \sim E) < \frac{1}{n}$. Let $G = \bigcap_n G_n$. Then G is a G_δ-set and $E \subset G$. Also $G \sim E \subset G_n \sim E$ and hence $\mu^*(G \sim E) \le \mu^*(G_n \sim E) < \frac{1}{n}$. Since n is arbitrary, $\mu^*(G \sim E) = 0$.

(iii) $\Rightarrow$ (i) Let (iii) hold. Since $\mu^*(G \sim E) = 0$, by Theorem 2.3(ii) $G \sim E$ is measurable and so $E = G \sim (G \sim E)$ is measurable. This completes the proof. □

Theorem 2.14 *The following statements are equivalent :*

(i) E is measurable.
(ii) For every $\epsilon > 0$ there is a closed set F such that $F \subset E$ and $\mu^*(E \sim F) < \epsilon$
(iii) There exists an F_σ set F such that $F \subset E$ and $\mu^*(E \sim F) = 0$.

Proof The proof follows from Theorem 2.13 by taking complement. In fact, if E is measurable then its complement $\widetilde{E}$ is also measurable and if G is an open set or a G_δ-set then its complement $\widetilde{G}$ is a closed set or an F_σ-set respectively and the result follows by considering $\widetilde{E}$ and $\widetilde{G}$. □

Theorem 2.15 *Let E be any measurable set. Then there is a sequence $\{G_n\}$ of open sets such that $E \subset G_{n+1} \subset G_n$ for all n and $\mu(E) = \mu(\cap G_n)$ and there is a sequence $\{F_n\}$ of closed sets such that $F_n \subset F_{n+1} \subset E$ for all n and $\mu(E) = \mu(\cup F_n)$.*

Proof By Theorem 2.13(ii) for every n there exists an open set G_n such that $E \subset G_n$ and $\mu(G_n \sim E) < \frac{1}{n}$ i.e $\mu(G_n) < \mu(E) + \frac{1}{n}$. We may suppose that $G_{n+1} \subset G_n$ for all n. Clearly $E \subset \bigcap_{n=1}^{\infty} G_n$ and so $\mu(E) \leq \mu(\bigcap_{n=1}^{\infty} G_n)$. On the other hand $\mu(\bigcap_{n=1}^{\infty} G_n) \leq \mu(G_n) < \mu(E) + \frac{1}{n}$ for all n and hence $\mu(\bigcap_{n=1}^{\infty} G_n) \leq \mu(E)$. The proof of the second part is similar by applying Theorem 2.14(ii). $\square$

Definition 2.3 For any set E and any $\alpha \in \mathbb{R}$, define $E + \alpha = \{x + \alpha : x \in E\}$ and $\alpha E = \{\alpha x : x \in E\}$

Theorem 2.16 *For any set E and any $\alpha \in \mathbb{R}$, $\mu^*(E + \alpha) = \mu^*(E)$ and $\mu^*(\alpha E) = |\alpha| \mu^*(E)$. If E is measurable then $E + \alpha$ and αE are measurable.*

Proof We may suppose that $\mu^*(E) < \infty$. Let $\epsilon > 0$ be arbitrary. Then there is a countable collection of open intervals $\{I_n\}$ such that $E \subset \bigcup_n I_n$ and $\sum_n |I_n| < \mu^*(E) + \epsilon$. Clearly $E + \alpha \subset \bigcup_n (I_n + \alpha)$. So

$$\mu^*(E + \alpha) \leq \sum_n |I_n + \alpha| = \sum_n |I_n| < \mu^*(E) + \epsilon.$$

Since ϵ is arbitrary. $\mu^*(E + \alpha) \leq \mu^*(E)$. Also for $\epsilon > 0$ there exists a sequence of open intervals $\{J_n\}$ such that $E + \alpha \subset \bigcup_n J_n$ and $\sum_n |J_n| < \mu^*(E + \alpha) + \epsilon$. Clearly $E \subset \bigcup_n (J_n - \alpha)$ and so

$$\mu^*(E) \leq \sum_n |J_n - \alpha| = \sum_n |J_n| < \mu^*(E + \alpha) + \epsilon.$$

Since $\epsilon > 0$ is arbitrary. $\mu^*(E) \leq \mu^*(E + \alpha)$. So, the first part is proved.

For the second part, note that if $\alpha = 0$ then $\alpha E = \{0\}$ and so $\mu^*(\alpha E) = \mu^*(\{0\}) = 0 = 0.\mu^*(E)$. Hence the result is proved. So, we suppose that $\alpha \neq 0$. Let $\epsilon > 0$ be arbitrary. Then there is a countable collection of open intervals $\{I_n\}$ such that $E \subset \bigcup_n I_n$ and $\sum_n |I_n| < \mu^*(E) + \epsilon$. So, $\alpha E \subset \bigcup_n (\alpha I_n)$ and hence

$$\mu^*(\alpha E) \leq \sum_n |\alpha I_n| = |\alpha| \sum_n |I_n| < |\alpha|(\mu^*(E) + \epsilon).$$

Since ϵ is arbitrary, $\mu^*(\alpha E) \leq |\alpha| \mu^*(E)$. Similarly for every $\epsilon > 0$ there is a countable collection of open intervals $\{J_n\}$ such that $\alpha E \subset \bigcup_n J_n$ and $\sum_n |J_n| < \mu^*(\alpha E) + \epsilon$. Since $E \subset \bigcup_n (\frac{1}{\alpha} J_n)$. $\mu^*(E) \leq \sum_n |\frac{1}{\alpha} J_n| = \frac{1}{|\alpha|} \sum_n |J_n| < \frac{1}{|\alpha|}(\mu^*(\alpha E) + \epsilon)$.

2.2 Measurable Sets and Their Properties

So $|\alpha|\mu^*(E) \le \mu^*(\alpha E) + \epsilon$ and since ϵ is arbitrary, $|\alpha|\mu^*(E) \le \mu^*(\alpha E)$. So the second part is proved.

Finally, suppose that E is measurable. Let $\epsilon > 0$ be arbitrary. Then there is a countable collection of open intervals $\{I_n\}$ such that $E \subset \bigcup_n I_n$ and $\sum_n |I_n| < \mu(E) + \epsilon$. So, $\bigcup_n I_n \sim E$ being measurable.

$$\mu(\bigcup_n I_n \sim E) = \mu(\bigcup_n I_n) - \mu(E) \le \sum_m |I_n| - \mu(E) < \epsilon \qquad (2.3)$$

Therefore since $\cup I_n \sim E$ is measurable, by the first part.

$$\mu^*((\bigcup_n I_n \sim E) + \alpha) = \mu^*(\bigcup_n I_n \sim E) = \mu(\bigcup_n I_n \sim E)$$

and so by (2.3), $\mu^*((\bigcup_n I_n \sim E) + \alpha) < \epsilon$. Since

$$\bigcup_n (I_n + \alpha) \sim (E + \alpha) = (\bigcup_n I_n \sim E) + \alpha, \ \mu^*(\bigcup_n (I_n + \alpha) \sim (E + \alpha)) < \epsilon.$$

Since ϵ is arbitrary, $\mu^*(\bigcup_n (I_n + \alpha) \sim (E + \alpha)) = 0$ and therefore $\bigcup_n (I_n + \alpha) \sim (E + \alpha)$ is measurable. Since $\bigcup_n (I_n + \alpha)$ is measurable and $E + \alpha \subset \bigcup_n (I_n + \alpha)$, $E + \alpha$ is measurable. For the measurability of αE note that $\bigcup_n (\alpha I_n \sim \alpha E) = \alpha(\bigcup_n I \sim E)$. Therefore by the above

$$\mu^*(\bigcup_n (\alpha I_n \sim \alpha E)) = \mu^*(\alpha(\bigcup_n I \sim E)) = |\alpha|\mu^*((\bigcup_n I \sim E)).$$

Since E is measurable, by (2.3)

$$\mu^*(\bigcup_n (\alpha I_n \sim \alpha E)) = |\alpha|\mu^*((\bigcup_n I_n \sim E)) < |\alpha|\epsilon.$$

Since ϵ is arbitrary, $\mu^*(\bigcup_n (\alpha I_n \sim \alpha E)) = 0$ and so $\bigcup_n (\alpha I_n \sim \alpha E)$ is measurable. Since $\bigcup_n (\alpha I_n \sim \alpha E) = \bigcup_n (\alpha I_n) \sim \alpha E$ and $\bigcup_n (\alpha I_n)$ is measurable, αE is measurable. $\square$

Theorem 2.17 *Let $\mu^*(E) < \infty$. Then E is measurable if and only if for every $\epsilon > 0$ there is a finite collection of open intervals $\{I_k : k = 1, 2, .., n\}$ such that the set $S = \bigcup_{k=1}^n I_k$ satisfies $\mu^*(S \Delta E) < \epsilon$.*

Proof Let E be measurable and let $\epsilon > 0$ be arbitrary. There exists a countable collection $\{I_k\}$ of open intervals such that

$$E \subset \bigcup_{k=1}^\infty I_k \text{ and } \sum_{k=1}^\infty |I_k| < \mu(E) + \tfrac{\epsilon}{2}.$$

Since $\sum_{k=1}^{\infty} |I_k|$ is convergent there is n such that $\sum_{k=n+1}^{\infty} |I_k| < \frac{\epsilon}{2}$.

$$\text{Let } S = \bigcup_{k=1}^{n} I_k \text{ and } G = \bigcup_{k=1}^{\infty} I_k.$$

Then $E \sim S \subset \bigcup_{k=n+1}^{\infty} I_n$ and so

$$\mu(E \sim S) \leq \sum_{k=n+1}^{\infty} |I_k| < \frac{\epsilon}{2}.$$

Also since $S \sim E \subset G \sim E$, we have

$$\mu(S \sim E) \leq \mu(G \sim E) = \mu(G) - \mu(E) \leq \sum_{k=1}^{\infty} |I_k| - \mu(E) < \frac{\epsilon}{2}.$$

Hence

$$\mu(E \triangle S) = \mu((E \sim S) \cup (S \sim E)) \leq \mu(E \sim S) + \mu(S \sim E) < \epsilon.$$

Conversely, suppose that the condition holds. Let $\epsilon > 0$ be arbitrary. By hypothesis these is a finite collection of open intervals $\{I_k : k = 1, 2, ..., n\}$ such that $S = \bigcup_{k=1}^{n} I_k$ satisfies $\mu^*(S \triangle E) < \epsilon$. Then since $\mu^*(E \sim S) \leq \mu^*(S \triangle E) < \epsilon$ there exits an open set G such that $E \sim S \subset G$ and $\mu(G) < \mu^*(E \sim S) + \epsilon < 2\epsilon$. The set $S \cup G$ is open and if $T = S \cup G$ then since

$$E = (E \cap S) \cup (E \sim S) \subset S \cup G = T, \; T \sim E \subset (S \sim E) \cup (G \sim E)$$

and so

$$\mu^*(T \sim E) \leq \mu^*(S \sim E) + \mu^*(G \sim E).$$

Therefore since

$$\mu^*(S \sim E) \leq \mu^*(S \triangle E) < \epsilon, \quad \mu^*(T \sim E) < \epsilon + \mu(G) < 3\epsilon.$$

Since S and G are open, T is open and so by Theorem 2.13(ii) E is measurable. □

Theorem 2.18 *Let A and B be such that $A \subset B$.*

(i) *If A is measurable and $\mu^*(B)$ is finite and $\mu^*(A) = \mu^*(B)$ then B is measurable.*
(ii) *If B is measurable and $\mu(B)$ is finite then A is measurable if and only if $\mu(B) = \mu^*(A) + \mu^*(B \sim A)$.*

Proof

(i) Since A is measurable
$\mu^*(B) = \mu^*(B \cap A) + \mu^*(B \sim A) = \mu(A) + \mu^*(B \sim A)$. Since $\mu(A) = \mu^*(B)$. $\mu^*(B \sim A) = 0$ so $B \sim A$ is measurable and therefore $A \cup (B \sim A) = B$ is measurable.

(ii) If A is measurable then
$$\mu(B) = \mu(B \cap A) + \mu(B \sim A) = \mu(A) + \mu(B \sim A).$$
Conversely, suppose that the condition holds. By Theorem 2.12 there is a measurable set E_1 such that $A \subset E_1$ and $\mu^*(A) = \mu(E_1)$. Let $E = E_1 \cap B$. Then E is measurable, $A \subset E$, $B \sim E \subset B \sim A$. and so $\mu(E) \geq \mu^*(A)$. Also $\mu^*(A) = \mu(E_1) \geq \mu(E)$ and hence $\mu(E) = \mu^*(A)$. Since E is measurable, by the given condition.
$$\mu^*(A) + \mu^*(B \sim A) = \mu(B) = \mu(B \cap E) + \mu(B \sim E) = \mu(E) + \mu(B \sim E)$$
and hence $\mu^*(B \sim A) = \mu(B \sim E)$. Since $B \sim E \subset B \sim A$ and $B \sim E$ is measurable and $\mu^*(B \sim A)$ is finite, applying part (i), $B \sim A$ is measurable. And so $A = B \sim (B \sim A)$ is measurable, completing the proof. □

2.3 Nonmeasurable Sets

To construct nonmeasurable sets we have to apply the axiom of choice which is as follows:

Axiom of choice, Given any non-void family of non-void sets $\mathcal{A} = \{A_i : i \in I\}$ it is possible to choose exactly one element from each set A_i.
In other words, there exists a function $c : \mathcal{A} \to \bigcup_{i \in I} A_i$ such that $C(A_i) \in A_i$ for each $i \in I$. The function c is called the choice function.

It may appear that this is obvious, but it cannot be proved independently. For this reason, it is called axiom of choice.

Theorem 2.19 *Every measurable set of positive measures contains a nonmeasurable subset.*

Proof Let E be any set of positive measures. We may suppose that E is bounded and that by Theorem 2.16 $E \subset [0, 1]$. For any two points $x, y \in E$, we say x is equivalent to y if and only if $x - y$ is a rational number. Clearly this gives an equivalence relation between points in E and so the set E is partitioned into distinct disjoint equivalence classes $\{E_\alpha : \alpha \in I, I$ is an index set$\}$ such that any two points x, y in a class E_α are equivalent and if $x \in E_\alpha$ and $y \in E_\beta$ where $\alpha, \beta \in I, \alpha \neq \beta$, then x is not equivalent to y. So, $E = \bigcup_{\alpha \in I} E_\alpha$. By axiom of choice, we choose exactly one point from each set E_α and form the set A by these points. We show that A is nonmeasurable.

Let $\{r_n\}$ be an enumeration of the set of rationals in $[-1, 1]$. For each n let $A_n = A + r_n$. Then the sets A_n are disjoint. For if $y \in A_n \cap A_m$, $n \neq m$, then since $y \in A_n$, $y = x + r_n$ for some $x \in A$ and by definition of A there is a point x_α in

E_α such that $x = x_\alpha$ and so $y = x_\alpha + r_n$. Similarly, since $y \in A_m$ there is $x_\beta \in E_\beta$ such that $y = x_\beta + r_m$. So, $x_\alpha - x_\beta = r_m - r_n$ which is a rational number. Showing that x_α and x_β are equivalent and hence they belong to the same class, which is a contradiction., since E_α and E_β are disjoint. Therefore $A_n \cap A_m = \emptyset$ for $n \neq m$. If possible, suppose that A is measurable. Then by Theorem 2.16, A_n is measurable for each n and $\mu(A) = \mu(A_n)$. Since $A_n \cap A_m = \emptyset$ for $n \neq m$,

$$\mu\left(\bigcup_{n=1}^{\infty} A_n\right) = \sum_{n=1}^{\infty} \mu(A_n) = \sum_{n=1}^{\infty} \mu(A). \tag{2.4}$$

Also $E \subset \bigcup_{n=1}^{\infty} A_n \subset [-1, 2]$. For, if $x \in E$ then $x \in E_\alpha$ for some $\alpha \in I$. If x_α is the member of E_α which is chosen to form A then x is equivalent to x_α and so $x - x_\alpha$ is a rational number. Let $x - x_\alpha = r_n$ then $x = x_\alpha + r_n$ and so $x \in A_n$ and therefore $E \subset \bigcup_{n=1}^{\infty} A_n$. The second inclusion is obvious. So

$$\mu(E) \leq \mu\left(\bigcup_{n=1}^{\infty} A_n\right) \leq 3 \tag{2.5}$$

From (2.4) and (2.5) $\mu(E) \leq \sum_{n=1}^{\infty} \mu(A) \leq 3$. Now if $\mu(A) = 0$ then $\mu(E) = 0$ which is a contradiction since E is of positive measure and if $\mu(A) > 0$ then $\infty \leq 3$ which is also a contradiction. This shows that A is not measurable. $\square$

Corollary 2.20 *Every set E of positive measure contains two disjoint nonmeasurable subsets E_1 and E_2 such that $E_1 \cup E_2 = E$.*

Proof By the above Theorem, the set E contains a nonmeasurable set E_1. Let $E_2 = E \sim E_1$. Then E_1 and E_2 are disjoint nonmeasurable sets and $E_1 \cup E_2 = E$. $\square$

Example 2.3 If F is measurable and $\mu^*(F \Delta G) = 0$ then G is measurable.

Since $F \Delta G = (F \sim G) \cup (G \sim F)$, $\mu^*(F \sim G) \leq \mu^*(F \Delta G) = 0$ and $\mu^*(G \sim F) \leq \mu^*(F \Delta G) = 0$. Hence by Theorem 2.3(ii) all of $F \Delta G$, $F \sim G$ and $G \sim F$ are measurable. Since $F \cap G = F \sim (F \sim G)$, $F \cap G$ is measurable. Since $G = (F \cap G) \cup (G \sim F)$, G is measurable.

2.4 Further Properties of Measurable Sets

Theorem 2.21 *For any set E, $\mu^*(E) = \inf\{\mu(T) : E \subset T; T \text{ is measurable}\}$ where the infimum is taken over the collection of all measurable super sets of E.*

2.4 Further Properties of Measurable Sets

Proof Let $\{I_n\}$ be any countable collection of open intervals such that $E \subset I_n$. Since by Theorems 2.10 and 2.4, $\cup I_n$ is measurable, considering the collection of all such countable collections $\{I_n\}$ and the collection of all measurable sets T such that $E \subset T$ we have

$$\{\mu(\bigcup I_n) : E \subset \bigcup I_n\} \subset \{\mu(T) : E \subset T; T \text{ is measurable}\}.$$

Hence

$$\inf\{\mu(\bigcup I_n) : E \subset \bigcup I_n\} \geq \inf\{\mu(T) : E \subset T; T \text{ is measurable }\}.$$

Therefore

$$\mu^*(E) = \inf\{\sum |I_n| : E \subset \cup I_n\} \geq \inf\{\mu(\cup I_n) : E \subset \cup I_n\}$$

$$\geq \inf\{\mu(T) : E \subset T; \ T \text{ is measurable}\}. \quad (2.6)$$

Also, if T is any measurable set such that $E \subset T$ then $\mu^*(E) \leq \mu(T)$ and hence

$$\mu^*(E) \leq \inf\{\mu(T) : E \subset T; T \text{ is measurable}\}. \quad (2.7)$$

The relations (2.6) and (2.7) complete the proof □

Remark Starting from an outer measure of sets, we get measurable sets and a measure. The above theorem shows that if measurable sets and a measure are given then outer measure of sets can be defined. In a latter chapter this will be discussed in detail.

Theorem 2.22 *If A is any closed set of positive measure then there is $\delta > 0$ such that the interval $(-\delta, \delta)$ is contained in the difference set $\{x - y : x, y \in A\}$.*

Proof We may suppose that $A \subset [0, 1]$. For each n let $G_n = \bigcup_{a \in A} (a - \frac{1}{n}, a + \frac{1}{n})$.

Then each G_n is open. Also $A \subset G_n$ for each n and $A = \overline{A} = \bigcap_{n=1}^{\infty} G_n$, where $\overline{A}$ is the closure of A. Since $G_{n+1} \subset G_n$, by Theorem 2.9 $\mu(A) = \lim_{n \to \infty} \mu(G_n)$. Let k be a positive integer such that $\mu(G_k) < \frac{3}{2}\mu(A)$. Choose δ such that $0 < \delta < \frac{1}{k}$. Let $z \in (-\delta, \delta)$ and $B = A - z$. So, if $\xi \in B$ then $\xi = a - z$ for some $a \in A$. Hence $a - \frac{1}{k} < a - \delta < a - z = \xi < a + \delta < a + \frac{1}{k}$ and so $\xi \in G_k$. Therefore $B \subset G_k$. Also by Theorem 2.16 $\mu(A) = \mu(B)$. Hence

$$\mu(G_k \sim (A \cap B)) \leq \mu(G_k \sim A) + \mu(G_k \sim B)$$
$$= \mu(G_k) - \mu(A) + \mu(G_k) - \mu(B)$$
$$= 2\mu(G_k) - 2\mu(A)$$
$$< 2\mu(G_k) - 2.\frac{2}{3}\mu(G_k) = \frac{2}{3}\mu(G_k).$$

Therefore $A \cap B \neq \emptyset$. Let $y \in A \cap B$. Since $y \in B$ there is $x \in A$ such that $y = x - z$. So, $z = x - y \in \{x - y : x, y \in A\}$. Hence $(-\delta, \delta) \subset \{x - y : x, y \in A\}$. □

Theorem 2.23 *If A is any measurable set of positive measure then there is $\delta > 0$ such that the interval $(-\delta, \delta)$ is contained in the difference set $\{x - y : x, y \in A\}$.*

Proof Let A be any measurable set of positive measure and let $0 < \epsilon < \mu(A)$. Then by Theorem 2.14 (ii) there is a closed set F such that $F \subset A$ and $\mu(A \sim F) < \epsilon$. So $\mu(A) - \mu(F) < \mu(A)$. Hence $\mu(F) > 0$. So by Theorem 2.22 there is $\delta > 0$ such that the interval $(-\delta, \delta)$ is contained in $\{x - y : x, y \in F\}$. Since $\{x - y : x, y \in F\} \subset \{x - y : x, y \in A\}$ the proof follows. □

Theorem 2.24 *Let E be any measurable set such that $\mu(E) < \infty$. Then for each real number $\alpha, 0 < \alpha < 1$, there is a measurable set $A \subset E$ such that $\mu(A) = \alpha\mu(E)$.*

Proof If $\mu(E) = 0$ this is obvious. So we suppose $\mu(E) > 0$. For each $x \geq 0$ let $f(x) = \mu(E \cap (-x, x))$. So, if $0 \leq x_1 < x_2$ then $f(x_2) - f(x_1) = \mu(E \cap (-x_2, -x_1)) + \mu(E \cap (x_1, x_2))$ and so $0 \leq f(x_2) - f(x_1) \leq 2(x_2 - x_1)$. Therefore f is a non-negative non-decreasing and uniformly continuous function on $[0, \infty)$ such that $f(0) = 0$. Also if $n < x < n + 1$ and $E_n = E \cap (-n, n)$ then $E_n \subset E \cap (-x, x) \subset E_{n+1}$ and so $\mu(E_n) \leq \mu(E \cap (-x, x)) \leq \mu(E_{n+1})$. So by Theorem 2.8

$$\lim_{x \to \infty} f(x) = \lim_{x \to \infty} \mu(E \cap (-x, x)) = \lim_{n \to \infty} \mu(E_n) = \mu(\lim_{n \to \infty} E_n) = \mu(E).$$

So, if $0 < \alpha < 1$ then $0 < \alpha\mu(E) < \mu(E)$ and hence by intermediate value theorem there exists a point z, $0 < z < \infty$ such that $f(z) = \alpha\mu(E)$. So putting $A = E \cap (-z, z)$, $\mu(A) = \alpha\mu(E)$. □

Remark From the above theorem it follows that if E is a measurable set such that $0 < \mu(E) < \infty$ then as x increases from 0 to ∞, $\mu(E \cap (-x, x))$ increases continuously from 0 to $\mu(E)$.

Theorem 2.25 *The measure μ defined on the class of all measurable sets of finite measure is a continuous completely additive set function.*

Proof From Theorems 2.5 and 2.6 it follows that μ is completely additive and from Theorem 1.12 the continuity of μ follows. □

Theorem 2.26 *Let A be a nonmeasurable set. If $\mu^*(A) < \infty$ then there exists $\delta > 0$ such that for every measurable set $E \subset A$, $\mu(E) < \mu^*(A) - \delta$.*

Proof Suppose that there is no such $\delta > 0$. Then for each positive integer n there is a measurable set E_n such that $E_n \subset A$ and $\mu(E_n) \geq \mu^*(A) - \frac{1}{n}$. Let $E = \bigcup_{n=1}^{\infty} E_n$. Then E is measurable and so $\mu(E) \geq \mu(E_n) \geq \mu^*(A) - \frac{1}{n}$ and hence $\mu(E) \geq \mu^*(A)$. So, $\mu(E) = \mu^*(A)$, since $E \subset A$. Therefore by Theorem 2.18 (i), A is measurable which is a contradiction. □

Example 2.4 If $\{E_n\}$ is a sequence of measurable sets then

(i) $\mu(\liminf_{n \to \infty} E_n) \leq \liminf_{n \to \infty} \mu(E_n)$, and

(ii) $\mu(\limsup_{n\to\infty} E_n) \geq \limsup_{n\to\infty} \mu(E_n)$, provided $\bigcup_{k=n}^{\infty} E_k$ has finite measure for some n. For (i) choose n and fix it. Then since $\bigcap_{k=n}^{\infty} E_k \subset E_n$, $\mu(\bigcap_{k=n}^{\infty} E_k) \leq \mu(E_n)$. Now letting $n \to \infty$,

$$\liminf_{n\to\infty} \mu(\bigcap_{k=n}^{\infty} E_k) \leq \liminf_{n\to\infty} \mu(E_n)$$

The sequence $\{\bigcap_{k=n}^{\infty} E_k\}$ is increasing and so by Theorem 2.8,

$$\lim_{n\to\infty} \mu(\bigcap_{k=n}^{\infty} E_k) = \mu(\bigcup_{n=1}^{\infty}\bigcap_{k=n}^{\infty} E_k) = \mu(\liminf_{n\to\infty} E_n)$$

completing the proof if (i). The proof if (ii) is similarly obtained by applying Theorem 2.9.

2.5 Vitali's Covering

Definition 2.4 Let $\mathcal{M}$ be a collection of intervals and E be any set. Then E is said to be covered by $\mathcal{M}$ in the sense of Vitali if for each $x \in E$ and each $\epsilon > 0$ there is an interval $I \in \mathcal{M}$ of length less than ϵ such that $x \in I$.

Theorem 2.27 (Vitali's Covering Theorem) *Let E be a set of finite outer measure and let $\mathcal{M}$ be a collection of intervals which covers E in the since of Vitali. Then there is a countable sub-collection of $\mathcal{M}$ of pairwise disjoint intervals $\{I_1, I_2, ...I_n, ...\}$ such that for every $\epsilon > 0$ there is N such that*

$$\mu^*(E \sim \bigcup_{n=1}^{N} I_n) < \epsilon$$

where μ^* is Lebesgue outer measure.

Proof We may suppose that all the intervals of $\mathcal{M}$ are closed. Let O be an open set of finite measure such that $E \subset O$. Let $\mathcal{M}_0$ be the collection of all intervals of $\mathcal{M}$ which lie completely within O. Clearly $\mathcal{M}_0$ also covers E in the sense of Vitali. We choose a countable collection of disjoint intervals from $\mathcal{M}_0$ as follows :
We choose an intervals say, I_1 from $\mathcal{M}_0$ such that I_1 contains some point of E. If $E \subset I_1$ then the proof is complete. Otherwise, we choose the intervals by induction. Suppose that the intervals

$$I_1, I_2, ..., I_n \tag{2.8}$$

are chosen and that they are pairwise disjoint. If $E \subset \bigcup_{k=1}^{n} I_k$ the proof is complete. If

$$E \sim \bigcup_{k=1}^{n} I_k \neq \emptyset \tag{2.9}$$

put $F_n = \bigcup_{k=1}^{n} I_n$, $G_n = O \sim F_n$. By (2.9) there exits intervals in $\mathcal{M}_0$ which are contained in G_n. Let d_n be the supremum of the lengths of the intervals in $\mathcal{M}_0$ which are contained in G_n. Then $d_n < \mu(O) < \infty$. So, there is an interval in $\mathcal{M}_0$ which is contained in G_n and whose length is greater then $\frac{1}{2}d_n$. Call this interval I_{n+1}. It is clear that I_{n+1} is disjoint from all the intervals in (2.8).

If the process of choosing these intervals terminates after a finite number of steps the proof is complete. Otherwise, we get the countable collection of pairwise disjoint intervals

$$I_1, I_2, ..., I_n, ... \tag{2.10}$$

For every n choose a closed interval J_n having the middle point as that of I_n and length of J_n is five times the length of I_n. Since $I'_n s$ in (2.10) are pairwise disjoint and are contained in the open set O

$$\sum \mu(I_n) \leq \mu(O) < \infty \tag{2.11}$$

Hence

$$\sum \mu(J_n) \leq 5\mu(O) < \infty.$$

So $\sum \mu(J_n)$ is convergent. Therefore given $\epsilon > 0$ there is N such that

$$\sum_{n=N+1}^{\infty} \mu(J_n) < \epsilon \tag{2.12}$$

We shall show that

$$E \sim \bigcup_{n=1}^{N} I_n \subset \bigcup_{n=N+1}^{\infty} J_n \tag{2.13}$$

Let $x \in E \sim \bigcup_{n=1}^{N} I_n$. Then $x \in G_N$. Since G_N is open, there is an interval $I \in \mathcal{M}_0$ such that $x \in I \subset G_N$. But I cannot be a subset of G_n for all n. For, in that case $\mu(I) \leq d_n < 2\mu(I_{n+1})$ for all n and hence by (2.11), $\mu(I) = 0$, which is a contradiction. Hence there is at least one n for which $I \not\subset G_n$ i.e.

$$I \cap F_n \neq \emptyset \tag{2.14}$$

Let n_o be the smallest integer for which (2.14) holds. Since $F_1 \subset F_2 \subset ...$ and $I \cap F_N = \emptyset, n_o > N$. Since $I \cap F_{n_o-1} = \emptyset$, from (2.14)

$$I \cap I_{n_o} \neq \emptyset. \tag{2.15}$$

2.5 Vitali's Covering

Since $I \subset G_{n_0-1}$
$$\mu(I) \leq d_{n_0-1} < 2\mu(I_{n_0}) \tag{2.16}$$

From (2.15) and (2.16) $I \subset J_{n_0}$ and hence $I \subset \bigcup_{n=N+1}^{\infty} J_n$ i.e $x \in \bigcup_{n=N+1}^{\infty} J_n$. So (2.13) is established. Hence from (2.13) and (2.12)

$$\mu^*(E \sim \bigcup_{n=1}^{N} I_n) \leq \sum_{n=N+1}^{\infty} \mu(J_n) < \epsilon,$$

completing the proof. □

Corollary 2.28 *Under the hypothesis of the above theorem there is a countable sub-collection of $\mathcal{M}$ of pairwise disjoint intervals $\{I_1, I_2, ..., I_n, ...\}$ such that*

$$\mu^*(E \sim \bigcup_{n=1}^{\infty} I_n) = 0.$$

This follows from Theorem 2.27.

Corollary 2.29 *Let $\{I_\alpha : \alpha \in \mathcal{A}\}$ be any collection of non-degenerate intervals and let $E = \bigcup_{\alpha \in \mathcal{A}} I_\alpha$. Then E is measurable.*

Proof We first suppose that E is bounded. Let $\mathcal{M} = \{I : I$ is an interval such that $I \subset I_\alpha$ for some $\alpha \in \mathcal{A}\}$. Then $\mathcal{M}$ is a Vitali cover of E. So by Corollary 2.28 there is a countable sub-collection of $\mathcal{M}$ of pairwise disjoint intervals $\{I_1, I_2, ...I_n...\}$ such that

$$\mu^*(E \sim \bigcup_{n=1}^{\infty} I_n) = 0.$$

So, by Theorem 2.3 (ii), $E \sim \bigcup_{n=1}^{\infty} I_n$ is measurable. Since

$$E = (E \sim \bigcup_{n=1}^{\infty} I_n) \cup (\bigcup_{n=1}^{\infty} I_n)$$

and since $\bigcup_{n=1}^{\infty} I_n$ is measurable by Theorems 2.4 and 2.10, the set E is measurable.

For the general case, for any positive integer k let $E_k = E \cap (-k, k)$. Then $E_k = \bigcup_{\alpha \in \mathcal{A}} (I_\alpha \cap (-k, k))$. So, by the above E_k is measurable. Since $E = \bigcup_{k=1}^{\infty} E_k$, the result follows. □

2.6 Exercises

1. Show that the set of rational numbers in [0, 1] is of outer measure zero. Find the outer measure of the set of all irrational numbers in [0, 1].
2. If A is the set of all rational numbers in [0, 1] and B is the set of all irrational numbers in [0, 1], find $\mu^*(A \triangle B)$.
3. Let $I_n = (\frac{1}{2n-1}, \frac{1}{2n})$ and let $A = \bigcup_{n=1}^{\infty} I_n$. Find $\mu^*(A)$.
4. If A and B are two sets such that $\sup A < \inf B$ then show that $\mu^*(A \bigcup B) = \mu^*(A) + \mu^*(B)$. Is the condition $\sup A < \inf B$ necessary for this result?
5. Show that for any set E and every $\epsilon > 0$ there is an open set G such that $E \subset G$ and $\mu(G) < \mu^*(E) + \epsilon$.
6. Show that every countable set is measurable and is of measure 0.
7. If $\mu^*(A) = 0$ show that $\mu^*(A \cup B) = \mu^*(B)$ for any set B.
8. If $\mu^*(A) = 0$ show that A cannot have an interior point.
9. For every bounded measurable set E prove that there is an F_σ set F and a G_δ set G such that $F \subset E \subset G$ and $\mu(F) = \mu(E) = \mu(G)$.
10. If E_1 and E_2 are measurable subsets of [0, 1] and $\mu(E_1) = 1$ prove that $\mu(E_1 \cap E_2) = \mu(E_2)$.
11. If $\{E_n\}$ is an increasing sequence of measurable sets and $E = \bigcup_{n=1}^{\infty} E_n$ then show that for every $\epsilon > 0$ there is n such that $\mu(E) < \mu(E_n) + \epsilon$.
12. If E and F are measurable sets then prove that $\mu(E) + \mu(F) = \mu(E \bigcup F) + \mu(E \bigcap F)$.
13. If F is measurable and $\mu^*(F \triangle G) = 0$ then show that G is measurable.
14. Let f be defined on [0, 1] by $f(0) = 0$, $f(x) = x \sin \frac{1}{x}$ for $x > 0$. Find the measure of the set $\{x : f(x) \geq 0\}$.
15. Let g be defined on [0, 1] by $g(0) = 0$, $g(x) = \frac{1}{x} \cos \frac{1}{x}$ for $x > 0$. Find the measure of the set $\{x : g(x) \geq 0\}$.
16. Show that there exist sets of Lebesgue measure zero which is not a Borel set. [Hint. Apply Theorem 2.12] [full answer is the consequence of Theorem 4.19]
17. Show that for every Lebesgue measurable set E there is a Borel set F such that $F \subset E$ and $\mu(F) = \mu(E)$ and there is a Borel set G such that $E \subset G$ and $\mu(E) = \mu(G)$.

Chapter 3
Measurable Functions

Measurable sets being introduced in the previous chapter, and the notion of measurable functions is now introduced. Before defining measurable function we prove

Theorem 3.1 *Let $E \subset \mathbb{R}$ be a measurable set and let $f : E \to \overline{\mathbb{R}}$. Then the following are equivalent:*

(i) *the set $\{x \in E : f(x) > r\}$ is measurable for each r*
(ii) *the set $\{x \in E : f(x) \geq r\}$ is measurable for each r*
(iii) *the set $\{x \in E : f(x) < r\}$ is measurable for each r*
(iv) *the set $\{x \in E : f(x) \leq r\}$ is measurable for each r*

where $\overline{\mathbb{R}}$ is the extended real number system, i.e. $\overline{\mathbb{R}} = \mathbb{R} \cup \{\infty, -\infty\}$.

Proof We show that $(i) \implies (ii) \implies (iii) \implies (iv) \implies (i)$. Suppose that (i) holds. Then since

$$\{x \in E : f(x) \geq r\} = \bigcap_{n=1}^{\infty} \{x \in E : f(x) > r - \frac{1}{n}\},$$

by condition (i) and Theorem 2.4, the right-hand side is measurable and so the left-hand side is measurable, proving (ii). Next suppose that (ii) holds. Then by (ii) and Theorem 2.3(i), (iii) holds. Now suppose that (iii) holds. Since

$$\{x \in E : f(x) \leq r\} = \bigcap_{n=1}^{\infty} \{x \in E : f(x) < r + \frac{1}{n}\}$$

by (iii) and Theorem 2.4 , (iv) holds. Finally Suppose (iv) holds, then by taking complement and using Theorem 2.3 (i) follows. □

3.1 Measurable Functions

Now we come to the definition of measurable functions.

Definition 3.1 Let E be a measurable subset of $\mathbb{R}$ and f be extended real-valued function on E. Then f is said to be measurable if the set $\{x \in E : f(x) > r\}$ is measurable for each $r \in \mathbb{R}$.

So by Theorem 3.1 f is measurable if any one of (i), (ii), (iii) and (iv) holds.

Note In this chapter measurable function means Lebesgue measurable function. But this definition of measurable function and all properties of measurable function which will be proved subsequently, will suffice for all other measures which will be introduced latter. In line with Definition 3.1 we introduce

Definition 3.2 The function $f : E \to \overline{\mathbb{R}}$ is said to be Borel measurable or a Borel function if $\{x : x \in E; f(x) > r\}$ is a Borel set for all $r \in \mathbb{R}$. (Note that Borel ses are also called Borel measurable sets).

3.2 Properties of Measurable Function

Theorem 3.2 If $f : E \to \overline{\mathbb{R}}$ is measurable then the set $\{x \in E : f(x) = \alpha\}$ is measurable for each $\alpha \in \overline{\mathbb{R}}$.

Proof If α is finite then since $\{x \in E : f(x) = \alpha\} = \{x \in E : f(x) \geq \alpha\} \cap \{x \in E : f(x) \leq \alpha\}$ the result follows from (ii) and (iv) of Theorem 3.1 and from the definition of measurability of f. Also since

$$\{x \in E : f(x) = \infty\} = \bigcap_{n=1}^{\infty} \{x \in E : f(x) > n\} \text{ and}$$
$$\{x \in E : f(x) = -\infty\} = \bigcap_{n=1}^{\infty} \{x \in E : f(x) < -n\}$$

the result follows for $\alpha = \infty$ and $\alpha = -\infty$. $\square$

Theorem 3.3 *Every function defined on a set of measure zero is measurable and every constant function defined on a measurable set is measurable.*

The proof is easy

Theorem 3.4 *Let E be any set. The characteristic function $\chi_E : \mathbb{R} \to \mathbb{R}$ is measurable if and only if E is measurable.*

Proof Let a be any real member. Then since

$$\{x : \chi_E(x) > a\} = \begin{cases} \emptyset, & \text{if } a \geq 1 \\ E, & \text{if } 0 \leq a < 1 \\ \mathbb{R} & \text{if } a < 0 \end{cases}$$

3.2 Properties of Measurable Function

if E is measurable then $\{x : \chi_E(x) > a\}$ is measurable and so χ_E is measurable and conversely, if χ_E is measurable then $\{x : \chi_E(x) > \frac{1}{2}\}$ is measurable which gives the result since $\{x : \chi_E(x) > \frac{1}{2}\} = E$. □

Theorem 3.5 *Let $f : E \to \mathbb{R}$, where E is measurable. Then f measurable if and only if the inverse image $f^{-1}(B)$ of every Borel set B is measurable.*

Proof Suppose that f is measurable. Let C be the class of all subsets A of $\mathbb{R}$ such that the inverse image $f^{-1}(A)$ of A is measurable. Since for all $A \subset \mathbb{R}$, $f^{-1}(\sim A) = \sim f^{-1}(A)$, where $\sim$ denotes complement and since for any countable collection $\{A_i\}$ of subsets of $\mathbb{R}$, $f^{-1}(\cup A_i) = \cup f^{-1}(A_i)$, C is a σ - algebra. Let (a, b) be any open interval. Let $E_1 = \{x \in E : f(x) \leq a\}$, $E_2 = \{x \in E : a < f(x) < b\}$, $E_3 = \{x \in E : f(x) \geq b\}$. Then since f is measurable E_1 and E_3 are measurable. Since $E_2 = E \sim E_1 \sim E_3$, the set E_2 is measurable and so $(a, b) \in C$. Thus C contains all open intervals. Since the Borel σ - algebra $\mathcal{B}$ is the smallest σ - algebra containing all open intervals, $\mathcal{B} \subset C$. So, if B is a Borel set then $B \in C$ and hence $f^{-1}(B)$ is measurable. Conversely, suppose that $f^{-1}(B)$ is measurable for every Borel set B. Then since (a, ∞) is a Borel set for every $a \in \mathbb{R}$, the set $\{x \in E : f(x) > a\}$ is measurable for every a and so, f is measurable. □

Note This result is not true if Borel set is replaced by Lebesgue measurable set. Counter example will be given latter. (See Theorem 4.19, some consequences).

Theorem 3.6 *If f and g are measurable functions on a measurable set E then the following functions are measurable on E.*
(i) kf, $k \in \mathbb{R}$; (ii) $f + g$; (iii) f^2; (iv) fg; (v) $|f|$; (vi) $\frac{1}{f}$; (vii) $\max[f, g]$ and $\min[f, g]$,
provided that in (ii) $f(x) + g(x)$ is not of the form $\infty - \infty$ for $x \in E$ and that in (vi) $f(x) \neq 0$ for $x \in E$.

Proof Let $a \in \mathbb{R}$.

(i) If $k = 0$ then $kf = 0$ on E and so kf is a constant function and so is measurable. (Note that if $f(x) = \infty$ or $-\infty$ for some $x \in E$ then $0 \cdot f(x)$ is undefined, but in integration theory $0 \cdot \infty$ and $0 \cdot -\infty$ are assumed to be 0 as convention). If $k \neq 0$ then

$$\{x \in E : kf(x) > a\} = \left\{x \in E : f(x) > \frac{a}{k}\right\} \text{ if } k > 0$$
$$= \left\{x \in E : f(x) < \frac{a}{k}\right\} \text{ if } k < 0$$

and so (i) is proved.

(ii) Let $\{r_i\}$ be an enumeration of the set of rational numbers in $\mathbb{R}$. Let

$$E_i = \{x \in E : f(x) > r_i\} \cap \{x \in E : g(x) > a - r_i\}$$
$$F = \{x \in E : f(x) = \infty\} \cap \{x \in E : g(x) \neq -\infty\}$$

$$G = \{x \in E : g(x) = \infty\} \cap \{x \in E : f(x) \neq -\infty\}$$
$$H = \{x \in E : f(x) + g(x) > a\}.$$

Clearly $\bigcup_{i=1}^{\infty} E_i \cup F \cup G \subset H$. To prove the reverse inclusion let $x \in H$. Then $f(x) + g(x) > a$. If both $f(x)$ and $g(x)$ are finite then $f(x) > a - g(x)$ and so there is r_i such that $f(x) > r_i > a - g(x)$ and hence $x \in E_i \subset \bigcup_{i=1}^{\infty} E_i$. If $f(x) = \infty$ and $-\infty < g(x) \leq \infty$ then $x \in F$ and if $g(x) = \infty$ and $-\infty < f(x) \leq \infty$ then $x \in G$. The other cases do not arise since $\infty - \infty$ is not allowed and $f(x) + g(x) > a$. Therefore $x \in \bigcup_{i=1}^{\infty} E_i \cup F \cup G$. Hence $H = \bigcup_{i=1}^{\infty} E_i \cup F \cup G$. Since f and g are measurable, E_i is measurable for each i. Also since by Theorem 3.2 the set $\{x \in E : f(x) = -\infty\}$ is measurable, its complement $\{x \in E : f(x) \neq -\infty\}$ is measurable. Similarly for the function g. Hence by Theorem 3.2 the sets F and G are measurable. Therefore H is measurable, proving (ii).

(iii) We have

$$\{x \in E : f^2(x) > a\} = E \text{ if } a < 0$$
$$= \{x \in E : f(x) > \sqrt{a}\} \cup \{x \in E : f(x) < -\sqrt{a}\} \text{ if } a \geq 0$$

and since f is measurable, (iii) follows.

(iv) Since $fg = \frac{1}{4}[(f+g)^2 - (f-g)^2]$, by (i), (ii) and (iii) the right-hand side is measurable and so fg is measurable.

(v) Since

$$\{x \in E : |f(x)| > a\} = E \text{ if } a < 0$$
$$= \{x \in E : f(x) > a\} \cup \{x \in E : f(x) < -a\} \text{ if } a \geq 0$$

the result follows.

(vi) We have

$$\left\{x \in E : \frac{1}{f(x)} > a\right\} = \{x \in E : f(x) > 0\} \text{ if } a = 0$$
$$= \{x \in E : f(x) > 0\} \cap \{x \in E : f(x) < \frac{1}{a}\} \text{ if } a > 0$$
$$= \{x \in E : f(x) > 0\} \cup (\{x \in E : f(x) < 0\} \cap \{x \in E : f(x) < \frac{1}{a}\}) \text{ if } a < 0$$

and so the result follows.

3.2 Properties of Measurable Function

(vii) Finally since

$$\{x \in E : max[f(x), g(x)] > a\} = \{x \in E : f(x) > a\} \cup \{x \in E : g(x) > a\}$$

and

$$\{x \in E : min[f(x), g(x)] < a\} = \{x \in E : f(x) < a\} \cup \{x \in E : g(x) < a\}$$

the proofs of (vii) are clear. $\square$

Example 3.1 Sum and product of two nonmeasurable functions may be measurable
Let E be a measurable set of positive measures. Then E contains a nonmeasurable set $E_1 \subset E$. Clearly $E \sim E_1$ is nonmeasurable. Define $f(x) = 1$ for $x \in E_1$ and $f(x) = 0$ for $x \in E \sim E_1$ and $g(x) = 0$ for $x \in E_1$ and $g(x) = 1$ for $x \in E \sim E_1$. Then $f(x) + g(x) = 1$ for all $x \in E$ and $f(x)g(x) = 0$ for all $x \in E$. So f and g are nonmeasurable but $f + g$ and fg are measureable.

Theorem 3.7

(i) If f is measurable on E_1 and on E_2 then f is measurable on $E_1 \cup E_2$.
(ii) If f is measurable on E and E_1 is a measurable subset of E then f is measurable on E_1.
(iii) If f is measurable on E and if g is defined on E such that $f(x) = g(x)$ except for a subset of E of measure zero, then g is measurable.
(iv) If f and g are measurable on E then the sets $\{x \in E : f(x) > g(x)\}$ and $\{x \in E : f(x) \neq g(x)\}$ are measurable.

Proof

(i) Since $\{x \in E_1 : f(x) > a\} \cup \{x \in E_2 : f(x) > a\} = \{x \in E_1 \cup E_2 : f(x) > a\}$, the result follows.
(ii) Since $\{x \in E_1 : f(x) > a\} = \{x \in E : f(x) > a\} \cap E_1$, the proof is clear.
(iii) Let $E_1 = \{x \in E : f(x) \neq g(x)\}$. Then E_1 is of measure zero. Since $f(x) = g(x)$ for $x \in E \cap \widetilde{E_1}$, we have

$$\{x \in E : g(x) > a\} = \{x \in E_1 : g(x) > a\} \cup \{x \in E \cap \widetilde{E_1} : g(x) > a\}$$
$$= \{x \in E_1 : g(x) > a\} \cup \{x \in E \cap \widetilde{E_1} : f(x) > a\}.$$

Since E_1 is of measure 0, the first set in the right side is measurable and since f is measurable and $E \cap \widetilde{E_1} \subset E$ the second set in the right side is also measurable by (ii). Hence g is measurable.

(iv) Let $\{r_i\}$ be an enumeration of the set of rational numbers in $\mathbb{R}$. Then since

$$\{x \in E : f(x) > g(x)\} = \bigcup_{i=1}^{\infty} \left(\{x \in E : f(x) > r_i\} \cap \{x \in E : g(x) < r_i\}\right)$$

the first part follows. Similarly the set $\{x \in E : f(x) < g(x)\}$ is measurable. Hence their union $\{x \in E : f(x) > g(x)\} \cup \{x \in E : f(x) < g(x)\}$ is measurable. So, $\{x \in E : f(x) \neq g(x)\}$ is measurable. $\square$

Example 3.2 For any measurable set E of positive measure there is a nonmeasurable function on E.

The function f defined in Example 3.1 is a nonmeasurable function on E.

Theorem 3.8 *If g is measurable and defined on a measurable set E and if f is continuous on an interval containing $g(E)$ then the composition function $f \circ g$ is measurable on E.*

Proof We have for any $a \in \mathbb{R}$

$$\{x : (f \circ g)(x) > a\} = (f \circ g)^{-1}(a, \infty)$$
$$= g^{-1} \circ f^{-1}(a, \infty)$$
$$= g^{-1}(f^{-1}(a, \infty))$$

Since f is continuous, $f^{-1}(a, \infty)$ is an open set and so there is a countable collection of disjoint open intervals $\{I_n\}$ such that $f^{-1}(a, \infty) = \bigcup_{n=1}^{\infty} I_n$. Now since g is measurable $g^{-1}(I_n)$ is measurable for each n and hence $g^{-1}(\bigcup_{n=1}^{\infty} I_n) = \bigcup_{n=1}^{\infty} g^{-1}(I_n)$ is measurable and so by the above equality, $\{x : f \circ g(x) > a\}$ is measurable, which completes the proof. □

Theorem 3.9 *If $\{f_n\}$ is a sequence of measurable functions on E then each of the following functions is measurable :*
(i) $\sup_n f_n$; *(ii)* $\inf_n f_n$; *(iii)* $\limsup_n f_n$; *(iv)* $\liminf_n f_n$. *(v) If $\{f_n\}$ converges then the limit function $\lim_n f_n$ is measurable.*

Proof Let $a \in \mathbb{R}$. Since

$$\{x \in E : \sup_n f_n(x) > a\} = \bigcup_{n=1}^{\infty} \{x \in E : f_n(x) > a\}$$

and

$$\{x \in E : \inf_n f_n(x) < a\} = \bigcup_{n=1}^{\infty} \{x \in E : f_n(x) < a\}$$

and since the right-hand side is measurable in both cases, the proofs of (i) and (ii) follow. To prove (iii) note that for any sequence of real numbers $\{r_n\}$, $\limsup_n r_n = \inf_k \sup_{k \leq n} r_n$. For, let $\limsup_n r_n = L$ and let $\epsilon > 0$ be arbitrary. Then there is k_0 such that $r_n < L + \epsilon$ for $n \geq k_0$ and so $\sup_{n \geq k_0} r_n \leq L + \epsilon$ which gives $\inf_k \sup_{k \leq n} r_n \leq L + \epsilon$. Again, there are an infinite number of values of n such that $r_n > L - \epsilon$ and so for each k, $\sup_{k \leq n} r_n > L - \epsilon$ and hence $\inf_k \sup_{k \leq n} r_n \geq L - \epsilon$. Since ϵ is arbitrary, $\inf_k \sup_{k \leq n} r_n = L = \limsup_n r_n$. Hence $\limsup_n f_n = \inf_k \sup_{k \leq n} f_n$ and so the proof of (iii) follows from (i) and (ii). Since $\liminf_n f_n = -\limsup_n (-f_n)$ the proof of (iv) follows from (iii). Finally, if f_n converges then $\lim_n f_n = \liminf_n f_n = \limsup_n f_n$ and so the proof of (v) follows from (iii) or (iv). □

Remark Regarding (i) and (ii) note that $\sup_a f_a$ and $\inf_a f_a$ need not be measurable for an arbitrary collection of measurable functions $\{f_a : a \in I\}$ unless the index set I is countable. For let E be any set of positive measure. Then by Theorem 2.18 E contains a subset A which is not Lebesgue measurable. For each $a \in A$ define $f_a(x) = 1$ if $x = a$ and $f_a(x) = 0$ if $x \in E \sim \{a\}$. Then each f_a is measurable on E and $\sup_a f_a = 1$ for $x \in A$ and $= 0$ for $x \in E \sim A$. So $\sup_a f_a = \chi_A$ the characteristic function of A. Since A is not measurable, by Theorem 3.4 χ_A is not measurable and hence $\sup_a f_a$ is not measurable.

Theorem 3.10 *The set of all points on which a sequence of measurable functions $\{f_n\}$ converges is measurable.*

Proof The set of points where $\{f_n\}$ converges is the set $\{x \in E : \limsup_n f_n(x) = \liminf_n f_n(x)\}$. So, the proof follows from Theorems 3.9 and 3.7 (iv). □

Theorem 3.11 *If f is measurable on each E_n, $n = 1, 2, \ldots$ then f is measurable on $\bigcup_{n=1}^{\infty} E_n$.*

Proof If $a \in \mathbb{R}$ then since

$$\{x \in \bigcup_{n=1}^{\infty} E_n : f(x) > a\} = \bigcup_{n=1}^{\infty} \{x \in E_n : f(x) > a\}$$

the result follows. □

3.3 Class of Measurable Functions

Here we consider functions which are measurable. It is clear that every function which is defined on a set of measure zero is measurable. However, it is not true that every function defined on a set of positive measure is measurable. For, if E is a measurable set of positive measure then by Theorem 2.18 the set E contains a nonmeasurable set say A. The characteristic function f of A i.e $f(x) = 1$ for $x \in A$ and $f(0) = 0$ for $x \in E \sim A$, is not measurable. We prove

Theorem 3.12 *If E is a measurable set and $f : E \to \mathbb{R}$ is continuous then f is measurable.*

Proof Since f is continuous on E, for each $\xi \in E$ and for every $\epsilon > 0$ there is $\delta > 0$ such that

$$|f(x) - f(\xi)| < \epsilon \text{ if } x \in E \cap (\xi - \delta, \xi + \delta).$$

Let $a \in \mathbb{R}$ and let $\xi \in \{x \in E : f(x) > a\}$. Then $\xi \in E$ and $f(\xi) > a$. Choosing $\epsilon > 0$ such that $f(\xi) - \epsilon > a$ we get an open interval $I_\xi = (\xi - \delta, \xi + \delta)$ such that

$x \in E \cap I_\xi$ implies $f(x) > a$ and so $E \cap I_\xi \subset \{x \in E : f(x) > a\}$. So, considering all points $\xi \in E$ such that $f(\xi) > a$ we have

$$\bigcup_{f(\xi)>a} (E \cap I_\xi) \subset \{x \in E : f(x) > a\}.$$

On the other hand if $\eta \in \{x \in E : f(x) > a\}$ then there is an open interval I_η such that $\eta \in E \cap I_\eta$ and so $\eta \in \bigcup_{f(\xi)>a} (E \cap I_\xi)$. Therefore

$$\bigcup_{f(\xi)>a} (E \cap I_\xi) = \{x \in E : f(x) > a\}$$

Let $G = \bigcup_{f(\xi)>a} I_\xi$. Since arbitrary union of open sets is an open set, G is an open set and so G is measurable. Since E is measurable $E \cap G$ is measurable. But $E \cap G = E \cap (\bigcup_{f(\xi)>a} I_\xi) = \bigcup_{f(\xi)>a} (E \cap I_\xi)$ and so $E \cap G = \{x \in E : f(x) > a\}$. Hence $\{x \in E : f(x) > a\}$ is measurable. So, f is measurable. □

Theorem 3.13 *If E is Borel measurable and $f : E \to \mathbb{R}$ is continuous then f is Borel measurable.*

Proof In the proof of the above theorem measurability of E is assumed to conclude that $E \cap G$ is measurable, where G is an open set. So, if E is Borel measurable, then $E \cap G$ is Borel measurable, since G is Borel measurable. The rest of the argument being valid for Borel measurable set, f is Borel measurable. □

Definition 3.3 Let E be any set. If a property holds on E except on a subset of E of measure zero then we say that the property holds almost everywhere, written a.e. on E.

Corollary 3.14 *Let f be defined on an interval I and let f be continuous a.e. on I. Then f is measurable.*

Proof Let E be the set of all points of I where f is continuous. Then $\mu(I \sim E) = 0$ and so E is measurable and the proof now follows from Theorem 3.12 □

Remark The measurability of E in Theorem 3.12 is necessary. For, if E is nonmeasurable and if $f(x) = 1$ for $x \in E$ then f is continuous on E but f is not measurable since $\{x \in E : f(x) > \frac{1}{2}\} = E$ which is not measurable.

Theorem 3.15 *Let $f : [a, b] \to \mathbb{R}$ be Riemann integrable. Then f is measurable.*

Proof By Lebesgue-Vitali Theorem for Riemann integral (see Theorem 5.19) f is continuous *a.e.* on $[a, b]$. So, the proof follows from Corollary 3.14. □

Theorem 3.16 *Let $f : [a, b] \to \mathbb{R}$ be measurable and let the derivative f' of f exist finite or infinite. Then the derivative function f' is measurable on $[a, b]$.*

3.3 Class of Measurable Functions

Proof Let $f(x) = f(b)$ for $x > b$. Then f is measurable on $[a, \infty)$. Hence using Theorem 3.6 the function $\phi_n(x) = n[f(x + \frac{1}{n}) - f(x)]$ is measurable for all n. Since $f'(x)$ exists, $\lim\limits_{n \to \infty} \phi_n(x) = f'(x)$ for $[a, b)$. So, by Theorem 3.9(v) the function f' is measurable on $[a, b)$ and hence on $[a, b]$. $\square$

Note If $f'(x) = \infty$ or $-\infty$ then f may not be continuous at x. For consider $f(x) = -1$ if $x < 0$, $f(0) = 0$, $f(x) = 1$ if $x > 0$ Then $f'(0) = \infty$. So measurability of f is assumed. However the measurability condition on f can be removed by applying a theorem of Denjoy [Theorem 6.6] who proved that the set of points x where $f'(x) = \pm \infty$ is of measure 0 and so, by this theorem f is continuous *a.e.* and hence is measurable, by Corollary 3.14

Theorem 3.17 *Let E be measurable and let $f : E \to \mathbb{R}$ be monotone. Then f is measurable.*

Proof If f is constant on E then the proof is clear. Suppose that f is non-constant and non-decreasing. Let $a \in \mathbb{R}$. Suppose $\inf\{f(x) : x \in E\} \le a < \sup\{f(x) : x \in E\}$. Then there is $x_0 \in \mathbb{R}$ such that

$$\{x \in E : f(x) > a\} = E \cap (x_0, \infty)$$

Also

$$\{x \in E : f(x) > a\} = \emptyset \text{ if } a \ge \sup\{f(x) : x \in E\}$$
$$= E \text{ if } a < \inf\{f(x) : x \in E\}.$$

So, f is measurable. If f is non-increasing the proof is similar. $\square$

Corollary 3.18 *If f is defined on a measurable set E and if f is strictly monotone on E, then the inverse function f^{-1} of f exists and is strictly monotone and hence is measurable.*

Proof It is clear that f is one - one from E onto $f(E)$. So, $f^{-1} : f(E) \to E$ exists and is strictly monotone and so the proof follows from the above theorem. $\square$

Theorem 3.19 *If $g : E \to \mathbb{R}$ is measurable and $f : \mathbb{R} \to \overline{\mathbb{R}}$ is Borel measurable then the composite function $f \circ g : E \to \overline{\mathbb{R}}$ is measurable.*

Proof Let $r \in \mathbb{R}$. Since f is Borel measurable $\{x : f(x) > r\}$ is a Borel set *i.e* $f^{-1}(r, \infty)$ is a Borel set. Since g is measurable, by Theorem 3.5 $g^{-1}(f^{-1}(r, \infty))$ is measurable. Since

$$\{x : x \in E; (f \circ g)(x) > r\} = (f \circ g)^{-1}(r, \infty) = g^{-1}(f^{-1}(r, \infty))$$

the set $\{x \in E : (f \circ g)(x) > r\}$ is measurable and hence $f \circ g$ is measurable. $\square$

Example 3.3 For any measurable set of positive measures there are an uncountable infinite number of nonmeasurable functions.

Let E be any set of positive measure. Then there is a nonmeasurable set $E_1 \subset E$. Clearly $E \sim E_1$ is nonmeasurable and so outer measure $\mu^*(E \sim E_1) > 0$ and hence $E \sim E_1$ contains uncountable infinite number of elements. For each $x \in E \sim E_1$ the set $E_1 \cup \{x\}$ is nonmeasurable. Define f_x on E such that $f_x(t) = 1$ if $t \in E_1 \cup \{x\}$ and $= 0$ if $t \in E \sim E_1 \cup \{x\}$. Then each f_x is nonmeasurable and $\{f_x : x \in E \sim E_1\}$ contains uncountable infinite number of nonmeasurable functions.

3.4 Further Properties of Measurable Functions. Simple Functions

Definition 3.4 Let E be any set. If $E = \bigcup_{i=1}^{n} E_i$ where the sets $E_1, E_2, ..., E_n$ are disjoint then the collection $\{E_1, E_2, ..., E_n\}$ is called a dissection of E. If all the sets $E_1, E_2, ..., E_n$ are measurable (obviously then E is also measurable) then $\{E_1, E_2, ..., E_n\}$ is called a measurable dissection or a measurable partition of E

Definition 3.5 Let $f : E \to \mathbb{R}$. Then f is said to be a simple function if there is a dissection $\{E_1, E_2, ..., E_n\}$ of E and real numbers $a_1, a_2, ..., a_n$ such that

$$f(x) = \sum_{i=1}^{n} a_i \chi_{E_i}(x)$$

where χ_{E_i} is the characteristic function of E_i, $i = 1, 2, ..., n$. That is, f has constant value a_i in E_i. If $\{E_1, E_2, ..., E_n\}$ is a measurable dissection of E then f is called a measurable simple function. It is clear that every constant function f on a measurable set is a measurable simple function.

If in particular E is an interval and the sets E_i are also intervals then f is called a step function. So, every step function on an interval is a measurable simple function but the converse is not true.

Example 3.4 There exists a nonmeasurable simple function.

In the previous Example (Example 3.3) each function f_x is a nonmeasurable simple function.

Theorem 3.20 *If f and g are measurable simple functions on E then $f + g$ and fg are also measurable simple functions on E.*

Proof Let $f = \sum_{i=1}^{n} a_i \chi_{E_i}$ and $g = \sum_{j=1}^{m} b_j \chi_{F_j}$, $E = \bigcup_{i=1}^{n} E_i = \bigcup_{j=1}^{m} F_j$. Let $H_{ij} = E_i \cap F_j$. Then the sets H_{ij} are disjoint and measurable and $E = \bigcup_{i=1}^{n} \bigcup_{j=1}^{m} H_{ij}$. Also

3.4 Further Properties of Measurable Functions. Simple Functions

$$f+g = \sum_{i=1}^{n}\sum_{j=1}^{m}(a_i+b_j)\chi_{H_{ij}}, \; fg = \sum_{i=1}^{n}\sum_{j=1}^{m}a_i b_j \chi_{H_{ij}}$$

So, $f+g$ and fg are measurable simple functions on E. □

Theorem 3.21 *If $f : E \to \overline{\mathbb{R}}$ is a non-negative measurable function then there is a non-decreasing sequence $\{f_n\}$ of non-negative measurable simple functions such that $\lim_n f_n(x) = f(x)$ for all $x \in E$. If f is bounded then this limit is uniform.*

Proof Define for each n

$$f_n(x) = \frac{i-1}{2^n} \text{ if } \frac{i-1}{2^n} \le f(x) < \frac{i}{2^n}, i = 1, 2, \ldots, n2^n$$
$$= n \text{ if } f(x) \ge n.$$

If $E_{in} = \{x \in E : \frac{i-1}{2^n} \le f(x) < \frac{i}{2^n}\}, i = 1, 2, \ldots, n2^n$ and $E_n = \{x \in E : f(x) \ge n\}$ then the sets E_{in}, E_n are all measurable and disjoint and

$$E = (\bigcup_{i=1}^{n2^n} E_{in}) \cup E_n.$$

Also $f_n(x) = \frac{i-1}{2^n}$ if $x \in E_{in}$ and $f_n(x) = n$ if $x \in E_n$. Therefore for each n f_n is a non-negative measurable simple function. To show that $f_n(x) \le f_{n+1}(x)$ for each n and each $x \in E$, let $x \in E$. Then $x \in E_{in}$ for some $i = 1, 2, \ldots n2^n$ or $x \in E_n$. Let $x \in E_{in}$. Then $\frac{i-1}{2^n} \le f(x) < \frac{i}{2^n}$ and so $\frac{2i-2}{2^{n+1}} \le f(x) < \frac{2i}{2^{n+1}}$. Hence either $\frac{2i-2}{2^{n+1}} \le f(x) < \frac{2i-1}{2^{n+1}}$ or $\frac{2i-1}{2^{n+1}} \le f < \frac{2i}{2^{n+1}}$. In the first case $f_n(x) = \frac{i-1}{2^n} = f_{n+1}(x)$ and in the second case $f_n(x) = \frac{i-1}{2^n} < \frac{2i-1}{2^{n+1}} = f_{n+1}(x)$. If $x \in E_n$ then $f(x) \ge n$. So, if $f(x) \ge n+1$ then $f_n(x) = n < n+1 = f_{n+1}(x)$ and if $n \le f(x) < n+1$ then $\frac{n2^{n+1}}{2^{n+1}} \le f(x) < \frac{(n+1)2^{n+1}}{2^{n+1}}$ and so $f(x)$ will lie in one of the intervals

$$\left[\frac{n2^{n+1}}{2^{n+1}}, \frac{n2^{n+1}+1}{2^{n+1}}\right), \left[\frac{n2^{n+1}+1}{2^{n+1}}, \frac{n2^{n+1}+2}{2^{n+1}}\right), \ldots, \left[\frac{n2^{n+1}+2^{n+1}-1}{2^{n+1}}, \frac{(n+1)2^{n+1}}{2^{n+1}}\right)$$

and then also $f_n(x) \le f_{n+1}(x)$, which shows that $\{f_n\}$ is non-decreasing. Finally, let $\xi \in E$. If $f(\xi) < \infty$, then for $n > f(\xi)$ and for some i, $i = 1, 2, \ldots, n2^n$, $f(\xi) \in E_{in}$ and hence

$$0 \le f(\xi) - f_n(\xi) < \frac{1}{2^n}$$

and so $\lim_n f_n(\xi) = f(\xi)$. If $f(\xi) = \infty$ then $f_n(\xi) = n$ for all n and hence $\lim_n f_n(\xi) = \infty = f(\xi)$. If f is bounded then there is N such that $f(x) < n$ for all $n \ge N$ and all $x \in E$ and so $0 \le f(x) - f_n(x) < \frac{1}{2^n}$ for all $n \ge N$ and all $x \in E$ showing that the limit is uniform. □

Theorem 3.22 *If $f : E \to \overline{\mathbb{R}}$ is measurable then there is a sequence $\{f_n\}$ of measurable simple functions such that $\lim_n f_n(x) = f(x)$ for $x \in E$. If f is bounded then this limit is uniform.*

Proof Let $f_+(x) = \max[f(x), 0]$ and $f_-(x) = \max[-f(x), 0]$. Then $f(x) = f_+(x) - f_-(x)$ for all x. Since each of f_+ and f_- are non-negative measurable function, by Theorem 3.21 there are sequences $\{g_n\}$ and $\{h_n\}$ of measurable simple functions such that $\lim_n g_n(x) = f_+(x)$ and $\lim_n h_n(x) = f_-(x)$ for $x \in E$. Putting $f_n = g_n - h_n$, $\{f_n\}$ is a sequence of measurable simple functions such that

$$\lim_n f_n(x) = \lim_n g_n(x) - \lim_n h_n(x) = f_+(x) - f_-(x) = f(x)$$

If f is bounded then f_+ and f_- are bounded and so by Theorem 3.21, the limits $\lim g_n$, $\lim h_n$ are uniform and so $\lim f_n$ is uniform. $\square$

Theorem 3.23 (Egorov) *Let $\{f_n\}$ be a sequence of measurable functions which converges to f a.e. on a measurable set E of finite measure. Then for every $\delta > 0$ there is a measurable subset $E_\delta \subset E$ such that $\mu(E \sim E_\delta) < \delta$ and $\{f_n\}$ converges to f uniformly on E_δ.*

Proof For each pair of positive integer (k, n) let

$$E_{kn} = \bigcap_{m=n}^{\infty} \{x \in E : |f_m(x) - f(x)| < \frac{1}{k}\}$$

$$H = \{x \in E : \lim_m f_m(x) = f(x)\}$$

Then the sets E_{kn} and H are measurable and $H \subset \bigcup_{n=1}^{\infty} E_{kn}$ for each k. Since $\{E_{kn}\}$ is increasing for fixed k by Theorem 2.8

$$\mu(E) = \mu(H) \leq \mu(\bigcup_{n=1}^{\infty} E_{kn}) = \mu(\lim_{n \to \infty} E_{kn}) = \lim_{n \to \infty} \mu(E_{kn})$$

and so

$$\lim_{n \to \infty} \mu(E \sim E_{kn}) = 0 \text{ for every fixed } k.$$

So, for every $\delta > 0$ there is $n = n_k$ such that $\mu(E \sim E_{kn_k}) < \frac{\delta}{2^k}$. Let $E_\delta = \bigcap_{k=1}^{\infty} E_{kn_k}$. then E_δ is measurable and

$$\mu(E \sim E_\delta) = \mu(\bigcup_{k=1}^{\infty}(E \sim E_{kn_k})) \leq \sum_{k=1}^{\infty} \mu(E \sim E_{kn_k}) < \sum_{k=1}^{\infty} \frac{\delta}{2^k} = \delta.$$

3.4 Further Properties of Measurable Functions. Simple Functions

Now we are to show that $\{f_n\}$ converges to f uniformly on E_δ.

Let $\epsilon > 0$ be arbitrary. Then there is k such that $\frac{1}{k} < \epsilon$. Since $E_\delta \subset E_{kn_k}$, for every, k if $x \in E_\delta$ then

$$|f_m(x) - f(x)| < \frac{1}{k} \text{ for all } m \geq n_k.$$

So,

$$|f_m(x) - f(x)| < \epsilon \text{ for all } m \geq n_k \text{ and for all } x \in E_\delta.$$

Since n_k depends on ϵ only and not on x, $\{f_n\}$ converges to f uniformly on E_δ. $\square$

Theorem 3.24 (Luzin) *Let f be a measurable function on a set E of finite measure. Then for every $\epsilon > 0$ there is a closed set $F \subset E$ such that $\mu(E \sim F) < \epsilon$ and the restriction f/F of f to F is continuous.*

Proof We first suppose that f is a measurable simple function. So there are disjoint measurable sets $E_1, E_2, \ldots, E_k$ and real numbers $a_1, a_2, \ldots, a_k$ such that

$$f = \sum_{i=1}^{k} a_i \chi_{E_i}, \quad E = \bigcup_{i=1}^{k} E_i,$$

where χ_{E_i} is the characteristic function of E_i. Since each E_i is measurable, there is a closed set $F_i \subset E_i$ such that $\mu(E_i \sim F_i) < \frac{\epsilon}{k}$. Let $F = \bigcup_{i=1}^{k} E_i$. Then F is closed, $F \subset E$ and $\mu(E \sim F) = \mu(\bigcup_{i=1}^{k}(E_i \sim F_i)) = \sum_{i=1}^{k} \mu(E_i \sim F_i) < \epsilon$. Since f is constant on each F_i, f/F is continuous on each F_i and so it is continuous on F. Thus the result is true for measurable simple functions.

Now let f be any measurable function. Then by Theorem 3.22 there is a sequence of measurable simple functions $\{f_n\}$ on E such that

$$f(x) = \lim_n f_n(x), \quad x \in E \tag{3.1}$$

Since the result is true for measurable simple functions, for each n there is a closed set $F_n \subset E$ such that $\mu(E \sim F_n) < \frac{\epsilon}{2^{n+1}}$ and the restriction f_n/F_n is continuous. Let $F_0 = \bigcap_{n=1}^{\infty} F_n$. Then

$$\mu(E \sim F_0) = \mu(\bigcup_{n=1}^{\infty}(E \sim F_n)) \leq \sum_{n=1}^{\infty} \mu(E \sim F_n) < \sum_{n=1}^{\infty} \frac{\epsilon}{2^{n+1}} = \frac{\epsilon}{2}. \tag{3.2}$$

Since $F_0 \subset E$, (3.1) is true on F_0. Hence by Theorem 3.23 there is $F_\epsilon \subset F_0$ such that

$$\mu(F_0 \sim F_\epsilon) < \frac{\epsilon}{4} \qquad (3.3)$$

and the limit in (3.1) is uniform on F_ϵ. Since f_n/F_n is continuous and $F_\epsilon \subset F_n$ for each n, the restriction function f_n/F_ϵ is continuous for each n and so f/F_ϵ is continuous. Choose a closed set $F \subset F_\epsilon$ such that

$$\mu(F_\epsilon \sim F) < \frac{\epsilon}{4} \qquad (3.4)$$

Since $F \subset F_\epsilon \subset F_0 \subset E$, from (3.2), (3.3) and (3.4)

$$\mu(E \sim F) = \mu\big((E \sim F_0) \cup (F_0 \sim F_\epsilon) \cup (F_\epsilon \sim F)\big)$$
$$\leq \mu(E \sim F_0) + \mu(F_0 \sim F_\epsilon) + \mu(F_\epsilon \sim F) < \epsilon.$$

Since f/F_ϵ is continuous, f/F is also continuous. □

Theorem 3.25 (Luzin) *If $f; [a,b] \to \mathbb{R}$ is measurable then for every $\epsilon > 0$ there is a continuous function $g_\epsilon; [a,b] \to \mathbb{R}$ such that $\mu(\{x \in [a,b] : f(x) \neq g_\epsilon(x)\}) < \epsilon$ and $\sup_{x \in [a,b]} g_\epsilon(x) = \sup_{x \in [a,b]} f(x)$, $\inf_{x \in [a,b]} g_\epsilon(x) = \inf_{x \in [a,b]} f(x)$*

Proof Let $\epsilon > 0$ be arbitrary. Then by Theorem 3.24 there is a closed set $F \subset [a,b]$ such that $\mu([a,b] \sim F) < \epsilon$ and the restriction f/F of f to F is continuous. Let $c = \inf F$ and $d = \sup F$. Then $a \leq c < d \leq b$. Since F is closed $c, d \in F$. Define $g_\epsilon(x) = f(x)$ for $x \in F$ and $g_\epsilon(x) = f(c)$ for $x < c$ and $g_\epsilon(x) = f(d)$ for $x > d$. We are to define g_ϵ on $G = [c,d] \sim F$. Since G is an open set, there is a countable collection of open intervals $\{(a_n, b_n)\}$ such that $G = \bigcup_{n=1}^{\infty} (a_n, b_n)$. If $x \in G$ then $x \in (a_n, b_n)$ for some n. Define

$$g_\epsilon(x) = \frac{f(b_n) - f(a_n)}{b_n - a_n}(x - a_n) + f(a_n) \text{ for } x \in (a_n, b_n).$$

So, g_ϵ is defined everywhere. Since f/F is continuous, g_ϵ is continuous everywhere. Since

$$\{x \in [a,b] : f(x) \neq g_\epsilon(x)\} \subset [a,b] \sim F \text{ and } \mu([a,b] \sim F) < \epsilon,$$

the proof is complete, the rest being clear. □

Theorem 3.26 (Borel) *Let $f : [a,b] \to \mathbb{R}$ be measurable. Then for every $\sigma > 0$ and every $\epsilon > 0$ there is a continuous function $\psi : [a,b] \to \mathbb{R}$ such that*

$$\mu(\{x \in [a,b] : |f(x) - \psi(x)| \geq \sigma\}) < \epsilon.$$

The proof follows from Theorem 3.25. In fact by Theorem 3.25 there is a continuous function $g_\epsilon : [a,b] \to \mathbb{R}$ such that

$$\mu(\{x \in [a,b] : f(x) \neq g_\epsilon(x)\}) < \epsilon$$

and hence

$$\mu(\{x \in [a,b] : |f(x) - g_\epsilon(x)| \geq \sigma\}) < \epsilon$$

and so putting $\psi = g_\epsilon$ the result follows.

3.5 Convergence Theorems for Measurable Functions

Definition 3.6 A sequence of functions $\{f_n\}$ defined on a set E is said to converge almost everywhere (briefly *a.e.*) on E to a functions f if $f_n(x) \to f(x)$ as $n \to \infty$ for $x \in E \sim E_1$ where $E_1 \subset E$ and E_1 is of measure zero and we write $f_n \to f$ [*a.e.*] on E.

Definition 3.7 a sequence of functions $\{f_n\}$ defined on a set E is said to converge almost uniformly (briefly, *a.u.*) on E to a function f if for every $\epsilon > 0$ there is a set $E_\epsilon \subset E$ such that $\mu^*(E_\epsilon) < \epsilon$ and $\{f_n\}$ converges uniformly to f on $E \sim E_\epsilon$ and we write $f_n \to f$ [*a.u.*] on E.

The condition of measurability of E in the above definitions is not necessary. Note that if $\{f_n\}$ converges uniformly to f on a set E then $\{f_n\}$ converges almost uniformly on E but the converse is not true. For, define for each n $f_n(x) = nxe^{-nx^2}$ for $x \in E = [0,1]$. Then $\{f_n\}$ converges everywhere is E to zero But $\{f_n\}$ does not converge uniformly to zero in E. Also $\{f_n\}$ converges uniformly in $E \cap [\epsilon, 1]$ for every ϵ, $0 < \epsilon < 1$, and hence $\{f_n\}$ converges almost uniformly to zero in E.

Definition 3.8 Let $\{f_n\}$ be a sequence of almost everywhere finite measurable functions and let f be an almost everywhere finite measurable function on a measurable set E. Then $\{f_n\}$ is said to converge to f in measure on E if for every $\sigma > 0$

$$\lim_{n \to \infty} \mu(\{x \in E : |f_n(x) - f(x)| \geq \sigma\}) = 0.$$

In this case we write $f_n \to f$ [meas.] on E.

Note that if a sequence $\{f_n\}$ converges in measure to two functions f and g then f and g are equal almost everywhere. For, given any $\sigma > 0$

$$\{x \in E : |f(x) - g(x)| \geq \sigma\} \subset \{x \in E : |f_n(x) - f(x)| \geq \frac{\sigma}{2}\} \bigcup \{x \in E : |f_n(x) - g(x)| \geq \frac{\sigma}{2}\}$$

So, taking measure and then letting $n \to \infty$

$$\mu(\{x \in E : |f(x) - g(x)| \geq \sigma\}) = 0.$$

Since
$$\{x \in E : f(x) \neq g(x)\} = \bigcup_{k=1}^{\infty} \{x \in E : |f(x) - g(x)| \geq \frac{1}{k}\}$$
the result follows.

Theorem 3.27 *If $\{f_n\}$ is a sequence of measurable function which converges to f a.e. on a measurable set E of finite measure then $f_n \to f$ [a.u.] on E as $n \to \infty$*

This is Egorov's Theorem 3.23.
We now prove the converse.

Theorem 3.28 *If a sequence of functions $\{f_n\}$ converges to f almost uniformly on a set E then $f_n \to f$ as $n \to \infty$ almost everywhere on E.*

Proof Since $f_n \to f$ [a.u.], for each positive integer k there is $A_k \subset E$ such that $\mu^*(A_k) < \frac{1}{k}$ and $f_n \to f$ uniformly on $E \sim A_k$. Let $A = \bigcup_{k=1}^{\infty} (E \sim A_k) = E \sim \bigcap_{k=1}^{\infty} A_k$. Then

$$\mu^*(E \sim A) = \mu^*(\bigcap_{k=1}^{\infty} A_k) \leq \mu^*(A_k) \to 0 \text{ as } k \to \infty.$$

So $\mu(E \sim A) = 0$. If $\xi \in A$ then $\xi \in E \sim A_k$ for some k. Since $f_n \to f$ uniformly on $E \sim A_k$, $f_n(\xi) \to f(\xi)$ as $n \to \infty$. Since $\mu(E \sim A) = 0$, $f_n \to f$ a.e. on E. □

Theorem 3.29 *Let $\{f_n\}$ be a sequence of measurable functions on a set E and let f be a measurable function on E. If $\{f_n\}$ converges almost uniformly to f on E then $f_n \to f$ in measure on E.*

Proof Let $\epsilon > 0$ be arbitrary. Since $f_n \to f$ [a.u.], there is $A_\epsilon \subset E$ such that $\mu(A_\epsilon) < \epsilon$ and $f_n \to f$ uniformly on $E \sim A_\epsilon$. Let $\sigma > 0$ be arbitrary. Then there is N such that

$$|f_n(x) - f(x)| < \sigma \text{ for } n \geq N \text{ and for all } x \in E \sim A_\epsilon.$$

Hence
$$\{x \in E : |f_n(x) - f(x)| \geq \sigma\} \subset A_\epsilon \text{ for } n \geq N.$$

So
$$\mu(\{x \in E : |f_n(x) - f(x)| \geq \sigma\}) \leq \mu(A_\epsilon) < \epsilon \text{ for } n \geq N.$$

Hence
$$\lim_{n \to \infty} \mu(\{x \in E : |f_n(x) - f(x)| \geq \sigma\}) < \epsilon.$$

3.5 Convergence Theorems for Measurable Functions

Since ϵ is arbitrary,

$$\lim_{n\to\infty} \mu(\{x \in E : |f_n(x) - f(x)| \geq \sigma\}) = 0.$$

And so $\{f_n\}$ converges to f in measure. □

Theorem 3.30 (Lebesgue) *If $\{f_n\}$ is a sequence of almost everywhere finite measurable functions which converges almost everywhere on a measurable set E of finite measure to a function f which is finite almost everywhere on E then $\{f_n\}$ converges to f in measure on E.*

Proof Let $A_n = \{x \in E : |f_n(x)| = \infty\}$, $B = \{x \in E : \lim_{n\to\infty} f_n(x) = f(x); f(x) \neq \pm\infty\}$, $C = E \sim B$, $D = (\bigcup_{n=1}^{\infty} A_n) \bigcup C$. Then $\mu(D) = 0$. Let $\sigma > 0$ be arbitrary. Let $E_k(\sigma) = \{x \in E : |f_k(x) - f(x)| \geq \sigma\}$, $R_n(\sigma) = \bigcup_{k=n}^{\infty} E_k(\sigma)$, $M = \bigcap_{n=1}^{\infty} R_n(\sigma)$. Then these sets are measurable and $R_1(\sigma) \supset R_2(\sigma) \supset \ldots \supset R_n(\sigma) \supset \ldots$.
Therefore by Theorem 2.9

$$\lim_{n\to\infty} \mu(R_n(\sigma)) = \mu(M).$$

Clearly $M \subset D$. For, if $x \notin D$ then $\lim_{n\to\infty} f_n(x) = f(x) \neq \pm\infty$. So, there is N such that

$$|f_k(x) - f(x)| < \sigma \text{ for } k \geq N.$$

Hence $x \notin E_k(\sigma)$ for $k \geq N$ and so $x \notin R_N(\sigma)$ and therefore $x \notin M$. Since $M \subset D$, and $\mu(D) = 0$, $\mu(M) = 0$. Therefore $\lim_{n\to\infty} \mu(R_n(\sigma)) = 0$. Since $E_n(\sigma) \subset R_n(\sigma)$ for all n, $\lim_{n\to\infty} \mu(E_n(\sigma)) = 0$ and so $\lim_{n\to\infty} \mu(\{x \in E : |f_n(x) - f(x)| \geq \sigma\}) = 0$.
Since σ is arbitrary, the result follows.

The converse of the above theorem is not true. We give an example in the following theorem. □

Theorem 3.31 *There exists a uniformly bounded sequence of measurable functions $\{f_n\}$ which converges in measure on $[0, 1)$ but $\{f_n\}$ does not converges at any point in $[0, 1)$.*

Proof For each positive integer k define k functions $\phi_1^k, \phi_2^k, \ldots, \phi_k^k$ as follows:

$$\phi_i^k = 1 \text{ if } x \in \left[\frac{i-1}{k}, \frac{i}{k}\right)$$
$$= 0 \text{ if } x \in [0, 1) \sim \left[\frac{i-1}{k}, \frac{i}{k}\right), \text{ for } i = 1, 2, \ldots, k.$$

The functions ϕ_i^k, $i = 1, 2, ..., k$; $k = 1, 2, ...$ now being defined we take

$$f_1 = \phi_1^1, \ f_2 = \phi_1^2, \ f_3 = \phi_2^2, \ f_4 = \phi_1^3, \ ...$$

So, for each n there is k and i such that $f_n = \phi_i^k$. Let $\sigma > 0$ be arbitrary. Then for each n

$$\{x \in [0, 1) : f_n(x) \geq \sigma\} = \{x \in [0, 1) : \phi_i^k(x) \geq \sigma\} = [\frac{i-1}{k}, \frac{i}{k}) \text{ or}$$

$\emptyset$ according as $\sigma \leq 1$ or $\sigma > 1$. Hence

$$\mu(\{x \in [0, 1) : f_n(x) \geq \sigma\}) \leq \frac{1}{k} \to 0 \text{ as } n \to \infty.$$

So, $\{f_n\}$ converges in measure to 0 on $[0, 1)$.

To show that $\{f_n\}$ does not converge at any point in $[0, 1)$ let $x_0 \in [0, 1)$. Then for every k there is i, $1 \leq i \leq k$, such that $x_0 \in [\frac{i-1}{k}, \frac{i}{k})$ and so $\phi_i^k(x_0) = 1$. So there is a subsequence of $\{f_n(x_0)\}$ each member of which is 1. Also for every $k > 1$ there is i, $1 \leq i \leq k$, such that $x_0 \notin [\frac{i-1}{k}, \frac{i}{k})$ and so $\phi_i^k(x_0) = 0$ and hence there is a subsequence of $\{f_n(x_0)\}$ each member of which is 0. Hence $\{f_n(x_0)\}$ does not converge. Since x_0 is any point in $[0, 1)$, $\{f_n\}$ converges nowhere in $[0, 1)$. $\square$

Theorem 3.32 (Riesz) *Let $\{f_n\}$ be a sequence of almost everywhere finite measurable functions and let f be an almost everywhere finite measurable functions on a measurable set E. If $\{f_n\}$ converges to f in measure on E then there exists a subsequence of $\{f_n\}$ which converges to f almost everywhere on E.*

Proof Consider the sequence $\{\frac{1}{2^k}\}$. Since $\{f_n\}$ converges to f in measure on E.

$$\lim_{n \to \infty} \mu(\{x \in E : |f_n(x) - f(x)| \geq \frac{1}{2}\}) = 0.$$

Hence there is n_1, such that

$$\mu(\{x \in E : |f_{n_1}(x) - f(x)| \geq \frac{1}{2}\}) < \frac{1}{2}.$$

By the same argument, there is $n_2 > n_1$ such that

$$\mu(\{x \in E : |f_{n_2}(x) - f(x)| \geq \frac{1}{2^2}\}) < \frac{1}{2^2}.$$

Applying this argument repeatedly we get a sequence of positive integers $\{n_k\}$, $n_k < n_{k+1}$, $k = 1, 2, ...$ such that

$$\mu(\{x \in E : |f_{n_k}(x) - f(x)| \geq \frac{1}{2^k}\}) < \frac{1}{2^k} \text{ for all } k, k = 1, 2, ...$$

3.5 Convergence Theorems for Measurable Functions

We shall show that $f_{n_k} \to f$ as $k \to \infty$ almost everywhere in E. Set

$$E_k = \{x : x \in E;\ |f_{n_k}(x) - f(x)| \geq \frac{1}{2^k}\}$$

$$R_n = \bigcup_{k=n}^{\infty} E_k,\quad M = \bigcap_{n=1}^{\infty} R_n.$$

Clearly all these sets are measurable and $R_1 \supset R_2 \supset ...$ and hence

$$\lim_{n \to \infty} \mu(R_n) = \mu(M).$$

Since $\mu(R_n) \leq \sum_{k=n}^{\infty} \mu(E_k) < \sum_{k=n}^{\infty} \frac{1}{2^k}$

and $\sum \frac{1}{2^k}$ is convergent, $\lim_{n \to \infty} \mu(R_n) = 0$ and so $\mu(M) = 0$. Let $x_0 \in E \sim M$. Then $x_0 \notin R_{n_0}$ for some n_0. So $x_0 \notin E_k$ for $k \geq n_0$. Hence

$$|f_{n_k}(x_0) - f(x_0)| < \frac{1}{2^k} \text{ for all } k \geq n_0.$$

Taking limit as $k \to \infty$

$$\lim_{k \to \infty} f_{n_k}(x_0) = f(x_0).$$

Since $x_0 \in E \sim M$ is arbitrary and $\mu(M) = 0$, the result follows. □

The following theorem gives a comparison of Theorems 3.30 and 3.32.

Theorem 3.33 *Let $\{f_n\}$ be a sequence of almost everywhere finite measurable functions and f be an almost everywhere finite measurable function on a measurable set E of finite measure. Then $\{f_n\}$ converges to f almost everywhere on E if and only if for every $\sigma > 0$*

$$\lim_{n \to \infty} \mu(\bigcup_{k=n}^{\infty} \{x : x \in E;\ |f_k(x) - f(x)| \geq \sigma\}) = 0.$$

Proof Let for $\sigma > 0$

$$B_{n,\sigma} = \{x \in E : |f_n(x) - f(x)| \geq \sigma\}\ \ B_\sigma = \bigcap_{n=1}^{\infty} \bigcup_{k=n}^{\infty} B_{k\sigma}.$$

Since $\bigcup_{k=n}^{\infty} B_{k\sigma}$ is a decreasing sequence of measurable sets converging to B_σ

$$\lim_{n\to\infty} \mu(\bigcup_{k=n}^{\infty} B_{k\sigma}) = \mu(B_\sigma) \qquad (3.5)$$

Also
$$\{x \in E : \lim_{n\to\infty} f_n(x) = f(x)\} = \bigcap_{\sigma>0}(E \sim B_\sigma) \qquad (3.6)$$

For, if $\lim_{n\to\infty} f_n(\xi) = f(\xi)$, $\xi \in E$, then for every $\sigma > 0$ there is n such that $|f_k(\xi) - f(\xi)| < \sigma$ for $k \geq n$ and so $\xi \notin \bigcup_{k=n}^{\infty} B_{k\sigma}$ and hence $\xi \notin B_\sigma$, showing that $\xi \in E \sim B_\sigma$. Since σ is arbitrary $\xi \in \bigcap_{\sigma>0} (E \sim B_\sigma)$. Again, if $\xi \in E \sim B_\sigma$ for every $\sigma > 0$ then for every σ there is n such that $|f_k(\xi) - f(\xi)| < \sigma$ for $k \geq n$ and hence $\lim_{n\to\infty} f_n(\xi) = f(\xi)$. So, (3.6) is proved.

Now, if $\sigma_1 < \sigma_2$ then for each k, $B_{k\sigma_2} \subset B_{k\sigma_1}$ and hence $B_{\sigma_2} \subset B_{\sigma_1}$. Hence

$$\bigcup_{\sigma>0} B_\sigma = \bigcup_{m=1}^{\infty} B_{\frac{1}{m}} \qquad (3.7)$$

and therefore from (3.6) and (3.7)

$$E \sim \{x \in E : \lim_{n\to\infty} f_n(x) = f(x)\} = \bigcup_{\sigma>0} B_\sigma = \bigcup_{m=1}^{\infty} B_{\frac{1}{m}} \qquad (3.8)$$

So, if $f_n \to f$ almost everywhere in E then $B_{\frac{1}{m}}$ is of measure 0 for all m. Let $\sigma > 0$ be arbitrary. Choose m such that $\frac{1}{m} < \sigma$ and then $B_\sigma < B_{\frac{1}{m}}$ and so $\mu(B_\sigma) = 0$. Hence from (3.5)

$$\lim_{n\to\infty} \mu(\bigcup_{k=n}^{\infty} B_{k\sigma}) = 0 \qquad (3.9)$$

Since $\sigma > 0$ is arbitrary, (3.9) holds for every $\sigma > 0$.

Conversely, suppose that (3.9) holds for every $\sigma > 0$.

Then by (3.5) $\mu(B_\sigma) = 0$ for every $\sigma > 0$ and hence $\mu(B_{\frac{1}{m}}) = 0$ for all m and so from (3.8) $f_n \to f$ almost everywhere in E. □

Definition 3.9 Let $\{f_n\}$ be a sequence of almost everywhere finite measurable functions defined on a measurable set E. Then $\{f_n\}$ is said to be a Cauchy sequence (or, a fundamental sequence) in measure if for arbitrary $\sigma > 0$ and $\epsilon > 0$ there is N such that
$$\mu(\{x : x \in E; \ |f_n(x) - f_m(x)| \geq \sigma\}) < \epsilon \quad \text{whenever} \quad n, m \geq N.$$

Theorem 3.34 *A sequence $\{f_n\}$ of measurable functions on a measurable set E converges in measure if and only if $\{f_n\}$ is a Cauchy sequence in measure on E.*

3.5 Convergence Theorems for Measurable Functions

Proof Let $\{f_n\}$ converge in measure to f and let $\sigma > 0$ and $\epsilon > 0$ be arbitrary. Then there is N such that

$$\mu(\{x \in E : |f_n(x) - f(x)| \geq \frac{\sigma}{2}\}) < \frac{\epsilon}{2} \text{ for } n \geq N.$$

Since

$$\{x \in E : |f_n(x) - f_m(x)| \geq \sigma\} \subset \{x \in E : |f_n(x) - f(x)| \geq \frac{\sigma}{2}\} \cup$$
$$\{x \in E : |f_m(x) - f(x)| \geq \frac{\sigma}{2}\}$$

we have

$$\mu(\{x \in E : |f_n(x) - f_m(x)| \geq \sigma\}) \leq \mu(\{x \in E : |f_n(x) - f(x)| \geq \frac{\sigma}{2}\}) +$$
$$\mu(\{x \in E : |f_m(x) - f(x)| \geq \frac{\sigma}{2}\}) < \frac{\epsilon}{2} + \frac{\epsilon}{2} = \epsilon$$
$$\text{whenever } n, m \geq N.$$

Hence $\{f_n\}$ is a Cauchy sequence in measure.

Conversely, suppose that $\{f_n\}$ is a Cauchy sequence in measure. Then for each k there is n_k such that

$$\mu(\{x : x \in E; |f_n(x) - f_m(x)| \geq \frac{1}{2^k}\}) < \frac{1}{2^k} \text{ for } n, m \geq n_k. \quad (3.10)$$

We may suppose that $n_1 < n_2 < \ldots < n_k < \ldots$. We first show that the subsequence $\{f_{n_k}\}$ converges almost uniformly on E.

Let

$$E_k = \{x \in E : |f_{n_k}(x) - f_{n_{k+1}}(x)| \geq \frac{1}{2^k}\}$$

$$R_k = \bigcup_{i=k}^{\infty} E_i.$$

Then by (3.10)

$$\mu(R_k) \leq \sum_{i=k}^{\infty} \mu(E_i) < \sum_{i=k}^{\infty} \frac{1}{2^i} = \frac{1}{2^{k-1}} \quad (3.11)$$

Let $\epsilon > 0$ be arbitrary. Choose k such that $\frac{1}{2^{k-1}} < \epsilon$. Then from (3.11) $\mu(R_k) < \epsilon$ and for all $x \in E \sim R_k$

$$|f_{n_i}(x) - f_{n_{i+1}}(x)| < \frac{1}{2^i} \text{ for } i \geq k.$$

Hence for $j > i \geq k$

$$|f_{n_j}(x) - f_{n_i}(x)| \leq \sum_{r=i}^{j-1} |f_{n_r}(x) - f_{n_{r+1}}(x)| < \sum_{r=i}^{j-1} \frac{1}{2^r} < \frac{1}{2^{k-1}} < \epsilon$$

for all $x \in E \sim R_k$. This shows that for every $\epsilon > 0$ there is a set $R_k \subset E$ such that $\mu(R_k) < \epsilon$ and $\{f_{n_i}\}$ converges uniformly on $E \sim R_k$. Hence $\{f_{n_i}\}$ converges almost uniformly on E.

Let g be the limit of $\{f_{n_k}\}$. Let $\epsilon > 0$ and $\sigma > 0$ be arbitrary. Since $\{f_{n_k}\}$ converges almost uniformly, there is as above a measurable set $E_\epsilon \subset E$, $\mu(E_\epsilon) < \frac{1}{2}\epsilon$ and k_0 such that

$$|f_{n_k}(x) - g(x)| < \frac{1}{2}\sigma \text{ for } k \geq k_0 \text{ and for all } x \in E \sim E_\epsilon. \tag{3.12}$$

Since $\{f_n\}$ is a Cauchy sequence in measure, there is $N \geq n_{k_0}$ such that

$$\mu(\{x \in E : |f_n(x) - f_{n_k}(x)| \geq \frac{1}{2}\sigma\}) < \frac{1}{2}\epsilon \text{ for } n, n_k \geq N. \tag{3.13}$$

Let $n \geq N$. Choose $n_k > N$. Then since

$$\{x \in E : |f_n(x) - g(x)| \geq \sigma\}$$
$$\subset \{x \in E : |f_n(x) - f_{n_k}(x)| \geq \frac{\sigma}{2}\} \cup \{x \in E : |f_{n_k}(x) - g(x)| \geq \frac{\sigma}{2}\}$$

we have

$$\mu(\{x \in E : |f_n(x) - g(x)| \geq \sigma\})$$
$$\leq \mu(\{x \in E : |f_n(x) - f_{n_k}(x)| \geq \frac{\sigma}{2}\}) + \mu(\{x \in E : |f_{n_k}(x) - g(x)| \geq \frac{\sigma}{2}\}). \tag{3.14}$$

Since by (3.12) $\{x \in E : |f_{n_k}(x) - g(x)| \geq \frac{\sigma}{2}\} \subset E_\epsilon$ we have from (3.13) and (3.14) and from the fact that $\mu(E_\epsilon) < \frac{\epsilon}{2}$

$$\mu(\{x \in E : |f_n(x) - g(x)| \geq \sigma\}) < \frac{\epsilon}{2} + \frac{\epsilon}{2} = \epsilon.$$

This being true for all $n \geq N$, since ϵ is arbitrary

$$\lim_{n \to \infty} \mu(\{x \in E : |f_n(x) - g(x)| \geq \sigma\}) = 0.$$

So, $\{f_n\}$ converges to g in measure on E. $\square$

3.6 Exercises

1. Show that the function $f(x) = 1$ if x is rational and $f(x) = 0$ if x is irrational is measurable in every interval $[a, b]$.
2. If $f(x) = x \cos \frac{1}{x}$ if $x \neq 0$ and $f(0) = 0$ show that f is measurable. Find the measure of the set $\{x \in [0, 1] : f(x) \geq 0\}$.
3. Give an example of a function f such that $|f|$ is measurable but f is not.
4. Show that there are functions f and g such that $f + g$ and $f.g$ are measurable but f and g are nonmeasurable.
5. Let $f(x) = \sin x$ if x is rational and $f(x) = 0$ if x is irrational show that f is measurable but nowhere continuous in $[0, 1]$.
6. Let $\{f_n\}$ be a sequence of functions defined on a set E such that for each n, f_n is finite $a.e.$ on E. Show that for almost all x in E, the function $f_n(x)$ is finite for all n.
7. Let $f(x) = n$ if $\frac{1}{n+1} < x \leq \frac{1}{n}$ and $f(0) = 0$. Show that f is measurable in $[0, 1]$.
8. Let $\{E_n\}$ be a sequence of disjoint measurable sets and let $f(x) = n$ if $x \in E_n$. Show that f is measurable on $\bigcup_{n=1}^{\infty} E_n$.
9. If $f : \mathbb{R} \to \mathbb{R}$ is continuous then show that f is Borel measurable.
10. If the sequences $\{f_n\}$ and $\{g_n\}$ converge to f and g in measure then prove that $\{f_n + g_n\}$ converges to $f + g$ in measure.
11. If the sequence $\{f_n\}$ converges to f in measure and g is a measurable almost everywhere finite function then prove that the sequence $\{f_n g\}$ converges to fg in measure.
12. Prove converse of Luzins theorem: Let E be a measurable set and let $f : E \to \mathbb{R}$. If for every $\epsilon > 0$ there is a closed set $F_\epsilon \subset E$ such that $\mu(E \sim F_\epsilon) < \epsilon$ and the restriction f/F_ϵ of f on F_ϵ is continuous then f is measurable. (Hint consider $\epsilon = \frac{1}{n}$, $n = 1, 2, ...$)
13. If $\{f_n\}$ converge to f in measure and α is a real number then show that $\{\alpha f_n\}$ converges to αf in measure.
14. If $\{f_n\}$ and $\{g_n\}$ are sequences of bounded measurable functions on a measurable set of finite measure and if $\{f_n\}$ and $\{g_n\}$ converge to f and g respectively in measure then show that $\{f_n g_n\}$ converges to fg in measure.
15. If $\{f_n\}$ converges to f in measure then show that $\{|f_n|\}$ converses to $|f|$ in measure.

Chapter 4
More About Sets and Functions

Properties of Lebesgue outer measure and Lebesgue measurable sets are discussed in Chap. 2. There are other special properties of Lebesgue measurable sets and Borel sets which will be discussed here. For this, we have to construct some sets and study them in the light of set theory. Since the Cardinality of a set is an essential property of it, we begin with Cardinal members.

4.1 Cardinal Members

The concept of counting elements was originated from the idea of making one-one correspondence between two sets and in this case, the two sets are said to have the same number of elements. In a later stage, positive integers are introduced calling them natural members which are used to count the elements of a set. The number system that we use is expressed with the help of ten digits, 0, 1, 2, 3, 4, 5, 6, 7, 8, and 9 as base and that is why we call it a decimal system ('deci' means 'ten'). If we wish to build up a system taking say, only two digits 0 and 1 or, only three digits 0, 1, and 2 as a base then also counting is possible and these systems are called binary systems or ternary systems respectively. In the ternary system the base being 0, 1, 2 the numbers are 0, 1, 2, 10, 11, 12, 100, ... and the other digits, 3, 4 ... , 9 are redundant. On the other hand, if some one wishes to consider more than ten digits, say eleven digits as base then we will have to import another new digit ξ different from 0, 1, ..., 9 and the numbers in this case are 0, 1, 2, 3, 4, 5, 6, 7, 8, 9, ξ, 10, 11, ... 19, 1ξ, 20, 21, ... Naturally, although the exact number of elements of a set remains fixed the expression of the actual numbers will vary for using different bases. We shall use binary and ternary systems somewhere. It may be noted that in any particular system, every digit has equal importance. For considering the decimal systems and decide not to use, say 1. Then

Proposition 4.1 *The set of points in $(-\infty, \infty)$ whose decimal representation does not contain the digit 1 is of measure zero.*

Proof Let E_1 be the set of points in $[0, 1]$ whose decimal representation does not contain 1. Divide $[0, 1]$ into ten equal sub-intervals $[0, 0.1]$ $[0.1, 0.2]$ ... $[0.9, 1]$. In the decimal representation of each point in $(0.1, 0.2)$ the first position is occupied by 1 and since the decimal representation of points of E does not contain 1, the interval $(0.1, 0.2)$ whose measure is $\frac{1}{10}$ lies outside E_1. In the second step divide each of the remaining 9 intervals $[0, 0.1], [0.2, 0.3], \ldots [0.9, 1]$ into ten equal sub-intervals. The sub-intervals of $[0, 0.1]$ are $[0, 0.01], [.01, 0.02] \ldots [0.9, 0.1]$. The decimal representation of each point of $(.01, 0.02)$ contains 1 in its second position and so the sub-interval $(0.01, 0.02)$ whose measure is $\frac{1}{10^2}$ lies outside E_1. Similarly dividing $(0.2, 0.3)$ into ten equal sub-intervals it is seen that $(0.21, 0.22)$ whose measure is $\frac{1}{10^2}$ lies outside E_1 and so considering all the 9 sub-intervals in this step, the sub-intervals which are not in E_1 is of total measure $\frac{9}{10^2}$. In the third step divide each of the remaining 81 sub-intervals into 10 equal sub-intervals and observe that the sub-intervals which contain points having 1 in the third position in its decimal representation is of total measure $\frac{9^2}{10^3}$ and these sub-intervals are not in E_1. This process being continued, the total measure of the sub-intervals which are not in E_1 is

$$\frac{1}{10} + \frac{9}{10^2} + \frac{9^2}{10^3} + \ldots = 1.$$

Hence $\mu(E_1) = 0$. Let E be the set of all points in $(-\infty, \infty)$ whose decimal representation does not contain the digit 1. Then $E = \bigcup_{n=-\infty}^{\infty} E \cap [n, n+1] = \bigcup_{n-\infty}^{\infty} E_n$ where $E_n = E \cap [n, n+1]$. Since $E_n = E_1 + n$ and since $\mu(E_1) = 0$, by Theorem 2.16 $\mu(E_n) = 0$. So $\mu(E) \leq \sum_{n=-\infty}^{\infty} \mu(E_n) = 0$, completing the proof. $\square$

The numbers that are used to determine the strength of a set are called cardinal numbers. The members $1, 2, 3, \ldots$ are cardinal members and they are sufficient to count the elements of a finite set. But for an infinite set, these numbers fail to assess the strength of that set and so infinite cardinal numbers are introduced.

Definition 4.1 Two non-empty sets A and B are said to be equivalent if there is a one-one correspondence between the elements of A and B. If A and B are equivalent then A and B are said to have the same cardinal numbers.

So, a cardinal number is associated with a class of equivalent set, all members of the class having the same cardinal number. If two sets A and B are equivalent we write $A \approx B$. The cardinal number of A is denoted by $\overline{\overline{A}}$. Thus $\overline{\overline{A}} = \overline{\overline{B}}$ if and only if $A \approx B$. The cardinal number of a finite set is the number of elements of that set since a set S containing n elements is such that $S \approx \{1, 2, \ldots n\}$. The cardinal number of the null set ϕ is 0. The cardinal number of the set of all positive integers is denoted by $\aleph_0$ or $\underline{a}$ while the cardinal number of the set of all real numbers is denoted by $\aleph_1$ or $\underline{c}$. The cardinal number $\underline{c}$ is called the power of the continuum.

4.1 Cardinal Members

Definition 4.2 Let u and v be cardinal numbers and let $\overline{\overline{A}} = u, \overline{\overline{B}} = v$. If A is equivalent to a subset of B we write $u \leq v$ and we write $u < v$ if $u \leq v$ but $u \neq v$.

It u, v and w are cardinal numbers then (i) $u \leq u$; (ii) if $u \leq v$ and $v \leq w$ then $u \leq w$ and (iii) if $u \leq v$ and $v \leq u$ then $u = v$.

The first two follow from the definition while the third follows from a theorem known as Schroder-Bernstein theorem.

Definition 4.3 Let u and v be any two cardinal numbers and let $\overline{\overline{A}} = u, \overline{\overline{B}} = v$, and $A \cap B = \emptyset$. (If $A \cap B \neq \emptyset$ take two distinct elements x, y and consider $A_1 = A \times \{x\}, B_1 = B \times \{y\}$., then $\overline{\overline{A_1}} = u, \overline{\overline{B_1}} = v$ and $A_1 \cap B_1 = \emptyset$) Then

(i) the sum $u + v$ is the cardinal numbers of the set $A \cup B$.
(ii) the product uv is the cardinal numbers of the set $A \times B$.
(iii) the exponential u^v is the cardinal number of A^B, that is the set of all functions from B to A.

It can be proved that cardinal numbers possess all the properties of addition, multiplication, and exponential as possessed by the set of positive integers.

Definition 4.4 If X is any set then the power set of X, denoted by $\mathcal{P}(X)$, is the set of all subsets of X.

It is clear that if X is the empty set $\emptyset$ then since $\emptyset$ is the only subset of $\emptyset$, $\mathcal{P}(\emptyset) = \{\emptyset\}$.

Theorem 4.1 *If a set X has cardinal numbers u then the cardinal number of $\mathcal{P}(X)$ is 2^u.*

Proof We show that $\mathcal{P}(X) \approx \{0, 1\}^X$ which is the set of all functions defined on X with values 0 or 1. Let F be the function on $\mathcal{P}(X)$ defined by $F(E) = \chi_E$, the characteristic function of $E \subset X$ i . e. $\chi_E(x) = 1$ if $x \in E$ and $= 0$ if $x \in X \sim E$. Then $\chi_E \in \{0, 1\}^X$. So F is a one-one function from $\mathcal{P}(X)$ onto $\{0, 1\}^X$. For, if $E_1, E_2 \in \mathcal{P}(X)$ and $E_1 \neq E_2$ then $\chi_{E_1} \neq \chi_{E_2}$ and so $F(E_1) \neq F(E_2)$. Again let $f \in \{0, 1\}^X$. Then $f : X \to \{0, 1\}$. Let $E = \{x : x \in X; f(x) = 1\}$. Then $E \in \mathcal{P}(X)$ and $f = \chi_E$ and hence $F(E) = f$. So, $\mathcal{P}(X) \approx \{0, 1\}^X$ which shows that $\overline{\overline{\mathcal{P}(X)}} = \overline{\overline{\{0, 1\}^X}} = \overline{\overline{\{0, 1\}}}^{\overline{\overline{X}}}$ by definition. Since cardinal number of $\{0, 1\}$ is 2 and that of X is u, $\overline{\overline{\mathcal{P}(X)}} = 2^u$. □

Theorem 4.2 (Cantor) *For any set X, $\overline{\overline{X}} < \overline{\overline{\mathcal{P}(X)}}$.*

Proof If $X = \emptyset$, the null set, then since $\emptyset \subset \emptyset$ and $\emptyset$ has no other subset, $\mathcal{P}(X) = \{\emptyset\}$ and so $\overline{\overline{X}} = 0 < 1 = \overline{\overline{\mathcal{P}(X)}}$. Hence the theorem is proved in this case. So, we supposed that $X \neq \emptyset$. Let $\overline{\overline{X}} = u$ and $\overline{\overline{\mathcal{P}(X)}} = v$. The function $f : X \to \mathcal{P}(X)$ defined by $f(x) = \{x\}$ for each $x \in X$, where $\{x\}$ is the set consisting of the single element x, gives a one-one correspondence from X onto a subset of $\mathcal{P}(X)$ and hence $u \leq v$. If possible, suppose that $u = v$. Then there is a one-one correspondence from X onto $\mathcal{P}(X)$. let h be the one-one function from X onto $\mathcal{P}(X)$ which gives this correspondence. Let $S = \{x \in X : x \notin h(x)\}$. Since $S \subset X$, $S \in \mathcal{P}(X)$. So, there is $x_0 \in X$ such that $h(x_0) = S$. If $x_0 \in S$ then $x_0 \notin h(x_0) = S$ which is a contradiction and if $x_0 \notin S$ then $x_0 \in h(x_0) = S$ which is also a contradiction. So, $u \neq v$. Hence $u < v$. □

Corollary 4.3 *If u is any cardinal number then $u < 2^u$.*

Proof Let X be a set such that $\overline{\overline{X}} = u$. Then by Theorem 4.1 $\overline{\overline{\mathcal{P}(X)}} = 2^u$ and by Theorem 4.2 $u < 2^u$. □

Theorem 4.4 *If u and v are cardinal numbers then $2^u 2^v = 2^{u+v}$.*

Proof (First) Suppose u and v are cardinal numbers of two disjoint sets X and Y so 2^u and 2^v are cardinal numbers of $\mathcal{P}(X)$ and $\mathcal{P}(Y)$. Hence we are to show $\mathcal{P}(X) \times \mathcal{P}(Y)$ and $\mathcal{P}(X \cup Y)$ are equivalent sets. Let $(A, B) \in \mathcal{P}(X) \times \mathcal{P}(Y)$. Then the correspondence $(A, B) \to A \cup B$ is one-one. For, if (A, B) and (A', B') are two members of $\mathcal{P}(X) \times \mathcal{P}(Y)$ then either $A \neq A'$ or $B \neq B'$ and so $A \cup B \neq A' \cup B'$. Thus distinct members of $\mathcal{P}(X) \times \mathcal{P}(Y)$ correspond to distinct members of $\mathcal{P}(X \cup Y)$. Finally, let $E \in \mathcal{P}(X \cup Y)$. Then $E \subset X \cup Y$. Since X and Y are disjoint, $E \cap X \subset X$ and $E \cap Y \subset Y$ and so $E \cap X \in \mathcal{P}(X)$ and $E \cap Y \in \mathcal{P}(Y)$. Clearly $(E \cap X, E \cap Y) \in \mathcal{P}(X) \times \mathcal{P}(Y)$ and $(E \cap X, E \cap Y) \to E$. So $\mathcal{P}(X) \times \mathcal{P}(Y)$ and $\mathcal{P}(X \cup Y)$ are equivalent. This completes the proof. □

Proof (Second) Let $X = \{0, 1\}$ are $\overline{\overline{A}} = u$, $\overline{\overline{B}} = v$, $A \cap B = \emptyset$. Then $X^A \times X^B \approx X^{A \cup B}$. For, let $(f, g) \in X^A \times X^B$ where $f \in X^A$ and $g \in X^B$. So, $f : A \to X$ and $g : B \to X$. Let $h : A \cup B \to X$ be such that $h = f$ on A and $h = g$ on B. So $h \in X^{A \cup B}$. Also if $(f_1, g_1) \in X^A \times X^B$ and $(f, g) \neq (f_1, g_1)$ and if (f_1, g_1) and (f, g) correspond to h_1 and h respectively then $h_1 \neq h$. Finally, if $h \in X^{A \cup B}$ then $h : A \cup B \to X$ and if f and g are restrictions of h on A and B respectively then $f : A \to X$ and $g : B \to X$ and so $(f, g) \in X^A \times X^B$. So $X^A \times X^B \approx X^{A \cup B}$. Hence $2^u 2^v = 2^{u+v}$. □

4.2 The Cardinal Numbers $\underline{a}$ and $\underline{c}$

The cardinal number of the set $\mathbb{N}$ of all positive integers and the cardinal number of the set $\mathbb{R}$ of all real numbers are denoted by $\underline{a}$ and $\underline{c}$ respectively. Since the set of all rational numbers $\mathbb{Q}$ is enumerable (i.e. countably infinite), the members of $\mathbb{Q}$ can be enumerated as $r_1, r_2, r_3, \ldots$ and so $\mathbb{Q}$ and $\mathbb{N}$ are equivalent and hence they have the same cardinal numbers. Therefore the cardinal number of $\mathbb{Q}$ is $\underline{a}$.

A real number r is called an algebraic number if r is a root of some polynomial equation $\sum_{i=0}^{n} a_i x^i = 0$ where the coefficients a_i are integers not all 0. Clearly every rational number is an algebraic number but the converse is not true. For, if r is a rational number $\frac{p}{q}$ then it is the root of $p - qx = 0$ and so r is an algebraic number

4.2 The Cardinal Numbers $\underline{a}$ and $\underline{c}$

but $\sqrt{2}$, which is not a rational number, is an algebraic numbers since $\sqrt{2}$ is the root of $2 - x^2 = 0$. So the set of rational numbers is a proper subset of the set of algebraic numbers. The real numbers which are not algebraic are called transcendental numbers.

Theorem 4.5 *The set of all real algebraic numbers is denumerable.*

Proof For each polynomial $f(x) = \sum_{i=0}^{n} a_i x^i$ define the weight of $f(x)$ to be the number $w(f) = n + \sum_{i=0}^{n} |a_i|$. Clearly every non-constant polynomial has weight ≥ 2. There are only a finite number of polynomials having a given weight. Arrange all polynomials first in the order of their weight and then in the order of their coefficients $a_0, a_1, a_2, \ldots$, taking first the non-negative a_i's and then the negative a_i's for a given i starting from $i = 0, 1, 2, \ldots$. That is, first in order of n then in order of a_0 then in order of a_1 and so on. Taking first the polynomials of weight 2 and then those of weight 3 and so on. Continuing this we get a sequence of polynomials $f_1, f_2, f_3 \ldots$ in which every polynomial appears just once. Each polynomial has at most a finite number of real roots. Number the roots of f_1 in the order in which they occur on the real line and then those of f_2 and so on, omitting those which are already numbered. In this way, we obtain a definite enumeration of all real algebraic numbers. □

Corollary 4.6 *The cardinal number of the set of all algebraic numbers is $\underline{a}$.*

Theorem 4.7 *Every interval $(a, b), [a, b], [a, b), (a, b]$ on the real line has the cardinal number $\underline{c}$.*

Proof Let $f : \mathbb{R} \to (-1, 1)$ be defined by

$$f(x) = \frac{x}{\sqrt{(1 + x^2)}}, \quad x \in \mathbb{R}$$

Then f gives a one-one correspondence between elements of $\mathbb{R}$ and $(-1, 1)$. Again let $g : (-1, 1) \to (a, b)$ be defined by

$$g(t) = a + \frac{(t + 1)}{2}(b - a), \quad t \in (-1, 1)$$

Then g gives a one-one correspondence between elements of $(-1, 1)$ and (a, b). So, the composite function $g \circ f : \mathbb{R} \to (a, b)$ gives a one-one correspondence between the elements of $\mathbb{R}$ and (a, b). Hence $\mathbb{R} \approx (a, b)$ and so $\underline{c} = \overline{\overline{\mathbb{R}}} = \overline{\overline{(a, b)}}$.

Since $(a, b) \subset [a, b] \subset \mathbb{R}$, $\overline{\overline{(a, b)}} \leq \overline{\overline{[a, b]}} \leq \overline{\overline{\mathbb{R}}}$ i.e. $\underline{c} \leq \overline{\overline{[a, b]}} \leq \underline{c}$ which gives $\overline{\overline{[a, b]}} = \underline{c}$. Similarly $\overline{\overline{[a, b)}} = \underline{c} = \overline{\overline{(a, b]}}$. □

Theorem 4.8 *If $\underline{a}$ and $\underline{c}$ are the cardinal numbers of the set of positive integers and of the set of real numbers then $2^{\underline{a}} = \underline{c}$.*

Proof Let $A = \{0, 1\}^{\mathbb{N}}$ where $\mathbb{N}$ is the set of positive integers. We show that $\overline{\overline{A}} = \underline{c}$. Let $t \in A$. Then t is a function from $\mathbb{N}$ with values 0 or 1. So, t is a sequence whose terms are either 0 or 1. That is $t = \{t_1, t_2, ...t_n, ...\}$ where $t_n = 0$ or 1 for all $n, n = 1, 2, ...$ Define $f : A \to \mathbb{R}$ by

$$f(t) = \sum_{n=1}^{\infty} \frac{t_n}{2^n}, \quad \text{if } t_n = 0 \text{ for infinite number of values of } n$$

$$= 1 + \sum_{n=1}^{\infty} \frac{t_n}{2^n} \text{ otherwise.}$$

We show that f is one-one. Let $t, t' \in A$, $t \neq t'$ and let $t = \{t_1, t_2, ..., t_n, ...\}$ and $t' = \{t'_1, t_2, ..., t'_n, ...\}$. Since $t \neq t'$, there is n such that $t_n \neq t'_n$. Let k be the smallest positive integer such that $t_k \neq t'_k$. Let $t_k = 0, t'_k = 1$. Then $t = \{t_1, t_2, ..., t_{k-1}, 0, t_{k+1}, t_{k+2}, ...\}$ and $t' = \{t_1, t_2, ..., t_{k-1}, 1, t'_{k+1}, t'_{k+2}, ...\}$.
Case-I: Let t contain infinite number of 0. Then

$$f(t) = \sum_{n=1}^{k-1} \frac{t_n}{2^n} + \frac{0}{2^k} + \sum_{n=k+1}^{\infty} \frac{t_n}{2^n}.$$

Since infinite number of term of $\sum_{n=k+1}^{\infty} \frac{t_n}{2^n}$ vanish,

$$\sum_{n=k+1}^{\infty} \frac{t_n}{2^n} < \sum_{n=k+1}^{\infty} \frac{1}{2^n} = \frac{1}{2^k} \text{ and so } f(t) < \sum_{n=1}^{k-1} \frac{t_n}{2^n} + \frac{1}{2^k}$$

Also since

$$f(t') = \sum_{n=1}^{k-1} \frac{t_n}{2^n} + \frac{1}{2^k} + \sum_{n=k+1}^{\infty} \frac{t'_n}{2^n}$$

Or

$$f(t') = 1 + \sum_{n=1}^{k-1} \frac{t_n}{2^n} + \frac{1}{2^k} + \sum_{n=k+1}^{\infty} \frac{t'_n}{2^n}$$

according as t' contains infinite number of 0 or not, $f(t) < f(t')$ and so $f(t) \neq f(t')$.

Case-II: Let t contain only finite number of 0. Then

$$f(t) = 1 + \sum_{n=1}^{k-1} \frac{t_n}{2^n} + \frac{0}{2^k} + \sum_{n=k+1}^{\infty} \frac{t_n}{2^n}$$

4.2 The Cardinal Numbers a and c

If t' contains only finite number 0 then an infinite number of terms of $\sum_{n=k+1}^{\infty} \frac{t'_n}{2^n}$ are $1/2^n$ and so

$$f(t') = 1 + \sum_{n=1}^{k-1} \frac{t_n}{2^n} + \frac{1}{2^k} + \sum_{n=k+1}^{\infty} \frac{t'_n}{2^n} > 1 + \sum_{n=1}^{k-1} \frac{t_n}{2^n} + \frac{1}{2^k} \geq f(t)$$

which gives $f(t) \neq f(t')$ and if t' contains an infinite number of 0 then

$$\sum_{n=k+1}^{\infty} \frac{t'_n}{2^n} < \sum_{n=k+1}^{\infty} \frac{1}{2^n} = \frac{1}{2^k}$$

and so

$$f(t') = \sum_{n=1}^{k-1} \frac{t_n}{2^n} + \frac{1}{2^k} + \sum_{n=k+1}^{\infty} \frac{t'_n}{2^n} < \sum_{n=1}^{k-1} \frac{t_n}{2^n} + \frac{2}{2^k} \leq \sum_{n=1}^{k-1} \frac{t_n}{2^n} + 1$$

$$< \sum_{n=1}^{k-1} \frac{t_n}{2^n} + 1 + \sum_{n=k+1}^{\infty} \frac{t_n}{2^n} = f(t)$$

which also gives $f(t) \neq f(t')$.

Thus in any case $f(t) \neq f(t')$ whenever $t, t' \in A$ and $t \neq t'$. So f gives a one-one correspondence between the elements of A and $f(A)$, where $f(A)$ is the image set of A for f. Hence $\overline{\overline{A}} = \overline{\overline{f(A)}}$. We show that $\overline{\overline{f(A)}} = c$. Let $\xi \in (0, 1)$. Expand ξ in the binary system so that

$$\xi = \sum_{n=1}^{\infty} \frac{\xi_n}{2^n} \text{ where } \xi_n = 0 \text{ or } 1 \text{ for all } n, n = 1, 2...$$

If $\xi_n = 0$ for infinite number of values of n then writing $t = \{\xi_1, \xi_2, ...\xi_n, ...\}$ we have

$$f(t) = \sum_{n=1}^{\infty} \frac{\xi_n}{2^n} = \xi.$$

and if $\xi_n = 0$ for only finite number of values of n then there is k such that $\xi_k = 0$ and $\xi_n = 1$ for all $n > k$ and so

$$\xi = \sum_{n=1}^{k-1} \frac{\xi_n}{2^n} + \frac{0}{2^k} + \sum_{n=k+1}^{\infty} \frac{\xi_n}{2^n} = \sum_{n=1}^{k-1} \frac{\xi_n}{2^n} + \frac{1}{2^k} = \sum_{n=1}^{k-1} \frac{\xi_n}{2^n} + \frac{1}{2^k} + \sum_{n=k+1}^{\infty} \frac{0}{2^n},$$

and so writing $t = \{\xi_1, \xi_2, ..\xi_{k-1}, 1, 0, 0, ..\}$ we have $f(t) = \xi$. So, $(0, 1) \subset f(A) \subset \mathbb{R}$. Hence $\underline{c} = \overline{\overline{(0, 1)}} \leq \overline{\overline{f(A)}} \leq \overline{\overline{\mathbb{R}}} = \underline{c}$ and so $\overline{\overline{f(A)}} = \underline{c}$. Since $\overline{\overline{A}} = \overline{\overline{f(A)}}$, $\overline{\overline{A}} = \underline{c}$. Also $A = \{0, 1\}^N$ and hence $\overline{\overline{A}} = 2^{\underline{a}}$. So $2^{\underline{a}} = \underline{c}$, completing the proof. □

From Corollary 4.3 we get $\underline{a} < 2^{\underline{a}}$. The natural question is whether there is any cardinal number between $\underline{a}$ and $2^{\underline{a}}$. The continuum hypothesis gives the answer.
Continuum hypothesis: There is no cardinal number between $\underline{a}$ and $2^{\underline{a}}$. That is, there is no cardinal number between $\underline{a}$ and $\underline{c}$.

Theorem 4.9 *If $\underline{a}$ and $\underline{c}$ are the cardinal numbers of the set of positive integers $\mathbb{N}$ and of the set of real number $\mathbb{R}$ respectively then*

(i) $\underline{a} + \underline{a} = \underline{a}$, $\underline{a}\underline{a} = \underline{a}$, $\underline{c} + \underline{c} = \underline{c}$, $\underline{c}\underline{c} = \underline{c}$
(ii) $n^{\underline{a}} = \underline{a}^{\underline{a}} = \underline{c}^{\underline{a}} = \underline{c}$, $n^{\underline{c}} = \underline{a}^{\underline{c}} = \underline{c}^{\underline{c}}$, n being a positive integer ≥ 2.
(iii) $\underline{a} + \underline{c} = \underline{c}$, $\underline{a}\underline{c} = \underline{c}$

Proof Since the union of two countable sets is countable and the Cartesian product of two countable sets is countable, the first two relations of (i) are clear. Since $(0, 1) \cup (2, 3) \subset \mathbb{R}$, $\underline{c} + \underline{c} \leq \underline{c} \leq \underline{c} + \underline{c}$ which gives the third of (i), and finally $\underline{c}\underline{c} = 2^{\underline{a}}2^{\underline{a}} = 2^{\underline{a}+\underline{a}} = 2^{\underline{a}} = \underline{c}$ which completes the proof of (i). To prove (ii), we have $\underline{c} = 2^{\underline{a}} \leq n^{\underline{a}} \leq \underline{a}^{\underline{a}} \leq \underline{c}^{\underline{a}} = (2^{\underline{a}})^{\underline{a}} = 2^{\underline{a}\underline{a}} = 2^{\underline{a}} = \underline{c}$ which gives the first relation of (ii) while $\underline{c}^{\underline{c}} = (2^{\underline{a}})^{\underline{c}} = 2^{\underline{a}\underline{c}} \leq n^{\underline{a}\underline{c}} \leq n^{\underline{c}\underline{c}} = n^{\underline{c}} \leq \underline{a}^{\underline{c}} \leq \underline{c}^{\underline{c}}$ which proves the second relation of (ii).
To prove (iii) let B be the set of rational numbers in $(1, 2)$. Then $[0, 1] \subset [0, 1] \cup B \subset [0, 1] \cup (1, 2] = [0, 2]$. So $\underline{a} \leq \underline{c} + \underline{a} \leq \underline{a}$ giving $\underline{a} + \underline{c} = \underline{a}$, proving the first relation of (iii). For the second, $\{\frac{1}{2}\} \times [0, 1] \subset B \times [0, 1] \subset (1, 2) \times [0, 1]$ where B is as above. Since $\{\frac{1}{2}\} \times [0, 1] \approx [0, 1]$ we have $\underline{c} \leq \underline{a}\underline{c} \leq \underline{c}\underline{c}$ and hence by the last of (i) $\underline{c} = \underline{a}\underline{c}$, completing the proof.

Example 4.1 The set of all sequences $\{x_n\}$ such that $x_n = 0$ or 1 for all n, $n = 1, 2, ...$ has the cardinal number $\underline{c}$.

A sequence $\{x_n\}$ is a function f defined on the set of positive integers $\mathbb{N}$ defined by $f(n) = x_n$. So if S is the set of all such sequences, $S = \{0, 1\}^{\mathbb{N}}$. Hence $\overline{\overline{S}} = \overline{\overline{\{0, 1\}^{\mathbb{N}}}} = 2^{\underline{a}} = \underline{c}$.

Example 4.2 The set of all sequences of rational numbers has a cardinal number is $\underline{c}$.

If $\{x_n\}$ is any sequence of rational numbers then $\{x_n\} \in Q^{\mathbb{N}}$ where Q is the set of rational numbers. So, $Q^{\mathbb{N}}$ is the set of all sequence $\{x_n\}$ such that $x_n \in Q$ for all n. Since the cardinal number of $Q^{\mathbb{N}}$ is $\underline{a}^{\underline{a}} = \underline{c}$, the result follows.

Example 4.3 The set of all sequences of real numbers has cardinal number $\underline{c}$.

As in Example 4.2, the set of such sequence is $\mathbb{R}^{\mathbb{N}}$ and the cardinal number of $\mathbb{R}^{\mathbb{N}}$ is $\underline{c}^{\underline{a}} = \underline{c}$.

4.3 Properties of Set

Definition 4.5 A set $A \subset \mathbb{R}$ is said to be dense in a set $B \subset \mathbb{R}$ if every point of B is either a point of A or a limit point of A. That is $B \subset \overline{A}$ where $\overline{A}$ is the closure of A.

A set $A \subset \mathbb{R}$ is said to be everywhere dense or simply dense if A is dense in every interval.

Definition 4.6 A set $A \subset \mathbb{R}$ is said to be dense in itself if every point of A is a limit point of A.

Definition 4.7 A set $A \subset \mathbb{R}$ is said to be perfect if A is closed and dense in itself.

Theorem 4.10 *Every non-void perfect set has the power of the continuum, i.e. has the cardinal number $\underline{c}$.*

Proof Let P be a non-void perfect set. We show that $\overline{\overline{P}} = \underline{c}$. Let $x \in P$. Let I be an open interval containing x. Since x is a limit point of P, the set $P \cap I$ is infinite. Choose two points x_0, x_1 in $P \cap I$ and disjoint open intervals I_0, I_1 having no common endpoints such that

$$x_{i_1} \subset I_{i_1} \subset I, |I_{i_1}| < 1, i_1 \in \{0, 1\}$$

were $|I_{i_1}|$ is the length of I_{i_1}.

Since x_0 and x_1 are limit points of P the sets $P \cap I_0$ and $P \cap I_1$ are infinite. Choose two points x_{00} and x_{01} from $P \cap I_0$ and two points x_{10} and x_{11} from $P \cap I_1$. Let I_{00}, I_{01}, I_{10} and I_{11} be the disjoint open intervals having no common endpoints such that

$$x_{i_1 i_2} \in I_{i_1 i_2} \subset I_{i_1}, |I_{i_1 i_2}| < \frac{1}{2}, i_k \in \{0, 1\}, k = 1, 2.$$

Continuing the process we get after nth step the set of points $\{x_{i_1 i_2 \ldots i_n}\}$ and disjoint open intervals having no common endpoints $\{I_{i_1 i_2 \ldots i_n}\}$, such that

$$x_{i_1 i_2 \ldots i_n} \in I_{i_1 i_2 \ldots i_n} \subset I_{i_1 i_2 \ldots i_{n-1}}, |I_{i_1 i_2 \ldots i_n}| < \frac{1}{n}, \ i_k \in \{0, 1\}, \ k = 1, 2, 3, \ldots, n.$$

This process being continued we get for each sequence $i_1, i_2, i_n, \ldots$ where $i_n \in \{0, 1\}$, a sequence of disjoint open intervals $I_{i_1}, I_{i_1 i_2}, \ldots, I_{i_1 i_2 \ldots i_n}, \ldots$ with the above property. So, by Cantor's theorem on nested intervals there is a unique point, say ξ, such that

$$\overline{I}_{i_1} \cap \overline{I}_{i_1 i_2} \cap \ldots \cap \overline{I}_{i_1 i_2 \ldots i_n \ldots} \cap \ldots = \{\xi\},$$

where $\overline{I}_{i_1}$ denotes the closure of I_{i_1}. Since ξ is the limit of the sequence $x_{i_1}, x_{i_1 i_2}, \ldots x_{i_1 i_2 \ldots i_n} \ldots$ each of which is a member of P, ξ is a limit point of P and since P is closed, $\xi \in P$. Let $i'_1, i'_2, \ldots i'_n, \ldots$, where $i'_n \in \{0, 1\}$ for each n, be a different sequence. Then there is k such that $i_k \neq i'_k$ and so $I_{i_1 i_2 \ldots i_n} \cap I_{i'_1 i'_2 \ldots i'_n} = \emptyset$ for $n \geq k$. So, if ξ' is the point corresponding to this sequence then $\xi \neq \xi'$, and $\xi' \in P$. So, the set

S of all sequence whose elements are either 0 or 1 is such that the elements of S are in one-one correspondence with the elements of a subset of P. Since S has the cardinal number $2^{\underline{a}} = \underline{c}$ (see Example 4.1), $\underline{c} \leq \overline{\overline{P}}$. Also $P \subset \mathbb{R}$ and hence $\overline{\overline{P}} \leq \overline{\overline{\mathbb{R}}} = \underline{c}$. Thus $\overline{\overline{P}} = \underline{c}$. □

Definition 4.8 A set $A \subset \mathbb{R}$ is said to be nowhere dense if the closure of A contains no interval. In other words A is nowhere dense if every intervals I containing points of A there is another interval $J \subset \mathbb{R}$ such that $J \subset I$ and $J \cap A = \emptyset$.

Clearly, every subset of a nowhere dense set is again a nowhere dense set. Note the difference between not dense and nowhere dense.

Definition 4.9 A set $A \subset \mathbb{R}$ is said to be of the first category if A is a countable union of nowhere dense sets.

Clearly, every subset of a set of the first category is again a set of the first category.

Definition 4.10 A set $A \subset \mathbb{R}$ is said to be of the second category if A is not a set of first category.

Definition 4.11 A set $A \subset \mathbb{R}$ is said to be residual if its complement $\widetilde{A}$ is of the first category.

Remark The notions described in Definitions 4.8 to 4.11 are also considered relative to a set instead of $\mathbb{R}$. For example, in Definition 4.8 if $A \subset B \subset \mathbb{R}$, then A is said to be nowhere dense in B every interval I containing points of A there is another interval J containing points of B such that $J \subset I$ and $J \cap A = \emptyset$.

Definition 4.11 needs the following theorem.

Theorem 4.11 *A set of the first category cannot contain any interval.*

Proof Let A be a set of the first category and let $A = \cup A_n$ where A_n is nowhere dense for each n. If possible, let A contain an interval (α, β). Since A_1 is nowhere dense there is a closed interval $[\alpha_1, \beta_1] \subset (\alpha, \beta)$ such that $[\alpha_1, \beta_1] \cap A_1 = \emptyset$ and $\beta_1 - \alpha_1 < 1$. Since A_2 is nowhere dense, there is a closed interval $[\alpha_2, \beta_2] \subset [\alpha_1, \beta_1]$ such that $[\alpha_2, \beta_2] \cap A_2 = \emptyset$ and $\beta_2 - \alpha_2 < \frac{1}{2}$. Continuing this process we get a sequence of closed intervals $\{[\alpha_n, \beta_n]\}$ such that for all n

$$[\alpha_{n+1}, \beta_{n+1}] \subset [\alpha_n, \beta_n] \text{ and } \beta_{n+1} - \alpha_{n+1} < \frac{1}{n+1}.$$

So, there is a point ξ which is such that $\bigcap_{n=1}^{\infty} [\alpha_n \beta_n] = \{\xi\}$. Now $\xi \in (\alpha, \beta)$ but $\xi \notin A_n$ for all n. Hence $\xi \notin A$ which is a contradiction. □

The above theorem shows that a residual set cannot be an empty set. We prove :

Theorem 4.12 *Every residual set has the power of the continuum.*

4.3 Properties of Set   71

Proof Let F be a residual set. Let $G = \tilde{F}$, the complement of F. Then G is of the first category. Let $G = \bigcup_{n=1}^{\infty} G_n$, where G_n is nowhere dense for all n. Since G_1 is nowhere dense, we choose two disjoint closed intervals I_0 and I_1 such that $I_{i_1} \cap G_1 = \emptyset, |I_{i_1}| < 1$ for $i_1 \in \{0, 1\}$, where $|I_{i_1}|$ is the length of I_{i_1}. Since G_2 is nowhere dense, we choose disjoint closed intervals I_{00}, I_{01}, I_{10} and I_{11} such that

$$I_{i_1 i_2} \subset I_{i_1}, I_{i_1 i_2} \cap G_2 = \emptyset, |I_{i_1 i_2}| < \frac{1}{2}, i_k \in \{0, 1\}, k = 1, 2$$

Continuing this process we get after nth step the set of disjoint closed intervals $\{I_{i_1, i_2,...i_n} : i_k \in \{0, 1\}; k = 1, 2, ..., n\}$ such that

$$I_{i_1 i_2..i_n} \subset I_{i_1 i_2..i_{n-1}}, I_{i_1 i_2..i_n} \cap G_n = \emptyset, |I_{i_1 i_2..i_n}| < \frac{1}{n}.$$

If this process is continued we get for each sequence $i_1, i_2, ..i_n, ..$ where $i_n \in \{0, 1\}$, a sequence of disjoint closed intervals $I_{i_1}, I_{i_1 i_2}, ..I_{i_1, i_2..i_n,...}$ with the above property. By Cantor's theorem there is a unique point ξ such that

$$I_{i_1} \cap I_{i_1 i_2} \cap ... \cap I_{i_1 i_2..i_n} \cap ... = \{\xi\}.$$

So, for each sequence $i_1, i_2, ...i_n, ...$ where $i_n \in \{0, 1\}$ there is a unique point ξ. Since ξ belongs to all the interval it cannot belong to any G_n and so $\xi \in F$. If $i'_1, i'_2, ...i'_n, ...$ where $i'_n \in \{0, 1\}$ for each n, is a different sequence then there is k such that $i_k \neq i'_k$ and so $I_{i_1 i_2...i_n} \cap I_{i'_1 i'_2...i'_n} = \emptyset$ for $n \geq k$. So, if ξ' is the point corresponding to this sequence then $\xi \neq \xi'$ and $\xi' \in F$. So, the set S of all sequence whose elements are either 0 or 1 is such that the elements of S are in one-one correspondence with the elements of a subset of F. Since the cardinal number of S is $\underline{c}$ (see Example 4.1), $\underline{c} \leq \overline{\overline{F}}$ and since $F \subset \mathbb{R}, \overline{\overline{F}} \leq \overline{\overline{\mathbb{R}}} = \underline{c}$. Hence $\overline{\overline{F}} = \underline{c}$. □

Theorem 4.13 (Baire Category Theorem) *A non-void closed set F is not a set of the first category in itself.*

Proof Suppose, if possible that F is of the first category in itself. So, $F = \bigcup_{n=1}^{\infty} A_n$ where each A_n is nowhere dense in F. Hence there is a point $x_1 \in F$ such that $x_1 \notin \overline{A_1}$, where $\overline{A_1}$ is the closure of A_1. So there is a positive number $\delta_1 < 1$ such that the interval $[x_1 - \delta_1, x_1 + \delta_1]$ contains no point of A_1. Since A_2 is nowhere dense in F there is $x_2 \in F \cap (x_1 - \delta_1, x_1 + \delta_1)$ such that $x_2 \notin \overline{A_2}$. So there is $\delta_2, 0 < \delta_2 < \frac{1}{2}$, such that $[x_2 - \delta_2, x_2 + \delta_2]$ contains no point of A_2 and $[x_2 - \delta_2, x_2 + \delta_2] \subset [x_1 - \delta_1, x_1 + \delta_1]$. Continuing this process we get a sequence of points $\{x_n\}$ and a sequence $\{\delta_n\}$ such that for all n

(i) $[x_{n+1} - \delta_{n+1}, x_{n+1} + \delta_{n+1}] \subset [x_n - \delta_n, x_n + \delta_n]$
(ii) $[x_n - \delta_n, x_n + \delta_n] \cap A_n = \emptyset$
(iii) $0 < \delta_n < \frac{1}{n}$.

By (i) and (iii) there is by Cantor's theorem one and only one point x_0 which is common to all intervals $[x_n - \delta_n, x_n + \delta_n]$. Clearly $\lim_{n \to \infty} x_n = x_0$. Since $x_n \in F$ for all n, $x_0 \in F$. But by (ii) $x_0 \notin A_n$ for all n and so $x_0 \notin \bigcup_{n=1}^{\infty} A_n$ This is a contradiction since $F = \bigcup_{n=1}^{\infty} A_n$. This completes the proof $\square$

The above theorem can also be stated as

Theorem 4.14 *If F is a closed set and if $F = \cup A_n$ then there is at least one A_n which is dense in a portion of F.*

Theorem 4.15 *The intersection of a countable number of dense G_δ—sets is again a dense G_δ—set.*

Proof Let $\{A_n\}$ be a sequence of dense G_δ—sets. Let $A_n = \bigcap_{k=1}^{\infty} V_{n,k}$ where $V_{n,k}$ is open for each pair n, k. Let $A = \bigcap_{n=1}^{\infty} A_n$. Then $A = \bigcap_{n,k=1}^{\infty} V_{n,k}$ and so A is a G_δ—set. Also $V_{n,k}$ is dense set for each pair n, k. For, suppose that there are n_0, k_0 such that V_{n_0,k_0} is not everywhere dense. Then there are x_0 and $r_0 > 0$ such that $[x_0 - r_0, x_0 + r_0] \cap V_{n_0,k_0} = \emptyset$ and so $[x_0 - r_0, x_0 + r_0] \cap A_{n_0} = \emptyset$ which is a contradiction since A_{n_0} is everywhere dense. To show A is dense, suppose, if possible, let there are x_0 and $r_0 > 0$ such that $[x_0 - r_0, x_0 + r_0] \cap A = \emptyset$. Thus

$$[x_0 - r_0, x_0 + r_0] \subset \tilde{A} = \bigcup_{n,k=1}^{\infty} \tilde{V}_{n,k} \tag{4.1}$$

Since $V_{n,k}$ is everywhere dense $\tilde{V}_{n,k}$ is nowhere dense for each pair n, k. Hence the right-hand side of (4.1) is a set of the first category. But this is a contradiction by Theorem 4.11. $\square$

4.4 Cantor Sets

Consider the closed interval $[0, 1]$ and any real number α, $0 < \alpha \leq 1$. Remove from $[0, 1]$ the open interval with center $\frac{1}{2}$ and of length $\frac{\alpha}{3}$ and call it I_1^1. Let J_1^1 and J_2^1 be the two remaining closed intervals each of length $\frac{1}{2}(1 - \frac{\alpha}{3})$, J_1^1 lying on the left of J_2^1. Again remove the central open intervals I_1^2 and I_2^2 each of length $\frac{\alpha}{3^2}$ from J_1^1 and J_2^1 respectively to get four closed intervals J_1^2, J_2^2, J_3^2 and J_4^2 each of length $\frac{1}{2}(\frac{1}{2}(1 - \frac{\alpha}{3}) - \frac{\alpha}{3^2}) = \frac{1}{4}(1 - \frac{\alpha}{3} - \frac{2\alpha}{3^2})$. Continuing the process remove at the nth step the open intervals I_k^n, $1 \leq k \leq 2^{n-1}$, each of length $\frac{\alpha}{3^n}$ and get the remaining closed intervals J_k^n, $1 \leq k \leq 2^n$ each of length $\frac{1}{2^n}(1 - \frac{\alpha}{3} - \ldots - \frac{2^{n-1}\alpha}{3^n})$. The process being continued indefinitely, the removed open intervals are $\{I_k^n : 1 \leq k \leq 2^{n-1}, n = $

1, 2, ...}. Let $A = \bigcup_{n=1}^{\infty} \bigcup_{k=1}^{2^{n-1}} I_k^n$. Then A is an open set and the open intervals I_k^n being disjoint.

$$\mu(A) = \sum_{n=1}^{\infty} \sum_{k=1}^{2^{n-1}} |I_k^n| = \sum_{n=1}^{\infty} \sum_{k=1}^{2^{n-1}} \frac{\alpha}{3^n} = \sum_{n=1}^{\infty} \frac{2^{n-1}\alpha}{3^n} = \frac{\alpha}{3} \sum_{n=1}^{\infty} (\frac{2}{3})^{n-1} = \alpha.$$

Let $C_\alpha = [0, 1] \sim A$. Then C_α is closed and $\mu(C_\alpha) = 1 - \alpha$. The set C_α is called the Cantor set corresponding to α.

From the construction, it follows that if

$$A_n = \bigcup_{k=1}^{2^{n-1}} I_k^n \text{ and } B_n = \bigcup_{k=1}^{2^n} J_k^n$$

then $A = \bigcup_{n=1}^{\infty} A_n$ and so $C_\alpha = [0, 1] \sim A = [0, 1] \sim \bigcup_{n=1}^{\infty} A_n = \bigcap_{n=1}^{\infty} B_n$.

The Cantor sets C_α have the following properties.

(*I*) For any fixed α, $0 < \alpha \le 1$, the set C_α is a nowhere dense perfect set.
For, since C_α is closed, we prove that C_α contains no interval. Suppose, if possible then $(a, b) \subset C_\alpha$. Then $(a, b) \subset \bigcap_{n=1}^{\infty} B_n$ and so $(a, b) \subset B_n$ for all n. So, $(a, b) \subset J_k^n$ for all n and for some k, $1 \le k \le 2^n$. Since J_k^n is of length $\frac{1}{2^n}(1 - \frac{\alpha}{3} - \dots - \frac{2^{n-1}\alpha}{3^n}) < \frac{1}{2^n}$, (a, b) is of length $< \frac{1}{2^n}$ for all n and hence $b - a = 0$ which is a contradiction. So C_α is nowhere dense.

To prove that C_α is perfect note that all the endpoints of the removed intervals I_k^n are in C_α and these endpoint are also the limit points of the endpoints. All other points of are also the limit points of the endpoints of the intervals I_k^n. Therefore C_α is dense in itself. Hence C_α is perfect.

(*II*) If $0 < \alpha_1 < \alpha_2 \le 1$, then $C_{\alpha_2} \subset C_{\alpha_1}$
For, each open interval which are removed corresponding to α_1 is a subset of the interval corresponding to α_2 and so with the above notation if $A_n(\alpha_1)$ and $A_n(\alpha_2)$ correspond to α_1 and α_2 respectively, then $A_n(\alpha_1) \subset A_n(\alpha_2)$ and hence $C_{\alpha_2} \subset C_{\alpha_1}$.

4.5 Cantor Ternary Set

If in the above construction we take $\alpha = 1$ then the corresponding set C_1 is called Cantor ternary set, because in this case all the closed intervals $[0, 1]$, J_k^n, $1 \le k \le 2^n$, $n = 1, 2, \dots$ are divided into three sub-intervals of equal length. From property *I* of the Cantor sets proved above it follows that the Cantor ternary set is a nowhere dense perfect set and so by Theorem 4.10 the Cantor set C_1 has the cardinal number $\underline{c}$. This can also be directly proved. Representing each point $x = 0.x_1x_2x_3\ldots$ in $[0, 1]$ in ternary scale, the points in each $I_k^n, 1 \leq k \leq 2^{n-1}, n = 1, 2, \ldots$ must have $x_i = 1$ for at lest one i and the points in $J_k^n, 1 \leq k \leq 2^n, n = 1, 2, \ldots$ are such that $x_i = 0$ or 2 for each i except that the endpoints of I_k^n have two representations namely for the interval $I_1^1, \frac{1}{3} = 0.1000\ldots$ or $0.0222\ldots$ and $\frac{2}{3} = 0.2000\ldots$ or $0.1222\ldots$ and so taking $\frac{1}{3} = 0.0222\ldots$ and $\frac{2}{3} = 0.2000\ldots$ we can avoid using 1 in the representation of the points in C_1. (Note that $0 = 0.000\ldots$ and $1 = .222\ldots$). So the points of C_1 are in one-one correspondence with the set of all sequences whose terms are either 0 or 2, and hence $\overline{\overline{C_1}} = 2^{\underline{a}} = \underline{c}$. (using Theorems 4.1, 4.8; see also Example 4.1). Thus the Cantor ternary set is a nowhere dense perfect set of measure zero having cardinal number $\underline{c}$.

Remarks

(i) By Theorem 2.1 (*iv*) and Theorem 2.3 (ii) it follows that every countable set is of measure zero. The Cantor ternary set shows that the converse is not true. That is, there exists a set of measure zero which is not countable.

(ii) Since every subset of a nowhere dense set is nowhere dense, every subset of a set first category is of the first category.

(iii) It should be noted that each Cantor set $C_\alpha, 0 < \alpha \leq 1$, being nowhere dense, is a set of the first category. But since it is closed it is not a set of the first category in itself, by Baire category theorem (Theorem 4.13).

(iv) A set of the first category may have full measure and a set of the second category may have measure zero. For, let $E_n = C_{\frac{1}{n}}$ where $C_{\frac{1}{n}}$ is the Cantor set C_α corresponding to $\alpha = \frac{1}{n}$, and let $E = \bigcup_{n=1}^{\infty} E_n$. Then E is of the first category. Since $0 < \alpha_1 < \alpha_2 \leq 1$ implies $C_{\alpha_2} \subset C_{\alpha_1}$, $\{E_n\}$ is an increasing sequence and hence $E = \lim_{n\to\infty} E_n$. So by Theorem 2.8

$$\mu(E) = \lim_{n\to\infty} \mu(E_n) = \lim_{n\to\infty} \mu(C_{\frac{1}{n}}) = \lim_{n\to\infty} (1 - \frac{1}{n}) = 1.$$

4.6 Cardinality of the σ-Algebra of All Borel Sets

Theorem 4.16 *The cardinality of the σ-algebra of all Borel sets in $[0, 1]$ is $\underline{c}$.*

To prove the theorem we consider the following steps:
Step I. The family of all open intervals in $[0, 1]$ has cardinality $\underline{c}$.

Proof Let $\mathcal{I}$ be the family of all open intervals in $[0, 1]$. Then the elements of $\mathcal{I}$ are in one-one correspondence with the elements of the set $E \subset [0, 1] \times [0, 1]$ where $E = \{(x, y) : x < y; x, y \in [0, 1]\}$. So, we are to show that $\overline{\overline{E}} = \underline{c}$. Since

4.6 Cardinality of the σ-Algebra of All Borel Sets

$E \subset [0, 1] \times [0, 1]$, $\overline{\overline{E}} \leq \underline{c} \cdot \underline{c} = \underline{c}$. Again consider the set $S = \{(0, y) : 0 < y < 1\}$. Then $S \subset E$ and so $\overline{\overline{S}} \leq \overline{\overline{E}}$. But the elements of S are in one-one correspondence with the points of the interval $(0, 1)$ and so by Theorem 4.7, $\overline{\overline{S}} = \underline{c}$. Hence $\underline{c} \leq \overline{\overline{E}}$. So, $\overline{\overline{E}} = \underline{c}$, completing the proof. □

Step II. The family of all open sets in $[0, 1]$ has cardinality $\underline{c}$, and the family of all closed sets in $[0, 1]$ has cardinality $\underline{c}$.

Proof Let $\mathcal{G}$ be the family of all open sets in $[0, 1]$ and let $G \in \mathcal{G}$. Then G is a countable union of disjoint open intervals in $[0, 1]$. Since the union of sets is independent of the order and since a definite order will be needed we are to arrange the constituent open intervals of G. We arrange these intervals in the decreasing order of there lengths and then their position on the real line. That is, if there is any interval I such that its length $|I| > \frac{1}{2}$ then call it I_1 and then consider the intervals I such that $\frac{1}{3} < |I| \leq \frac{1}{2}$ and then $\frac{1}{4} < |I| \leq \frac{1}{3}$ and so on. If there are two or more intervals of equal length then order them according to there position on $[0, 1]$. Under this arrangement we take $G = \bigcup_n I_n$ and so G corresponds to a unique sequence of intervals $\{I_1, I_2, I_3, \ldots\}$. Also each sequence of disjoint open intervals in $[0, 1]$ arranged in the decreasing order of their lengths and then there position on $[0, 1]$ corresponds to a member of $\mathcal{G}$ and this correspondence is one-one. So, if $\mathcal{I}$ is the family of all open intervals on $[0, 1]$ then $\mathcal{G}$ is equivalent to a subset of $\mathcal{I}^{\mathbb{N}}$ and so $\overline{\overline{\mathcal{G}}} \leq \overline{\overline{\mathcal{I}}}^{\mathbb{N}}$ and therefore by step I, $\overline{\overline{\mathcal{G}}} \leq \underline{c}^{\underline{a}} = \underline{c}$. But $\mathcal{I} \subset \mathcal{G}$ and so $\overline{\overline{\mathcal{I}}} \leq \overline{\overline{\mathcal{G}}}$ and so by step I, $\underline{c} \leq \overline{\overline{\mathcal{G}}}$, Hence $\overline{\overline{\mathcal{G}}} = \underline{c}$.

For the second part note that if $\mathcal{F}$ is the family of all closed sets in $[0, 1]$ then a set $G \in \mathcal{G}$ if and only if its complement $\tilde{G} \in \mathcal{F}$ and so $\mathcal{G}$ and $\mathcal{F}$ are equivalent and hence they have the same cardinal number. This completes the proof of step II. □

Proof of the theorem. Let $C_0 = \mathcal{G} \cup \mathcal{F}$ where $\mathcal{G}$ and $\mathcal{F}$ are respectively the family of all open sets and of all closed sets in $[0, 1]$. So $\overline{\overline{C_0}} = \overline{\overline{\mathcal{G}}} + \overline{\overline{\mathcal{F}}} = \underline{c} + \underline{c} = \underline{c}$ by step II. Let $\mathcal{A}_1$ be the family of all set which are countable union of members of C_0. Let $S \in \mathcal{A}_1$. Then S is a countable union of members of C_0. We shall show that $\mathcal{A}_1$ is equivalent to a subset of $C_0^{\mathbb{N}}$, the set of all sequences of members of C_0. Since the union does not depend on the order and since S should correspond to exactly one member of $C_0^{\mathbb{N}}$, we are to choose precisely one member of $C_0^{\mathbb{N}}$ for S. We call two sequences $\{E_1, E_2, \ldots\}$ and $\{F_1, F_2, \ldots\}$ of members of C_0 equivalent if for each $i \in \mathbb{N}$ there is $j \in \mathbb{N}$ such that $E_i = F_j$ and conversely. Then by a well known therefore of algebra $C_0^{\mathbb{N}}$ is divided into distinct disjoint classes of equivalent sequences. By axiom of choice we choose exactly one sequence, from each class. Since $S \in \mathcal{A}_1$, take $S = \bigcup_{i=1}^{\infty} E_i$, where $\{E_1, E_2, \ldots\}$ is such a chosen sequence. Then S corresponds to a unique sequence in $C_0^{\mathbb{N}}$. So, $\mathcal{A}_1$ is equivalent to a subset of $C_0^{\mathbb{N}}$ and hence $\overline{\overline{\mathcal{A}_1}} \leq \overline{\overline{C_0^{\mathbb{N}}}} = \overline{\overline{C_0}}^{\mathbb{N}} = \underline{c}^{\underline{a}} = \underline{c}$. Since each member of C_0 can be considered as a countable union of itself, $C_0 \subset \mathcal{A}_1$. Hence $\underline{c} = \overline{\overline{C_0}} \leq \overline{\overline{\mathcal{A}_1}} \leq \underline{c}$ and so $\overline{\overline{\mathcal{A}_1}} = \underline{c}$.

Again let $\mathcal{B}_1$ be the family of all sets F such that its $\tilde{F} \in \mathcal{A}_1$. So $\mathcal{A}_1$ and $\mathcal{B}_1$ are equivalent and hence $\overline{\overline{\mathcal{B}_1}} = \overline{\overline{\mathcal{A}_1}} = \underline{c}$. Let $C_1 = \mathcal{A}_1 \cup \mathcal{B}_1$. Then $\overline{\overline{C_1}} = \overline{\overline{\mathcal{A}_1}} + \overline{\overline{\mathcal{B}_1}} = \underline{c} + \underline{c} = \underline{c}$. Considering the family $\mathcal{A}_2$ of all sets which are countable union of members of C_1 and the family $\mathcal{B}_2$ of all sets whose complements are in $\mathcal{A}_2$ and writing $C_2 = \mathcal{A}_2 \cup \mathcal{B}_2$ we can show as above that $\overline{\overline{C_2}} = \underline{c}$. Continuing this process we get a sequence $C_0, C_1, C_2, \ldots, C_n, \ldots$ such that $\overline{\overline{C_n}} = \underline{c}$ for all n.

Let
$$C = C_0 \cup C_1 \cup C_2 \cup \ldots \cup C_n \cup \ldots$$

Clearly C is a σ-algebra and $\overline{\overline{C}} \le \underline{c} + \underline{c} + \ldots = \underline{c}$. Also the Borel σ-algebra $\mathcal{B}$ is the smallest σ-algebra containing open sets and hence $\mathcal{B} \subset C$. So, $\overline{\overline{\mathcal{B}}} \le \overline{\overline{C}} \le \underline{c}$. Since $C_0 \subset \mathcal{B}, \overline{\overline{C_0}} \le \overline{\overline{\mathcal{B}}}$. Since $C_0 = \mathcal{G} \cup \mathcal{F}$, by step II $\overline{\overline{C_0}} = \underline{c} + \underline{c} = \underline{c}$, So, $\underline{c} \le \overline{\overline{\mathcal{B}}}$. Hence $\overline{\overline{\mathcal{B}}} = \underline{c}$.

4.7 Cardinality of the σ-Algebra of All Lebesgue Measurable Sets

Theorem 4.17 *The cardinality of the σ-algebra of all Lebesgue measurable sets in $[0, 1]$ in $2^{\underline{c}}$.*

Proof Let $\mathcal{L}$ be the class of all Lebesgue measurable sets in $[0, 1]$. Let $\mathcal{P}$ be the Cantor ternary set in $[0, 1]$ considered in the last section. Then $\mathcal{P}$ has measure zero. So, every subset of $\mathcal{P}$ is measurable by Theorem 2.3 (ii). Since cardinal number of $\mathcal{P}$ is $\underline{c}$ and $\mathcal{P}$ has $2^{\underline{c}}$ subset by Theorem 4.1 all of which are measurable the cardinal number of $\mathcal{L}$ is greater then or equal to $2^{\underline{c}}$. On the other hand $\mathcal{L} \subset \mathbb{P}([0, 1])$ where $\mathbb{P}([0, 1])$ denotes the power set of $[0, 1]$ i.e. the family of all subsets of $[0, 1]$. Since the cardinal number of $[0, 1]$ is $\underline{c}$, the cardinal number of $\mathbb{P}([0, 1]) = 2^{\underline{c}}$. Since $\mathcal{L} \subset \mathbb{P}([0, 1]) \overline{\overline{\mathcal{L}}} \le 2^{\underline{c}}$. Hence $\overline{\overline{\mathcal{L}}} = 2^{\underline{c}}$. $\square$

Remark By Theorem 4.2 $\underline{c} < 2^{\underline{c}}$ and so by Theorems 4.16 and 4.17 there are uncountably many Lebesgue measurable sets which are not Borel measurable. In the next section we shall show that there are $2^{\underline{c}}$ subsets in $[0, 1]$ of Lebesgue measure zero which are not Borel measurable.

4.8 Cardinality of the Class of All Nonmeasurable Sets

Theorem 4.18 *The cardinality of the set of all subsets of $[0, 1]$ which are not Lebesgue measurable is $2^{\underline{c}}$.*

Proof Let C be the set of all subsets of $[0, 1]$ which are not Lebesgue measurable. We are to show that $\overline{\overline{C}} = 2^{\underline{c}}$.

Let $\mathcal{L}$ be the set of all subsets of $[0, 1]$ which are Lebesgue measurable and let $\mathcal{S}$ be the set of all subsets of $[0, 3]$ which are not Lebesgue measurable (such sets exists by Theorem 2.19). Choose a subset $E \subset [2, 3]$ which is not Lebesgue measurable and fix it. Then for each $F \in \mathcal{L}$, $E \cup F$ is not Lebesgue measurable. For, if $E \cup F$ is Lebesgue measurable then $[2, 3] \cap (E \cup F) = E$ would be Lebesgue measurable which is a contradiction. Hence $E \cup F \in \mathcal{S}$. Thus for each $F \in \mathcal{L}$ there is a set $E \cup F$ in $\mathcal{S}$ and if $F_1, F_2 \in \mathcal{L}$ and $F_1 \neq F_2$ then $E \cup F_1 \neq E \cup F_2$. So $\mathcal{L}$ is equivalent to a subset of $\mathcal{S}$. Hence $\overline{\overline{\mathcal{L}}} \leq \overline{\overline{\mathcal{S}}}$, and so by Theorem 4.17 $2^{\underline{c}} \leq \overline{\overline{\mathcal{S}}}$. On the other hand $\mathcal{S} \subset \mathcal{P}(\mathbb{R})$ where $\mathcal{P}(\mathbb{R})$ is the set of all subsets of $\mathbb{R}$ and so by Theorem 4.1 $\overline{\overline{\mathcal{S}}} \leq \overline{\overline{\mathcal{P}(\mathbb{R})}} = 2^{\underline{c}}$. Hence $\overline{\overline{\mathcal{S}}} = 2^{\underline{c}}$.

The function $f(t) = 3t$ is one-one from $[0, 1]$ onto $[0, 3]$ and so for each member $A \in C$ there is a unique member B defined by $B = \{3t; t \in A\} = 3A$ and applying Theorem 2.16, B is not Lebesgue measurable and so $B \in \mathcal{S}$. Also for each $B \in \mathcal{S}$ there is $A \in C$ such that A corresponds to B. Hence $\overline{\overline{C}} = \overline{\overline{\mathcal{S}}}$ and so $\overline{\overline{C}} = 2^{\underline{c}}$. □

4.9 Cantor Function

We consider the Cantor ternary set defined in Sect. 4.5 and denote this set by P. The set P is constructed step by step by dividing first the closed interval $[0, 1]$ into there equal parts and then removing the middle open interval $I_1^1 = (\frac{1}{3}, \frac{2}{3})$ from $[0, 1]$ and then in the second step dividing each of the remaining closed intervals $J_1^1 = [0, \frac{1}{3}]$ and $J_2^1 = [\frac{2}{3}, 1]$ into three equal parts and then removing the open intervals $I_1^2 = (\frac{1}{9}, \frac{2}{9})$ and $I_2^2 = (\frac{7}{9}, \frac{8}{9})$ respectively and so on. In the nth step the removed open intervals are I_k^n, $1 \leq k \leq 2^{n-1}$ and the remaining closed intervals are J_k^n, $1 \leq k \leq 2^n$. The process being continued indefinitely the remaining set is the Cantor ternary set P. Let $G = [0, 1] \sim P$.

We now define the Cantor function $\theta : [0, 1] \to [0, 1]$. Define θ step by step such that

$$\theta(x) = \frac{1}{2} \text{ if } x \in I_1^1 = (\frac{1}{3}, \frac{2}{3})$$

$$\theta(x) = \frac{1}{4} \text{ if } x \in I_1^2 = (\frac{1}{9}, \frac{2}{9}), \quad \theta(x) = \frac{3}{4} \text{ if } x \in I_2^2 = (\frac{7}{9}, \frac{8}{9})$$

and so on. In the nth step there are 2^{n-1} open intervals I_k^n, $1 \leq k \leq 2^{n-1}$, which are removed and define

$$\theta(x) = \frac{2k-1}{2^n} \text{ if } x \in I_k^n, 1 \leq k \leq 2^{n-1}.$$

Continuing this process θ is defined in $G = \bigcup_{n=1}^{\infty} \bigcup_{k=1}^{2^{n-1}} I_k^n$. To define θ on P we take $\theta(0) = 0$. and $\theta(1) = 1$ and for any other points $x \in P$ we define $\theta(x)$ as follows : Since θ is non-decreasing on G and since every point of P is a limit point of G, the left-hand limit $\theta(x - 0)$ and the right-hand limit $\theta(x + 0)$ at $x \in P$ with respect to G exist and $\theta(x - 0) \leq \theta(x + 0)$. We claim that $\theta(x - 0) = \theta(x + 0)$. For if $\theta(x - 0) < \theta(x + 0)$ then θ will take no value between $\theta(x - 0)$ and $\theta(x + 0)$ on G which is a contradiction since the values of θ taken on G is the set $\{\frac{2k-1}{2^n}; 1 \leq k \leq 2^{n-1}, n = 1, 2, ...\}$ which is everywhere dense in $[0, 1]$. Therefore $\theta(x - 0) = \theta(x + 0)$. We define $\theta(x) = \theta(x - 0) = \theta(x + 0)$ for $x \in P$. Thus θ is defined on $[0, 1]$. Clearly θ is continuous and non-decreasing on $[0, 1]$. The function θ is called Cantor function on $[0, 1]$ and has many interesting properties and also is helpful for constructing many counter examples.

4.10 Lebesgue Function

To define the Lebesgue function we consider the interval I_k^n as in Sect. 4.9. We need a sequence $\{f_n\}$ of functions on $[0, 1]$ For fixed n which is defined as follows :

$f_n(x) = 0$ if $x = 0$

$= \frac{1}{2}$ for $x \in \overline{I}_1^1$

$= \frac{1}{4}$ for $x \in \overline{I}_1^2$ and $= \frac{3}{4}$ for $x \in \overline{I}_2^2$

$= \frac{2k-1}{2^n}$ for $x \in \overline{I}_k^n, 1 \leq k \leq 2^{n-1}$

where $\overline{I}_k^r$ denotes the closure of I_k^r, $1 \leq k \leq 2^{r-1}$, $r = 1, 2, ...n$. Further, f_n is defined in the complementary intervals such that f_n is linear and continuous in each closed interval J_k^r, $1 \leq k \leq 2^r$, $r = 1, 2, ...n$. Clearly f_n is continuous and non-decreasing in $[0, 1]$ for all n. Also $|f_n(x) - f_m(x)| < \frac{1}{2^m}$ for $n > m$ and for all $x \in [0, 1]$. So, $\{f_n\}$ is a Cauchy sequence and hence the sequence $\{f_n\}$ converges uniformly to a continuous function f on $[0, 1]$. This function f is called the Lebesgue function. Since $f(x) = \theta(x)$ for all $x \in G = \bigcup_{n=1}^{\infty} \bigcup_{k=1}^{2^{n-1}} I_k^n$ and since $\mu(G) = 1$, $f = \theta$ almost everywhere on $[0, 1]$. Since f and θ are continuous in $[0, 1]$, $f(x) = \theta(x)$ for all $x \in [0, 1]$. So, the Cantor function θ and the Lebesgue function f are same.

4.11 Properties of Cantor Function

The following interesting properties of Cantor function may be noted :
I Although $\mu(G) = 1$ the image of G under θ i.e. $\theta(G) = \{\theta(x) : x \in G\}$ is such that $\mu(\theta(G)) = 0$, since $\theta(G)$ is the set $\{\frac{2k-1}{2^n} : 1 \le k \le 2^{n-1}; n = 1, 2, ...\}$.
II Although $\mu(P) = 0$ the image of P under θ is the whole interval $[0, 1]$ and hence $\mu(\theta(P)) = 1$.
III The derivative θ' of θ exists and is 0 for all $x \in G$ and hence $\theta' = 0$ almost everywhere in $[0, 1]$ but θ is non-constant in $[0, 1]$ and

$$\int_0^1 \theta'(x)dx = 0 < 1 = \theta(1) - \theta(0)$$

where the integral is the Lebesgue integral which will be introduced in the next chapter.

4.12 Associated Cantor Function and Its Inverse Function and Some Consequences

We shall construct two functions from the Cantor function θ which will be helpful for many purposes.

The function θ assumes the constant value $\frac{2k-1}{2^n}$ in the open interval I_k^n, $1 \le k \le 2^{n-1}$, $n = 1, 2, ...$ and since θ is continuous, θ also assumes the same value $\frac{2k-1}{2^n}$ at the two endpoints of I_k^n. Since P contains both the endpoints of each I_k^n, the values assumed by θ on G are also assumed by θ on P. Let P_0 be the set of all points of P which are not the left-hand endpoints of the intervals I_k^n, $1 \le k \le 2^{n-1}$, $n = 1, 2, ...$. So, the values assumed by θ on G are assumed by θ on P_0 at the right-hand endpoint of the intervals I_k^n. Define $\phi : P_0 \to [0, 1]$ by $\phi(t) = \theta(t)$ for $t \in P_0$ i.e. ϕ is the restriction of θ on P_0. We shall show that ϕ is one-one, strictly increasing and continuous from P_0 onto $[0, 1]$.

Let $x_1, x_2, \in P_0$, $x_1 < x_2$. If x_2 is the right-hand endpoint of some interval I_k^n say $I_{k_0}^{n_0}$ then $\phi(x_2) = \theta(x_2) = \frac{2k_0-1}{2^{n_0}}$ and since x_1 is not the left-hand endpoint of $I_{k_0}^{n_0}$ there is an interval, say $I_{k_1}^{n_1}$, $n_1 > n_0$ between x_1 and the left-hand endpoint of $I_{k_0}^{n_0}$ such that $\frac{2k_1-1}{2^{n_1}} < \frac{2k_0-1}{2^{n_0}}$. Since $\theta(x_1) \le \frac{2k_1-1}{2^{n_1}}$, $\phi(x_1) = \theta(x_1) \le \frac{2k_1-1}{2^{n_1}} < \phi(x_2)$. If x_1 is the right-hand endpoint of some interval I_k^n say $I_{k_2}^{n_2}$ then $\phi(x_1) = \theta(x_1) = \frac{2k_2-1}{2^{n_2}}$ and there is an interval, say $I_{k_3}^{n_3}$, $n_3 > n_2$, between x_1 and x_2 such that $\frac{2k_2-1}{2^{n_2}} < \frac{2k_3-1}{2^{n_3}}$. Since $\frac{2k_3-1}{2^{n_3}} < \theta(x_2)$, $\phi(x_1) < \theta(x_2) = \phi(x_2)$. If none of x_1 and x_2 is the right-hand endpoint of intervals of G there are intervals I_k^n whose endpoints lie strictly between x_1 and x_2 and so choosing such interval I_k^n, $\phi(x_1) = \theta(x_1) < \frac{2k-1}{2^n} < \theta(x_2) = \phi(x_2)$.

So, ϕ is one - one and strictly increasing on P_0. Since θ is continuous and $\phi = \theta$ on P_0, ϕ is continuous on P_0. Finally, θ assumes every value of $[0, 1]$ of which $\frac{2k-1}{2^n}, 1 \leq k \leq 2^{n-1}, n = 1, 2, \ldots$ are assumed by θ on G. Since θ is continuous, these values are also assumed by θ at the right-hand endpoints of all the open intervals of G and since these endpoints are in P_0, θ assumes every value of $[0, 1]$ on P_0. Hence $\theta(P_0) = [0, 1]$ and so $\phi(P_0) = [0, 1]$.

Definition 4.12 Let P be the Cantor ternary set in $[0, 1]$ and let P_0 be the set of all points of P except the left-hand endpoints of all its complementary open intervals. The restriction of the Cantor function θ on P_0 will be called Associated Cantor Function on P_0 and will be denoted by ϕ.

Since the associated Cantor function ϕ is one-one, strictly increasing and continuous on P_0 taking every value of $[0, 1]$, ϕ has an inverse function $\phi^{-1} : [0, 1] \to P_0$ which is also one-one, strictly increasing and continuous on $[0, 1]$ taking every value of P_0. Both the associated Cantor function and its inverse function have interesting application.

Definition 4.13 The inverse function of the associated Cantor function is called the inverse Cantor function.

We now have

Theorem 4.19 *There exist a nowhere dense set $P_0 \subset [0, 1]$ which is dense in itself and of measure 0 and there are two functions $\phi : P_0 \to [0, 1]$ and $\psi : [0, 1] \to P_0$ such that ϕ is one-one, strictly increasing and continuous on P_0 taking every value of $[0, 1]$ and ψ is one-one, strictly increasing and continuous on $[0, 1]$ taking every value of P_0.*

From the above discussion, the associated Cantor function ϕ and its inverse function $\psi = \phi^{-1}$ satisfy the requirements of the theorem and so the proof is complete.

Some consequences Clearly the functions ϕ and ψ are measurable. We shall use these function for counter examples.

I. Let $E \subset [0, 1]$ be any set which is not Lebesgue measurable. Let $S = \{y : y \in P_0; y = \psi(x) \text{ for some } x \in E\}$. Then S is Lebesgue measurable since $S \subset P_0$ and $\mu(P_0) = 0$. Since $\psi^{-1}(S) = E$ which is not Lebesgue measurable, the Borel set in Theorem 3.5 (which states that f is measurable if and only if $f^{-1}(B)$ is measurable for every Borel set) cannot be replaced by Lebesgue measurable set.

II. There are $2^{\underline{c}}$ Lebesgue measurable subset of Lebesgue measure 0 which are not Borel sets. Let $\mathcal{C}$ be the class of all subsets of $[0, 1]$ which are not Lebesgue measurable. Let $S \in \mathcal{C}$ and let $E = \psi(S)$, the image of S under ψ. Since ψ is one-one $\psi^{-1}(E) = S$. Since $E \subset P_0$ $\mu(E) = 0$. We claim that E is not a Borel set. For, if E is a Borel set then since ψ is measurable, by Theorem 3.5, $\psi^{-1}(E) = S$ would be measurable which is a contradiction. Since $\overline{\overline{\mathcal{C}}} = 2^{\underline{c}}$ by Theorem 4.18, and since distinct members of $\mathcal{C}$ corresponds to distinct subsets of P_0, the class of all subsets of P_0 which are not Borel set has cardinality $2^{\underline{c}}$.

4.13 Functions Similar to Cantor Function

Instead of considering the Cantor ternary set consider for any real number α, $0 < \alpha < 1$ and the set C_α defined in Sect. 4.4. Then $\mu(C_\alpha) = 1 - \alpha$. Define a function on [0, 1] with its values in [0, 1] just like the Cantor function θ is defined on [0, 1]. Then this function is also continuous and non-decreasing in [0, 1] taking the constant value $\frac{2k-1}{2^n}$ on each interval I_k^n, $1 \leq k \leq 2^{n-1}$, $n = 1, 2, ...$ and takes all the values in [0, 1]. This function differs from the Cantor function θ in the sense that its derivative vanishes on the set $[0, 1] \sim C_\alpha$ which is of measure α whereas the derivative θ' of θ vanishes $a.e.$ on [0, 1]. This function can also serve for many counter examples.

4.14 Exercises

1. Prove that the set of all function $f : \mathbb{R} \to \mathbb{R}$ has cardinal number $2^{\underline{c}}$.
2. Prove that the set of all continuous function $f : \mathbb{R} \to \mathbb{R}$ has cardinal number $\underline{c}$. Hint : Since a continuous function is completely known if it is known at an everywhere dense set of points in its domain, the cardinal number is $\overline{\overline{\mathbb{R}^Q}} = \underline{c}^{\underline{a}} = \underline{c}$.
3. The set of all function $f : \mathbb{R}^m \to \mathbb{R}^n$ has cardinal number $2^{\underline{c}}$. Prove this.
4. Prove that the set of all differentiable function $f : \mathbb{R} \to \mathbb{R}$ has cardinal number $\underline{c}$.
5. Find the cardinal number of the set of all Lebesgue measurable sets.
6. Find the cardinal number of the set of all Lebesgue measurable function.
7. Find the cardinal number of the set of all intervals which have rational endpoints. Hint : If $E = \{(x, y) : x < y; x, y \text{ rational}\}$ then $E \subset Q \times Q$ and so $\overline{\overline{E}} \leq \underline{a}$ and if $S = \{(1, y) : 1 < y, y \text{ rational}\}$ then $\overline{\overline{S}} = \overline{\overline{Q}} = \underline{a}$. Also $S \subset E$ implies $\overline{\overline{E}} \geq \overline{\overline{S}} = \underline{a}$.
8. Show that the members of the set of all real numbers are in one-one correspondence with the members of the set of all complex numbers.
9. Define a function on [0, 1] with values in [0, 1] just like the Cantor function but considering the Cantor like set C_α, $0 < \alpha < 1$.
10. Consider the Cantor ternary set $P = C_1$ and the Cantor function θ. Let $\{I_k^n; 1 \leq k \leq 2^{n-1}, n = 1, 2, ...\}$ are the open intervals in [0, 1] complementary to P. Define for each n

$$f_n(x) = 0 \text{ if } x = 0$$
$$= 1 \text{ if } x = 1$$
$$= \frac{2k-1}{m} \text{ if } x \in I_k^n, 1 \leq k \leq 2^{m-1}, 1 \leq m \leq n$$

and f_n is linear on each closed interval of the set $[0, 1] \sim \bigcup_{m=1}^{n} (\bigcup_{k=1}^{2^m-1} I_k^m)$. Show that the function f_n is continuous for each n and the sequence $\{f_n\}$ converges uniformity to θ.

11. Show that the points of the Cantor ternary set P in $[0, 1]$ are of the form $x = \sum_{k=1}^{\infty} \frac{2}{3^k} e_k$ where each e_k is either 0 or 1 and this representation is unique and the Cantor function θ is given by $\theta(x) = \sum_{k=1}^{\infty} \frac{e_k}{2^k}$ for each $x = \sum_{k=1}^{\infty} \frac{2}{3^k} e_k$.

12. Show that the Cantor ternary set P and the Cantor function θ in $[0, 1]$ can also be defined by

$$P = \{x : x \in [0, 1]; x = \sum_{k=1}^{\infty} \frac{2}{3^k} e_k, e_k = 0 \text{ or } 1\}$$

and

$$\theta(t) = \sum_{k=1}^{\infty} \frac{e_k}{2^k} \text{ for } t \in P \text{ and } = \sup\{\theta(x), x \in P; x < t\} \text{ for } t \in [0, 1] \sim P.$$

13. Show that the set of all transcendental numbers has cardinal number $\underline{c}$.

Chapter 5
The Lebesgue Integral

The Riemann integral of a bounded function f on a closed interval $[a, b]$ (as modified by Darboux) is defined by partitioning $[a, b]$ into a finite number of sub-intervals by points $D : a = x_0 < x_1 < x_2 < \cdots < x_n = b$, which may be called 'Riemann partition', and considering the upper and lower Riemann-Darboux sums defined by

$$S(D, f) = \sum_{r=1}^{n} M_r(x_r - x_{r-1}) \text{ and } s(D, f) = \sum_{r=1}^{n} m_r(x_r - x_{r-1})$$

where M_r and m_r are the upper and lower bounds of f on $[x_{r-1}, x_r]$ and defining the upper and lower Riemann integrals

$$(R)\overline{\int_a^b} f(x)dx = \inf_D S(D, f) \text{ and } (R)\underline{\int_a^b} f(x)dx = \inf_D s(D, f),$$

the infimum and supremum being taken over all partitions D of [a, b]. If the upper and lower Riemann integrals are equal then f is said to be Riemann integrable in $[a, b]$ and the common value is called the Riemann integral of f which is denoted by $(R)\int_a^b f(x)dx$.

5.1 Lebesgue Integral of Bounded Functions

Let E be a measurable set of finite measures. By a Lebesgue partition τ of E we mean a finite collection $\tau = \{e_i : i = 1, 2, \ldots, n\}$ of measurable subsets of E such that $E = \bigcup_{i=1}^{n} e_i$, $e_i \cap e_j = \phi$ for $i \neq j$. If $\tau = \{e_i : i = 1, 2, \ldots, n\}$ and $\tau' = \{e'_j : i = 1, 2, \ldots, m\}$ are two partitions of E then τ' is said to be a refinement of τ if every e'_j is contained in some e_i; in other words the sets e'_j is obtained by partitioning some or all e_i into measurable subsets. Let f be a bounded function on E. For every partition $\tau = \{e_i : i = 1, 2, \ldots, n\}$ of E form the sum

$$S(\tau) = S(\tau, f) = \sum_{i=1}^{n} M_i \mu(e_i) \text{ where } M_i = \sup\{f(x) : x \in e_i\}$$

$$s(\tau) = s(\tau, f) = \sum_{i=1}^{n} m_i \mu(e_i) \text{ where } m_i = \inf\{f(x) : x \in e_i\}.$$

Then $S(\tau)$ and $s(\tau)$ are called upper and lower Lebesgue sums of f corresponding to τ.

Note that in the above definition of Lebesgue partition $\tau = \{e_i : i = 1, 2, \ldots, n\}$ of E the condition $e_i \cap e_j = \phi$ may be replaced by $\mu(e_i \cap e_j) = 0$, since this change will not affect $S(\tau)$ and $s(\tau)$.

Lemma 5.1 *If $\tau' = \{e'_j\}$ is refinement of $\tau = \{e_i\}$ then $s(\tau) \leq s(\tau')$ and $S(\tau') \leq S(\tau)$.*

Proof Since the sets e'_j are obtained by breaking some or all of e_i it is sufficient to prove the lemma for the case when τ' is obtained from τ by breaking one of e_i say e_r into two sets and e'_r and e''_r. Then $e_r = e'_r \cup e''_r$ and $e'_r \cap e''_r = \emptyset$. Putting $M_r = \sup\{f(x) : x \in e_r\}$, $M'_r = \sup\{f(x) : x \in e'_r\}$ and $M''_r = \sup\{f(x) : x \in e''_r\}$ we have $M'_r \leq M_r$, and $M''_r \leq M_r$ and $\mu(e_r) = \mu(e'_r) + \mu(e''_r)$ and so

$$\begin{aligned} S(\tau) - S(\tau') &= M_r \mu(e_r) - M'_r \mu(e'_r) - M''_r \mu(e''_r) \\ &= (M_r - M'_r)\mu(e'_r) + (M_r - M''_r)\mu(e''_r) \\ &\geq 0 \end{aligned}$$

Similarly $s(\tau') - s(\tau) \geq 0$. $\square$

Lemma 5.2 *For any two partitions τ_1 and τ_2, $s(\tau_1) \leq S(\tau_2)$.*

Proof Let $\tau = \{e_i \cap e_j : e_i \in \tau_1, e_j \in \tau_2 \text{ and } e_i \cap e_j \neq \emptyset\}$. So τ is a partition of E and is refinement of both τ_1 and τ_2. Hence by Lemma 5.1, $s(\tau_1) \leq s(\tau) \leq S(\tau) \leq S(\tau_2)$ $\square$

Lemma 5.3 *Considering all Lebesgue partitions τ of E, if $\{s(\tau)\}$ and $\{S(\tau)\}$ are the collections of all lower and upper sums respectively then $\sup_\tau \{s(\tau)\} \leq \inf_\tau \{S(\tau)\}$.*

The proof follows from Lemma 5.2

Definition 5.1 The lower and upper Lebesgue integral of f on E are defined by

$$\underline{\int_E} f dx = \sup_\tau \{s(\tau)\} \text{ and } \overline{\int_E} f dx = \inf_\tau \{S(\tau)\}$$

5.1 Lebesgue Integral of Bounded Functions

respectively. By Lemma 5.3, $\underline{\int_E} f dx \leq \overline{\int_E} f dx$. If $\underline{\int_E} f dx = \overline{\int_E} f dx$, then f is said to be Lebesgue integrable and the common value is called the Lebesgue integral of f on E and is denoted by $(L) \int_E f dx$, or simply $\int_E f$ if there is no confusion. If E is an interval, say $E = [a, b]$, then we write $\int_a^b f dx$ instead of $\int_{[a,b]} f dx$.

Theorem 5.4 *Let E be a measurable set of finite measures and $f : E \to \mathbb{R}$ be bounded. Then*

(i) *The upper and lower integrals are finite.*
(ii) *If $f = c$, a constant on E then f is integrable on E and $\int_E f dx = c\mu(E)$.*
(iii) *If for some partition τ of E, $s(\tau) = S(\tau)$ then f is integrable on E and the common value is the integral f on E.*
(iv) *If f is Riemann integrable on $[a, b]$ then f is Lebesgue integrable on $[a, b]$ and the integrals are equal, but the converse is not true.*
(v) *If f is measurable then f is Lebesgue integrable.*

Proof Since f is bounded on E, there are m and M such that $m \leq f \leq M$ on E and so $m\mu(E) \leq s(\tau) \leq S(\tau) \leq M\mu(E)$ for all partition τ on E. This proves (i) and (ii). The proof of (iii) follows from the definition of lower and upper integrals.
(iv) Suppose that f is Riemann integrable on $[a, b]$. Since every Riemann partition $D : a = x_0 < x_1 < x_2 < .. < x_n = b$ gives the Lebesgue partition $\tau = \{[x_{r-1}, x_r] : r = 1, 2, 3, \ldots, n\}$, every upper Riemann-Darboux sum $S(D, f) = \sum_{r=1}^{n} M_r(x_r - x_{r-1})$ gives the upper Lebesgue sum $S(\tau, f) = \sum_{r=1}^{n} M_r\mu[x_r, x_{r-1}]$. Considering all Riemann partition D and all Lebesgue partition τ the family $\{S(D, f)\}$ in contained in the family $\{S(\tau, f)\}$. Hence

$$(R)\overline{\int_a^b} f dx = \inf_D S(D, f) \geq \inf_\tau S(\tau, f) = (L)\overline{\int_a^b} f dx.$$

Similarly,

$$(R)\underline{\int_a^b} f dx = \sup_D s(D, f) \leq \sup_\tau s(\tau, f) = (L)\underline{\int_a^b} f dx.$$

So

$$(R)\underline{\int_a^b} f dx \leq (L)\underline{\int_a^b} f dx \leq (L)\overline{\int_a^b} f dx \leq (R)\overline{\int_a^b} f dx$$

Since f is Riemann integrable, f is Lebesgue integrable and the integrals are equal. To complete the proof consider the following example:
Let $f(x) = 1$ if x is a rational and $f(x) = 0$, if x is a irrational. Then for any Riemann partition D of $[a, b]$, $S(D, f) = b - a$ and $s(D, f) = 0$ and so f is not

Riemann integrable in $[a, b]$. To show that f is Lebesgue integrable in $[a, b]$, let $e_1 = [a, b] \cap Q$ and $e_2 = [a, b] \cap T$ where Q and T are the sets of rationals and irrationals respectively. Then $\tau = \{e_1, e_2\}$ in a Lebesgue partition of $[a, b]$. Since $f(x) = 1$ for $x \in e_1$ and $f(x) = 0$ for $x \in e_2$, $S(\tau, f) = 1.\mu(e_1) + 0.\mu(e_2) = 0 = s(\tau, f)$ and so by (iii) f is Lebesgue integrable on $[a, b]$ and the Lebesgue integral is 0.

(v) Suppose that f is measurable. Since f is bounded there are M and m such that $m < f(x) < M$ for $x \in E$. Let $\epsilon > 0$ be arbitrary, Take a partition $m = \eta_0 < \eta_1 < \cdots < \eta_n = M$ such that $max\{\eta_i - \eta_{i-1} : 0 < i \leq n\} < \frac{\epsilon}{\mu(E)+1}$. Let, $e_i = \{x \in E : \eta_{i-1} \leq f(x) < \eta_i\}$, for $i = 1, 2, \ldots n - 1$, and $e_n = \{x \in E : \eta_{n-1} \leq f(x) \leq \eta_n\}$. Then $\tau = \{e_i : 1 \leq i \leq n\}$ is a Lebesgue partition of E. Let $m_i = inf\{f(x) : x \in e_i\}$ and $M_i = \sup\{f(x) : x \in e_i\}$ for $i = 1, 2, \ldots, n$. So

$$S(\tau, f) - s(\tau, f) = \sum_{i=1}^{n}(M_i - m_i)\mu(e_i)$$

$$\leq \sum_{i=1}^{n}(\eta_i - \eta_{i-1})\mu(e_i)$$

$$\leq \frac{\epsilon}{\mu(E) + 1}\sum_{i=1}^{n}\mu(e_i) < \frac{\epsilon}{\mu(E) + 1}\mu(E) < \epsilon$$

Hence,

$$0 \leq \overline{\int_E} f dx - \underline{\int_E} f dx \leq S(\tau, f) - s(\tau, f) < \epsilon.$$

Since ϵ is arbitrary,

$$\overline{\int_E} f dx = \underline{\int_E} f dx,$$

and so f is integrable. □

Note In view of part (iv) of the above theorem Lebesgue integral of bounded function is more general than the Riemann integral.

Theorem 5.5 (i) Let E_1 and E_2 be disjoint measurable sets of finite measure and let $f : E \to \mathbb{R}$ be bounded where $E = E_1 \cup E_2$. Then

$$\underline{\int_{E_1}} f dx + \underline{\int_{E_2}} f dx = \underline{\int_E} f dx, \quad \overline{\int_{E_1}} f dx + \overline{\int_{E_2}} f dx = \overline{\int_E} f dx \quad (5.1)$$

Hence f is integrable on E if and only if f is integrable on each of E_1 and E_2. In either case

$$\int_{E_1} f dx + \int_{E_2} f dx = \int_E f dx \quad (5.2)$$

5.1 Lebesgue Integral of Bounded Functions

(ii) Let $\{E_n\}$ be a sequence of disjoint measurable sets and $E = \bigcup_{n=1}^{\infty} E_n$ be of finite measure. Let $f : E \to \mathbb{R}$ be bounded. Then

$$\sum_{n=1}^{\infty} \underline{\int_{E_n}} f\,dx = \underline{\int_E} f\,dx, \quad \sum_{n=1}^{\infty} \overline{\int_{E_n}} f\,dx = \overline{\int_E} f\,dx \tag{5.3}$$

Hence f is integrable on E if and only if f is integrable on each E_n. In either case

$$\sum_{n=1}^{\infty} \int_{E_n} f\,dx = \int_E f\,dx \tag{5.4}$$

Proof (i) Let $\tau = \{e_k : k = 1, 2, \ldots, l\}$ be any partition of E and let $e'_k = e_k \cap E_1$, $e''_k = e_k \cap E_2$. Then $E_1 = \bigcup_{k=1}^{l} e'_k$ and $E_2 = \bigcup_{k=1}^{l} e''_k$. So $\tau' = \{e'_k : e'_k \neq \emptyset, k = 1, 2, \ldots, l\}$ and $\tau'' = \{e''_k : e''_k \neq \emptyset, k = 1, 2, \ldots, l\}$ are Lebesgue partitions of E_1 and E_2 respectively. If m_k and M_k are lower and upper bounds of f on e_k and m'_k, M'_k and m''_k, M''_k are those on e'_k and e''_k respectively, then $m_k \leq m'_k$, $m_k \leq m''_k$, $M_k \geq M'_k$, $M_k \geq M''_k$. Since $e'_k \cup e''_k = e_k$ and $e'_k \cap e''_k = \emptyset$, $\mu(e'_k) + \mu(e''_k) = \mu(e_k)$ and so $m_k \mu(e_k) \leq m'_k \mu(e'_k) + m''_k \mu(e''_k)$ and $M_k \mu(e_k) \geq M'_k \mu(e'_k) + M''_k \mu(e''_k)$ for $k = 1, 2, \ldots, l$. Hence

$$s(\tau, f) \leq s(\tau', f) + s(\tau'', f) \leq S(\tau', f) + S(\tau'', f) \leq S(\tau, f) \tag{5.5}$$

and so $s(\tau, f) \leq \underline{\int_{E_1}} f\,dx + \underline{\int_{E_2}} f\,dx$ and hence

$$\underline{\int_E} f\,dx \leq \underline{\int_{E_1}} f\,dx + \underline{\int_{E_2}} f\,dx \tag{5.6}$$

Let $\epsilon > 0$ be arbitrary. Then there exist partitions τ_1 and τ_2 of E_1 and E_2 respectively such that $\underline{\int_{E_1}} f\,dx - \frac{\epsilon}{2} < s(\tau_1, f)$ and $\underline{\int_{E_2}} f\,dx - \frac{\epsilon}{2} < s(\tau_2, f)$ and so if τ is the partition of E defined by $\tau = \tau_1 \cup \tau_2$ then

$$\underline{\int_{E_1}} f\,dx + \underline{\int_{E_2}} f\,dx - \epsilon < s(\tau_1, f) + s(\tau_2, f) = s(\tau, f) \leq \underline{\int_E} f\,dx$$

Since ϵ is arbitrary

$$\underline{\int_{E_1}} f\,dx + \underline{\int_{E_2}} f\,dx \leq \underline{\int_E} f\,dx \tag{5.7}$$

The first relation of (5.1) follows from (5.6) and (5.7). The second is similar. Now suppose that f is integrable on E. Then from (5.5)

$$s(\tau, f) \leq \underline{\int_{E_1} f dx} + \underline{\int_{E_2} f dx} \leq \overline{\int_{E_1} f dx} + \overline{\int_{E_2} f dx} \leq S(\tau', f) + S(\tau'', f) \leq S(\tau, f).$$

Since f is integrable on E this gives

$$\int_E f dx = \underline{\int_{E_1} f dx} + \underline{\int_{E_2} f dx} = \overline{\int_{E_1} f dx} + \overline{\int_{E_2} f dx} = \int_E f dx$$

which shows that f is integrable on E_1 as well as on E_2.

Conversely suppose that f is integrable on E_1 and E_2. Then form (5.1) f is integrable on E and (5.2) holds.

(ii) For any positive integer n let $S_n = \bigcup_{k=1}^{n} E_k$, $R_n = \bigcup_{k=n+1}^{\infty} E_k$. Then $E = S_n \cup R_n$ and $S_n \cap R_n = \phi$. Applying (i) repeatedly

$$\int_E f dx = \int_{S_n} f dx + \int_{R_n} f dx = \sum_{k=1}^{n} \int_{E_k} f dx + \int_{R_n} f dx \quad (5.8)$$

Since f is bounded there are A, B such that $A \leq f(x) \leq B$ for $x \in E$. Let τ be any partition of R_n. Then

$$A\mu(R_n) \leq s(\tau, f) \leq \underline{\int_{R_n} f dx} \leq \overline{\int_{R_n} f dx} \leq S(\tau, f) \leq B\mu(R_n) \quad (5.9)$$

By Theorem 2.6 $\mu(E) = \sum_{k=1}^{\infty} \mu(E_n)$ and so $\sum_{k=1}^{\infty} \mu(E_k)$ is a convergent series of non-negative terms. Hence $\mu(R_n) = \sum_{k=n+1}^{\infty} \mu(E_k) \to 0$ as $n \to \infty$. Hence from (5.9)

$$\lim_{n \to \infty} \underline{\int_{R_n} f dx} = 0, \quad \lim_{n \to \infty} \overline{\int_{R_n} f dx} = 0 \quad (5.10)$$

So form (5.8) and (5.10)

$$\int_E f dx = \lim_{n \to \infty} \sum_{k=1}^{n} \int_{E_k} f dx = \sum_{k=1}^{\infty} \int_{E_k} f dx$$

proving the first relation of (5.3). The proof of the second is similar.

5.1 Lebesgue Integral of Bounded Functions

Finally suppose that f is integrable on E. Take n arbitrary. Let $F_n = E \sim E_n$. Then $E = E_n \cup F_n$ and $E_n \cap F_n = \phi$. So by first part (i), f is integrable on E_n. Since n is arbitrary, f is integrable on each E_n.

Conversely, suppose that f is integrable on each E_n. Then by (5.3) f is integrable on E. The relation (5.4) now follows from (5.3). □

Note By Theorems 5.4(ii) and 5.5, every measurable simple function on a bounded measurable set is integrable.

Theorem 5.6 *Let E be a measurable set of finite measures and let $f : E \to \mathbb{R}$ be bounded. Then*

(i) *for every partition τ of E there are measurable bounded simple functions ϕ_τ and ψ_τ on E such that $\phi_\tau \leq f \leq \psi_\tau$ on E which satisfy $s(\tau, f) = \int_E \phi_\tau dx$ and $S(\tau, f) = \int_E \psi_\tau dx$*

(ii) *Considering all partition τ of E*

$$\sup_\tau \{s(\tau, f)\} = \sup \left\{ \int_E \phi dx : \phi \text{ is a measurable simple function and } \phi \leq f \right\}$$

$$\inf_\tau \{S(\tau, f)\} = \inf \left\{ \int_E \psi dx : \psi \text{ is a measurable simple function and } \psi \geq f \right\}$$

Proof (i) Let $\tau = \{e_k : k = 1, 2, \ldots, n\}$ and $m_k = \inf f$ and $M_k = \sup f$ on e_k. Let $\phi_\tau = m_k$ and $\psi_\tau = M_k$ on e_k. Then ϕ_τ and ψ_τ one measurable bounded simple function such that $\phi_\tau \leq f \leq \psi_\tau$ on E. By Theorem 5.4(ii), $\int_{e_k} \phi_\tau dx = m_k \mu(e_k)$ and $\int_{e_k} \psi_\tau dx = M_k \mu(e_k)$ and so by Theorem 5.5, $s(\tau, f) = \int_E \phi_\tau dx$ and $S(\tau, f) = \int_E \psi_\tau dx$.

(ii) Let $\mathcal{A}$ be the family of all measurable simple function ϕ on E such that $\phi \leq f$ and let $\mathcal{B} = \{\int_E \phi dx : \phi \in \mathcal{A}\}$. Let C be the family of all lower sums $s(\tau, f)$. Let $s(\tau, f) \in C$. Then by (i) there is a measurable simple function ϕ_τ on E such that $\phi_\tau \leq f$ and $s(\tau, f) = \int_E \phi_\tau dx$. Hence $s(\tau, f) \in \mathcal{B}$ and so $C \subset \mathcal{B}$. Hence $\sup C \leq \sup \mathcal{B}$. If possible, suppose $\sup C < \sup \mathcal{B}$. Then there is λ such that $\sup C < \lambda$ and $\lambda \in \mathcal{B}$. So there is $\phi \in \mathcal{A}$ such that $\lambda = \int_E \phi dx$. Since ϕ is a measurable simple function there is a partition $\tau = \{e_i : i = 1, 2, \ldots, l\}$ of E and constants c_i, $i = 1, 2, \ldots, l$, such that $\phi(x) = c_i$ for $x \in e_i$, $i = 1, 2, \ldots, l$. Since $\phi \in \mathcal{A}, \phi \leq f$ and so $c_i \leq f(x)$ for $x \in e_i$. Hence $c_i \leq m_i = \inf\{f(x) : x \in e_i\}$ for $i = 1, 2, \ldots, l$. So

$$\lambda = \int_E \phi dx = \sum_{i=1}^l c_i \mu(e_i) \leq \sum_{i=1}^l m_i \mu(e_i) = s(\tau, f) \leq \sup C,$$

which is a contradiction, since $\sup C < \lambda$. Hence $\sup C = \sup \mathcal{B}$ and so the first part of (ii) is proved. The proof of the second part is similar.

In view of Theorem 5.6 (ii) we get the following definition which is equivalent to Definition 5.1. □

Definition 5.2 Let E be a measurable set of finite measures and let $f : E \to \mathbb{R}$ be bounded. Then the lower and upper Lebesgue integral of f on E are defined by

$$\underline{\int_E} f\,dx = \sup\left\{\int_E \phi\,dx : \phi \text{ is a measurable simple function and } \phi \leq f\right\}$$

and

$$\overline{\int_E} f\,dx = \inf\left\{\int_E \psi\,dx : \psi \text{ is a measurable simple function and } \psi \geq f\right\}$$

respectively. If they are equal then f is said to be Lebesgue integrable and the common value is the Lebesgue integral of f.

Theorem 5.7 Let E be a measurable set of finite measure and $f : E \to \mathbb{R}$ be bounded. Then the following statements are equivalent:

(i) f is integrable.
(ii) For every $\epsilon > 0$ there is a partition τ of E such that $S(\tau, f) - s(\tau, f) < \epsilon$ where $S(\tau, f)$ and $s(\tau, f)$ are the upper and lower sums corresponding to τ.
(iii) For every $\epsilon > 0$ there are measurable bounded simple functions ϕ_ϵ and ψ_ϵ on E such that $\phi_\epsilon \leq f \leq \psi_\epsilon$ on E and $\int_E \psi_\epsilon\,dx - \int_E \phi_\epsilon\,dx < \epsilon$.

Proof (i) $\Rightarrow$ (ii) Suppose that f is integrable on E. Then $\sup_\tau s(\tau, f) = \int_E f\,dx = \inf_\tau S(\tau, f)$.

So, for every $\epsilon > 0$ there is a partition $\tau_1 = \{e'_j\}$ and a partition $\tau_2 = \{e''_k\}$ of E such that

$$S(\tau_1, f) < \int_E f\,dx + \frac{\epsilon}{2} \text{ and } s(\tau_2, f) > \int_E f\,dx - \frac{\epsilon}{2}. \quad (5.11)$$

Let $\tau = \{e_{jk}\}$ where $e_{jk} = e'_j \cap e''_k$. Then τ is a partition of E which is a refinement of both τ_1 and τ_2.
Hence by Lamma 5.1,

$$S(\tau, f) \leq S(\tau_1, f) \text{ and } s(\tau, f) \geq s(\tau_2, f)$$

and so from (5.11)

$$S(\tau, f) < \int_E f\,dx + \frac{\epsilon}{2} \text{ and } s(\tau, f) > \int_E f\,dx - \frac{\epsilon}{2}$$

which gives

$$S(\tau, f) - s(\tau, f) < \epsilon$$

$(ii) \Rightarrow (iii)$.

Let $\epsilon > 0$ be arbitrary. Then there is a partition τ of E such that $S(\tau, f) - s(\tau, f) < \epsilon$. By Theorem 5.6$(i)$ there are bounded measurable simple functions ϕ_τ and ψ_τ on E such that $\phi_\tau \leq f \leq \psi_\tau$ on E which satisfy $s(\tau, f) = \int_E \phi_\tau dx$ and $S(\tau, f) = \int_E \psi_\tau dx$. Since τ depends on ϵ, the functions ϕ_τ and ψ_τ also depends on ϵ and so writing $\phi_\tau = \phi_\epsilon$ and $\psi_\tau = \psi_\epsilon$ we have

$$\int_E \psi_\epsilon dx - \int_E \phi_\epsilon dx < \epsilon$$

$(iii) \rightarrow (i)$.

Let $\epsilon > 0$ be arbitrary. Then applying Definition 5.2 and (iii)

$$0 \leq \overline{\int_E} f dx - \underline{\int_E} f dx \leq \int_E \psi_\epsilon dx - \int_E \phi_\epsilon dx < \epsilon$$

Since ϵ is arbitrary f is integrable. $\square$

5.2 Properties of the Lebesgue Integral of Bounded Functions

Theorem 5.8 *Let E be a measurable set of finite measures and let $f : E \to \mathbb{R}$ be such that $A \leq f \leq B$ on E. If f is integrable then*

$$A\mu(E) \leq \int_E f dx \leq B\mu(E)$$

Proof Let $\tau = \{e_i : i = 1, 2, \ldots, n\}$ be any partition of E and let $m_i = \inf\{f(x) : x \in e_i\}$, $M_i = \sup\{f(x) : x \in e_i\}$. Then $A \leq m_i \leq M_i \leq B$ and so

$$A\mu(E) = A \sum \mu(e_i) \leq \sum m_i \mu(e_i) = s(\tau, f) \leq \underline{\int_E} f dx = \int_E f dx = \overline{\int_E} f dx$$

$$\leq S(\tau, f) = \sum M_i \mu(e_i) \leq B \sum \mu(e_i) = B\mu(E)$$

$\square$

Corollary 5.9 *If f is a constant, say $f = c$ on E then*

$$\int_E f dx = c\mu(E).$$

Corollary 5.10 *If f is bounded and integrable and $f \geq 0$ on E then*

$$\int_E f\,dx \geq 0$$

Corollary 5.11 *If f is bounded on a measurable set E of measure zero then f is integrable on E and*

$$\int_E f\,dx = 0.$$

[Hint: In the above theorem it is proved that $A\mu(E) \leq \int_E f\,dx \leq B\mu(E)$].

Theorem 5.12 *If f and g are integrable on a measurable set E of finite measure and if $f = g$ a.e on E then*

$$\int_E f\,dx = \int_E g\,dx$$

Proof Let $A = \{x \in E : f(x) \neq g(x)\}$, $B = E \sim A$. Then $\mu(A) = 0$ and so by Theorem 5.5 and Corollary 5.11

$$\int_E f\,dx = \int_A f\,dx + \int_B f\,dx = \int_B f\,dx = \int_B g\,dx = \int_A g\,dx + \int_B g\,dx = \int_E g\,dx$$

$\square$

Theorem 5.13 *If f is integrable and non-negative on a measurable set E of finite measure and if E_1 is any measurable subset of E then*

$$\int_{E_1} f\,dx \leq \int_E f\,dx$$

Proof Let $E_2 = E \sim E_1$. Then applying Theorem 5.5 and Corollary 5.10

$$\int_E f\,dx = \int_{E_1} f\,dx + \int_{E_2} f\,dx \geq \int_{E_1} f\,dx.$$

$\square$

Theorem 5.14 *Let E be a measurable set of finite measures.*

(i) *If f is integrable on E and k is constant then kf is integrable on E and*

$$\int_E kf\,dx = k\int_E f\,dx$$

(ii) *If f and g are integrable on E then $f + g$ is also integrable on E and*

$$\int_E (f+g)\,dx = \int_E f\,dx + \int_E g\,dx$$

5.2 Properties of the Lebesgue Integral of Bounded Functions

(iii) If f and g are integrable on E and $f \geq g$ a.e on E then

$$\int_E f\,dx \geq \int_E g\,dx.$$

Proof (i) If $k = 0$ the proof is trivial. So we suppose that $k \neq 0$. Let $\tau = \{e_i\}$ be any partition of E and let m_i and M_i be the tower and upper bounds of f on e_i. Then when $k > 0$, km_i and kM_i are the lower and upper bounds of kf on e_i and so $ks(\tau, f) = s(\tau, kf)$ and $kS(\tau, f) = S(\tau, kf)$ and therefore

$$\underline{\int_E} kf\,dx = \sup_\tau s(\tau, kf) = k \sup_\tau s(\tau, f) = k \underline{\int_E} f\,dx$$

$$= k \inf_\tau S(\tau, f) = \inf_\tau S(\tau, kf) = \overline{\int_E} kf\,dx$$

proving (i) and when $k < 0$, kM_i and km_i are the lower and upper bounds of kf on e_i and so $ks(\tau, f) = S(\tau, kf)$ and $kS(\tau, f) = s(\tau, kf)$ and therefore

$$\overline{\int_E} kf\,dx = \inf_\tau S(\tau, kf) = \inf_\tau ks(\tau, f) = k \sup_\tau s(\tau, f) = k \underline{\int_E} f\,dx$$

$$= k \inf_\tau S(\tau, f) = \sup_\tau kS(\tau, f) = \sup_\tau s(\tau, kf) = \underline{\int_E} kf\,dx$$

completing the proof of (i).

(ii) Let $h = f + g$ and $\epsilon > 0$ be a arbitrary. By Theorem 5.7 there are partitions $\tau' = \{e_i'\}$ and $\tau'' = \{e_j''\}$ of E such that

$$S(\tau', f) - s(\tau', f) < \frac{\epsilon}{2} \text{ and } S(\tau'', g) - s(\tau'', g) < \frac{\epsilon}{2} \quad (5.12)$$

Let $\tau = \{e_k\}$ where $e_k = e_i' \cap e_j''$ and $e_k \neq \emptyset$. Let m_k', m_k'' and m_k be the lower bounds of f, g and h on e_k and let M_k', M_k'' and M_k be the upper bounds of f, g and h respectively on e_k. Then $m_k' + m_k'' \leq m_k \leq M_k \leq M_k' + M_k''$ for all k. So

$$s(\tau, f) + s(\tau, g) \leq s(\tau, h) \leq \underline{\int_E} h\,dx \leq \overline{\int_E} h\,dx \leq S(\tau, h) \leq S(\tau, f) + S(\tau, g) \quad (5.13)$$

Also

$$s(\tau, f) + s(\tau, g) \leq \int_E f\,dx + \int_E g\,dx \leq S(\tau, f) + S(\tau, g) \quad (5.14)$$

From (5.12) and (5.13), $0 \leq \overline{\int_E} h dx - \underline{\int_E} h dx < \epsilon$. Since ϵ is arbitrary h is integrable on E. This fact together with (5.12) and (5.13), (5.14), shows

$$\left| \int_E h dx - \int_E f dx - \int_E g dx \right| < \epsilon.$$

Since ϵ is arbitrary, the result follows.

(iii) Let $E_1 = \{x \in E : f(x) \geq g(x)\}$ and $E_2 = E \sim E_1$. Then for any partition τ of E_1, $s(\tau, f) \geq s(\tau, g)$ and so $\int_{E_1} f dx \geq \int_{E_1} g dx$. Since E_2 is of measure zero by Theorem 5.5 and Corollary 5.11

$$\int_E f dx = \int_{E_1} f dx + \int_{E_2} f dx = \int_{E_1} f dx \geq \int_{E_1} g dx$$

$$= \int_{E_1} g dx + \int_{E_2} g dx = \int_E g dx$$

$\square$

Theorem 5.15 *Let f be a bounded function on a measurable set E of finite measure. Then f is integrable on E if and only if f is measurable on E*

Proof The 'if' part in proved in Theorem 5.4 (v). We prove the 'only if' part. Let f be integrable on E. Then by Theorem 5.7 for each positive integer n there exist measurable bounded simple functions ϕ_n and ψ_n such that $\phi_n \leq f \leq \psi_n$ on E and $\int_E \psi_n dx - \int_E \phi_n dx < \frac{1}{n}$. Let $\phi(x) = \sup_n \phi_n(x)$ and $\psi(x) = \inf_n \psi_n(x)$ for $x \in E$. Then by Theorem 3.9, ϕ and ψ are measurable. Also $\phi \leq f \leq \psi$ on E. Let $S = \{x \in E : \psi(x) - \phi(x) > 0\}$ and for each positive integer k let $S_{nk} = \{x \in E; \psi_n(x) - \phi_n(x) > \frac{1}{k}\}$. Then $S \subset \bigcup_{k=1}^{\infty} \bigcap_{n=1}^{\infty} S_{nk}$. Fix k. Then for all n, by Theorem 5.14 and by Corollary 5.10

$$\frac{1}{n} > \int_E \psi_n dx - \int_E \phi_n dx = \int_E (\psi_n - \phi_n) dx$$

$$\geq \int_{S_{nk}} (\psi_n - \phi_n) dx$$

$$\geq \frac{1}{k} \mu(S_{nk}) \geq \frac{1}{k} \mu \left(\bigcap_{n=1}^{\infty} S_{nk} \right).$$

5.2 Properties of the Lebesgue Integral of Bounded Functions

Hence $\mu\left(\bigcap_{n=1}^{\infty} S_{nk}\right) < \frac{k}{n}$. Letting $n \to \infty$, $\mu\left(\bigcap_{n=1}^{\infty} S_{nk}\right) = 0$.

This being true for all k, $\mu(S) \leq \sum_{k=1}^{\infty} \mu\left(\bigcap_{n=1}^{\infty} S_{nk}\right) = 0$ which shows that $\mu(S) = 0$ i.e. $\phi = \psi$ a.e on E. Since $\phi \leq f \leq \psi$ and ϕ and ψ are measurable, f is measurable. □

Remarks

(i) From the above theorem, it follows that for every measurable set E such that $0 < \mu(E) < \infty$ there exists a bounded function f on E which is not Lebesgue integrable. For, by Theorem 2.19 there exists a nonmeasurable set $E_1 \subset E$. Let $f(x) = 1$ for $x \in E_1$ and $f(x) = 0$ for $x \in E \sim E_1$. Then f is bounded and nonmeasurable and so by the above theorem f is not Lebesgue integrable.

(ii) The class of all bounded functions which are not Lebesgue integrable on $[0, 1]$ has the cardinality $2^{\underline{c}}$. For, let $E \subset [0, 1]$ be a nonmeasurable set. Define $f(x) = 1$ for $x \in E$ and $f(x) = 0$ for $x \in [0, 1] \sim E$. Then f is bounded in $[0, 1]$ but is not measurable and so by the above theorem f is not Lebesgue integrable on $[0, 1]$. Since by Theorem 4.18 the class of all nonmeasurable subsets of $[0, 1]$ has the cardinality $2^{\underline{c}}$, the result follows.

Theorem 5.16 . *Let E be a measurable set of finite measure*
(i) If f and g are integrable on E then $max[f, g]$ and $min[f, g]$ are integrable E.
(ii) If f is integrable and $f \geq 0$ on E and if $\int_E f\, dx = 0$ then $f = 0$ a.e on E.
(iii) f is integrable on E if and only if $|f|$ is integrable on E. In either case

$$\left|\int_E f\, dx\right| \leq \int_E |f|\, dx$$

Proof (i), By Theorem 5.15, f and g are measurable and so by Theorem 3.6 $max[f, g]$ and $min[f, g]$ are measurable. Also they are bounded since f and g are bounded and so (i) is proved by Theorem 5.15
(ii) Let $A = \{x \in E : f(x) > 0\}$ and $A_n = \{x \in E : f(x) > \frac{1}{n}\}, n = 1, 2, \ldots$. Then $A = \bigcup_{n=1}^{\infty} A_n$. Since f is integrable, it is measurable and so the sets A, A_k are measurable and hence $\mu(A) \leq \sum_{n=1}^{\infty} \mu(A_n)$. If possible suppose that $\mu(A_n) > 0$ for some n. Then writing $E \sim A_n = B_n$ and using Corollary 5.10

$$\int_E f\, dx = \int_{A_n} f\, dx + \int_{B_n} f\, dx \geq \int_{A_n} f\, dx \geq \frac{1}{n}\mu(A_n) > 0$$

which is a contradiction. Hence $\mu(A_n) = 0$ for all n and so $\mu(A) = 0$ and hence $f = 0$ a.e on E.
(iii) Let f be integrable on E. Let $f_+ = max[f, 0]$ and $f_- = max[-f, 0]$. Then $f = f_+ - f_-$ and $|f| = f_+ + f_-$. By (i) f_+ and f_- are integrable on E and so $|f|$

is integrable on E. Conversely suppose that $|f|$ is integrable on E. Let $E_+ = \{x \in E : f(x) \geq 0\}$ and $E_- = \{x \in E : f(x) < 0\}$. Then $E_+ \cup E_- = E$ and $E_+ \cap E_- = \emptyset$. Since $|f|$ is integrable on E by Theorem 5.5(i), $|f|$ is integrable on E_+ and E_-. Since $|f| = f$ on E_+, f is integrable on E_+ and since $|f| = -f$ on E_-, $-f$ is integrable on E_-. Therefore f is integrable on $E_+ \cup E_- = E$. Finally, since $-|f| \leq f \leq |f|$, by Theorem 5.14(iii),

$$-\int_E |f|dx \leq \int_E f dx \leq \int_E |f|dx \text{ and so } \left|\int_E f dx\right| \leq \int_E |f|dx.$$

□

Note The part (iii) of the above theorem is special for Lebesgue integral. The Riemann integral has the property that if f is Riemann integrable then $|f|$ is Riemann integrable but the converse is not true. For, consider $f(x) = 1$ for x is rational and $f(x) = -1$ for x is irrational. For this reason, the Lebesgue integral is called absolute integral. Other more general integrals have not this property.

Theorem 5.17 (Bounded convergence theorem) *Let E be a measurable set of finite measure and let $\{f_n\}$ be a sequence of integrable function on E such that $|f_n(x)| \leq M$ for all n and for all $x \in E$. If $\lim_{n \to \infty} f_n(x) = f(x)$ for almost all $x \in E$ then f is integrable on E (assuming $f = 0$ where $\lim f_n$ does not exist) and*

$$\lim_{n \to \infty} \int_E f_n(x)dx = \int_E f(x)dx$$

Proof Since $|f_n(x)| \leq M$ for all n and for all $x \in E$, the functions f, f_n are all bounded on E. Also by Theorem 3.9, the limit function f is measurable and so by Theorem 5.15 the function f is integrable. Let $\epsilon > 0$ be arbitrary. Then by Egorov's Theorem (Theorem 3.23) there is a measurable set E_ϵ such that $\mu(E \sim E_\epsilon) < \epsilon$ and $\{f_n\}$ converges to f uniformly on E_ϵ. So, there is N such that

$$|f_n(x) - f(x)| < \epsilon \text{ for } n \geq N \text{ and } x \in E_\epsilon$$

So for $n \geq N$

$$\left|\int_E f_n(x)dx - \int_E f(x)dx\right| = \left|\int_E (f_n(x) - f(x))dx\right|$$
$$\leq \int_E |f_n(x) - f(x)|dx$$
$$= \int_{E_\epsilon} |f_n(x) - f(x)|dx + \int_{E \sim E_\epsilon} |f_n(x) - f(x)|dx$$
$$< \epsilon\mu(E_\epsilon) + 2M\mu(E \sim E_\epsilon)$$
$$< \epsilon\mu(E_\epsilon) + 2M\epsilon.$$

Since ϵ is arbitrary, the theorem is proved. □

5.3 Lebesgue's Criterion for Riemann Integrability

If a function f is Riemann integrable in $[a, b]$ then f may have discontinuity at an everywhere dense set of points in $[a, b]$. For, consider $f : [0, 1] \to \mathbb{R}$ defined by

$f(x) = 0$ if $x = 0$ or an irrational number in $(0, 1)$

$\quad\quad = \dfrac{1}{q}$ if x is rational in $(0,1]$ and $x = \dfrac{p}{q}$, $p \leq q$, p and q are relatively prime

Then f is discontinuous at each rational point in $(0, 1]$ but f is Riemann integrable in $[0,1]$. On the other hand the function $f(x) = 1$ for x rational and $f(x) = 0$ for x irrational is discontinuous everywhere and it is not Riemann integrable in $[0,1]$. Therefore, it is natural to ask what is the role of continuity of f for its Riemann integrability. The Lebesgue - Vitali Theorem will give an answer.

For any interval I and $f : I \to \mathbb{R}$ the oscillation of f on I is defined by

$$O(f; I) = \sup\{f(x) : x \in I\} - \inf\{f(x) : x \in I\}.$$

The oscillation of f at a point $x \in I$ is defined by

$$O(f; x) = \lim_{\delta \to 0} O(f; I_\delta) \text{ where } I_\delta = (x - \delta, x + \delta) \cap I$$

the limit exists since $O(f; I_\delta)$ decreases as δ decreases. The following theorem will be used.

Theorem 5.18 *Let $f : I \to \mathbb{R}$. Then*

(i) *f is continuous at $x \in I$ if and only if $O(f; x) = 0$*
(ii) *For every $\alpha > 0$ the set $\{x \in I : O(f; x) \geq \alpha\}$ is closed in I.*
(iii) *The set $\{x \in I : O(f; x) > 0\}$ is an F_σ set.*

The proofs of (i) and (ii) are not difficult since

$$\{x \in I : O(f; x) > 0\} = \bigcup_{k=1}^{\infty} \left\{x \in I : O(f; x) \geq \dfrac{1}{k}\right\}$$

the proof of (iii) follows from (ii).

The following theorem is known as the Lebesgues criterion for Riemann integrability.

Theorem 5.19 (Lebesgue-Vitali) *A bounded function f is Riemann integrable in $[a, b]$ if and only if the set of points of discontinuity of f in $[a, b]$ is of measure zero.*

Proof Let f be Riemann integrable in $[a, b]$. We shall show that the set $H = \{x \in [a, b] : O(f; x) > 0\}$ is of measure zero, which by Theorem 5.18(i) proves that f is continuous almost everywhere. Let k be any positive integer, since f is Riemann integrable, there is a partition $D : a = x_0 < x_1 < \cdots < x_n = b$ of $[a, b]$ such that

$$S(D, f) - s(D, f) = \sum_{r=1}^{n}(M_r - m_r)(x_r - x_{r-1}) < \frac{1}{k^2}$$

Where M_r and m_r one the upper and lower bounds of f in $I_r = [x_{r-1}, x_r]$. Hence

$$\sum_{r=1}^{n} O(f; I_r) \mid I_r \mid < \frac{1}{k^2} \qquad (5.15)$$

where $\mid I_r \mid$ is the length of I_r. Let $\{J_i\}$ be the subfamily of the family of intervals $\{I_r : r = 1, 2, \ldots, n\}$ such that there is some point x_i in the interior of J_i for which $O(f; x_i) \geq \frac{1}{k}$. Let $H_k = \{x \in [a, b] : O(f; x) \geq \frac{1}{k}\}$. Then all points of H_k. Except possibly the endpoints of the intervals $I_r, r = 1, 2, \ldots n$, are covered by the subfamily $\{J_i\}$. Since the set of endpoints of the intervals I_r has measure zero.

$$\mu(H_k) \leq \sum_{i} \mid J_i \mid \qquad (5.16)$$

where $\sum_{i}$ denotes the summation over all J_i having the above property, Also by (5.15)

$$\frac{1}{k^2} > \sum_{i} O(f; J_i) \mid J_i \mid \geq \sum_{i} O(f; x_i) \mid J_i \mid \geq \frac{1}{k}\sum_{i} \mid J_i \mid .$$

Therefore $\sum_{i} \mid J_i \mid < \frac{1}{k}$ and so by (5.16) $\mu(H_k) < \frac{1}{k}$. Hence

$$\lim_{k \to \infty} \mu(H_k) = 0 \qquad (5.17)$$

Since $H = \bigcup_{k=1}^{\infty} H_k$ and $\{H_k\}$ is increasing, by Theorem 2.8 $\mu(H) = \lim_{k \to \infty} \mu(H_k)$ and so by (5.17), $\mu(H) = 0$.

Conversely, suppose that the set of points of discontinuity of f in $[a, b]$ is of measure zero. Let $E \subset [a, b]$ be the set of points where f is discontinuous. Let $\epsilon > 0$ be arbitrary. Since $\mu(E) = 0$, there is a countable collection of open intervals $\{I_n\}$ such that $E \subset \cup I_n$ and

$$\sum \mid I_n \mid < \frac{\epsilon}{2(M - m) + 1} \qquad (5.18)$$

where $M = \sup\{f(x) : x \in [a, b]\}$ and $m = \inf\{f(x) : x \in [a, b]\}$. Let $x \in [a, b] \sim E$. Then f is continuous at x so $O(f, x) = 0$ and hence there is an open interval J_x containing x such that

$$O(f; J_x) < \frac{\epsilon}{2(b - a)} \qquad (5.19)$$

The collection of open intervals $\{I_n\} \cup \{J_x : x \in [a,b] \sim E\}$ is an open cover of $[a,b]$ and so there is a finite sub-collection of this collection which also covers $[a,b]$. Let

$$J_1, J_2, \ldots, J_p, J_{x_1}, J_{x_2}, \ldots, J_{x_q} \tag{5.20}$$

be the finite sub-collection which covers $[a,b]$. By cutting down the portions of the intervals in (5.20) which are overlapped or which are outside $[a,b]$, we get the set of non-overlapping intervals

$$J'_1, J'_2, \ldots J'_p, J'_{x_1}, J'_{x_2}, \ldots J'_{x_q} \tag{5.21}$$

such that $[a,b] = \left(\bigcup_{i=1}^{p} J'_i\right) \cup \left(\bigcup_{i=1}^{p} J'_{x_i}\right)$ and the endpoints of the intervals in (5.21) give a partition say D of $[a,b]$. Then

$$S(D, f) - s(D, f) = \sum_{i=1}^{p}(M_i - m_i)|J'_i| + \sum_{i=1}^{q}(M_{x_i} - m_{x_i})|J'_{x_i}| \tag{5.22}$$

where M_i and m_i are the upper and lower bounds of f in J'_i and M_{x_i} and m_{x_i} are the upper and lower bounds of f in J'_{x_i} respectively. Since $M_i - m_i \leq M - m$, and from (5.19) $M_{x_i} - m_{x_i} \leq \frac{\epsilon}{2(b-a)}$ we have from (5.22)

$$S(D, f) - s(D, f) = (M - m)\sum_{i=1}^{p}|J'_i| + \frac{\epsilon}{2(b-a)}\sum_{i=1}^{q}|J'_{x_i}| \tag{5.23}$$

Since every J_i in (5.20) is an I_n and $J'_i \subset J_i$ for $i = 1, 2, \ldots, p$ we get from (5.23) and (5.18)

$$S(D, f) - s(D, f) \leq (M - m)\frac{\epsilon}{2(M - m) + 1} + \frac{\epsilon}{2(b - a)}.(b - a) < \epsilon$$

and so f is Riemann integrable in $[a,b]$ $\qquad\square$

5.4 Riemann Integrability of Bounded Derivatives. Volterra Function

In the theory of Riemann integration, the Fundamental Theorem of Integral Calculus demands that if a function f is differentiable then the derivative f' should be Riemann integrable in order to satisfy the relation $\int_a^b f' dx = f(b) - f(a)$ on $[a,b]$. Since a derivative function may not be bounded the question arises whether a bounded derivative is Riemann integrable. The answer is negative. There are functions f

which have bounded derivative f' but f' is not Riemann integrable. An example of such a function is given by V. Volterra.

Theorem 5.20 *If $[a, b]$ is any closed interval then there is a function $\phi : [a, b] \to \mathbb{R}$ such that ϕ is differentiable in $[a, b]$ and*
(i) $\phi'_+(a) = 0 = \phi'_-(b) = \phi(a) = \phi(b)$
(ii) $|\phi'(x)| \leq 2(b - a) + 1$ for all $x \in [a, b]$
(iii) ϕ' is continuous in (a,b) but ϕ' oscillates between $+1$ and -1 infinitely many times in every right neighborhood of a and in every left neighborhood of b and hence ϕ' is discontinuous at a and at b.
(iv) $|\phi(x)| \leq (x - a)^2$ for $x \in [a, \frac{a+b}{2}]$ and $|\phi(x)| \leq (b - x)^2$ for $x \in [\frac{a+b}{2}, b]$.

Proof Let N be the smallest positive integer such that $a < a + \frac{2}{(4N+1)\pi} < \frac{a+b}{2}$. Define $\phi : [a, b] \to \mathbb{R}$ as follows

$$\phi(x) = (x - a)^2 \sin \frac{1}{x - a} \quad \text{for} \quad a < x \leq a + \frac{2}{(4N + 1)\pi}$$

$$= \left(\frac{2}{(4N + 1)\pi}\right)^2 \quad \text{for} \quad a + \frac{2}{(4N + 1)\pi} < x < b - \frac{2}{(4N + 1)\pi}$$

$$= (b - x)^2 \sin \frac{1}{b - x} \quad \text{for} \quad b - \frac{2}{(4N + 1)\pi} \leq x < b$$

$$= 0 \quad \text{for} \quad x = a \text{ and } x = b$$

Clearly $\phi'_+(a) = 0 = \phi'_-(b)$, and

$$\phi'(x) = 2(x - a) \sin \frac{1}{x - a} - \cos \frac{1}{x - a} \quad \text{for} \quad a < x \leq a + \frac{2}{(4N + 1)\pi}$$

$$= 0 \quad \text{for} \quad a + \frac{2}{(4N + 1)\pi} < x < b - \frac{2}{(4N + 1)\pi}$$

$$= -2(b - x) \sin \frac{1}{b - x} + \cos \frac{1}{b - x} \quad \text{for} \quad b - \frac{2}{(4N + 1)\pi} \leq x < b.$$

So ϕ satisfies all the conditions and the theorem is proved. $\square$

In the next theorem we consider a nowhere dense closed set (sec Definition 4.8). Note that a set $E \subset [a, b]$ is nowhere dense and closed if and only if its complement $[a, b] \sim E$ is a countable union of disjoint open intervals which are everywhere dense in $[a, b]$.

Theorem 5.21 *Let E be a nowhere dense closed subset of $[a, b]$. Then there is a function $f : [a, b] \to \mathbb{R}$ such that*
(i) f is differentiable in $[a, b]$
(ii) $f'(x) = 0$ if $x \in E$
(iii) f' is bounded in $[a, b]$
(iv) f' is continuous for every $x \in [a, b] \sim E$ and discontinuous for every $x \in E$

5.4 Riemann Integrability of Bounded Derivatives. Volterra Function

Proof We may suppose that $a = \inf E$ and $b = \sup E$. For, if not then we define $f(x) = 0$ for $x \in [a, \inf E] \cup [\sup E, b]$. Then $[a, b] \sim E$ is a countable union of disjoint open intervals $\{I_k\}$ which are everywhere dense in $[a, b]$. Let $[a, b] \sim E = \cup I_k$. Let $[a_k\ b_k]$ be the closure of the open interval I_k. Then by Theorem 5.20, there is, for each k, a function $\phi_k : [a_k, b_k] \to \mathbb{R}$ such that
(1k) $(\phi_k)'_+(a_k) = 0 = (\phi_k)'_-(b_k) = \phi_k(a_k) = \phi_k(b_k)$
(2k) $|\phi'_k(x)| \leq 2(b_k - a_k) + 1$ for all $x \in [a_k, b_k]$
(3k) ϕ'_k is continuous in (a_k, b_k) but ϕ'_k oscillates between $+1$ and -1 infinitely many times in every right and left neighborhood of a_k and b_k respectively
(4k) $|\phi_k(x)| \leq (x - a_k)^2$ for $x \in [a_k, \frac{a_k+b_k}{2}]$ and $|\phi_k(x)| \leq (b_k - x)^2$ for $x \in [\frac{a_k+b_k}{2}, b_k]$.
Define $f : [a, b] \to \mathbb{R}$ by

$$f(x) = 0 \text{ for } x \in E$$
$$= \phi_k(x) \text{ for } x \in I_k, k = 1, 2, \ldots$$

Then f is differentiable on each I_k. Let $\xi \in E$. If ξ is an isolated point of E from the right then ξ is the left endpoint of some interval I_k and hence $f'_+(\xi) = (\phi_k)'_+(\xi) = 0$ Let ξ be not isolated from the right. Then since the intervals $\{I_k : k = 1, 2, \ldots\}$ are everywhere dense in $[a, b]$ there is a subsequence $\{I_{k_n}\}$ of $\{I_k\}$ such that $\{I_{k_n}\}$ converges to ξ from the right side. Let $\delta > 0$ be arbitrary and let $\xi < x < \xi + \delta$. If $x \in E$ then

$$\left|\frac{f(x) - f(\xi)}{x - \xi}\right| = 0 \tag{5.24}$$

If $x \notin E$ then for some K_n, $x \in I_{k_n} = (a_{k_n}, b_{k_n})$ and so when $x \in (a_{k_n}, \frac{a_{k_n}+b_{k_n}}{2}]$

$$\left|\frac{f(x) - f(\xi)}{x - \xi}\right| = \left|\frac{f(x)}{x - \xi}\right| = \left|\frac{\phi_{k_n}(x)}{x - \xi}\right| < \frac{(x - a_{k_n})^2}{x - \xi} < x - a_{k_n} < \delta \tag{5.25}$$

and when $x \in (\frac{a_{k_n}+b_{k_n}}{2}, b_{k_n})$

$$\left|\frac{f(x) - f(\xi)}{x - \xi}\right| = \left|\frac{\phi_{k_n}(x)}{x - \xi}\right| \leq \frac{(b_{k_n} - x)^2}{x - \xi} < \frac{(x - a_{k_n})^2}{x - \xi} < x - a_{k_n} < \delta \tag{5.26}$$

Since δ is arbitrary, it follows from (5.24), (5.25) and (5.26) that $f'_+(\xi) = 0$. It can similarly be shown that the left-hand derivative exists at ξ and $f'_-(\xi) = 0$. So the derivative $f'(\xi)$ exists. Since ξ is any point of E, f' exists and $f' = 0$ on E. Also if $x \in I_k$ for some k then $|f'(x)| = |\phi'_k(x)| \leq 2(b_k - a_k) + 1 < 2(b - a) + 1$ by (2k) and hence f' exists and is bounded in $[a, b]$, proving (i), (ii), and (iii). Since $f' = \phi'_k$ in I_k, f' is continuous in I_k for each k. So f' is continuous in $[a, b] \sim E$. Let $\xi \in E$. If ξ is an isolated point of E from the right then ξ is the left endpoint of some interval I_k and since $f'(x) = \phi'_k(x)$ for all $x \in I_k$, by (3k) f' is not continuous at ξ. If ξ is not isolated from the right then also by (3k) every right neighborhood of ξ contains points where f' assumes the value 1 and -1 and so f' cannot be continuous at ξ. Since ξ is arbitrary, f' is discontinuous for every $x \in E$. □

Theorem 5.22 *For every interval $[a, b]$ and every α, $0 < \alpha \leq 1$ there is a nowhere dense closed set $E_\alpha \subset [a, b]$ of measure $(b - a)(1 - \alpha)$ and a function $f : [a, b] \to \mathbb{R}$ such that the derivative f' of f exists and is bounded in $[a, b]$ and f' is continuous for every $x \in [a, b] \sim E_\alpha$ and is discontinuous at each point of E_α.*

Proof Fix α and $[a, b]$. Let C_α be the Cantor set in $[0,1]$ corresponding to α, $0 < \alpha \leq 1$ (See Sect. 4.4). Applying the transformation $T(x) = (b - a)x + a$ which transforms the interval $[0, 1]$ to the interval $[a, b]$, the set C_α is transformed into the set $E_\alpha = \{(b - a)x + a : x \in C_\alpha\}$. Since C_α is nowhere dense and closed in $[0, 1]$, E_α is nowhere dense and closed in $[a, b]$ and since $\mu(C_\alpha) = 1 - \alpha$, by Theorem 2.16, $\mu(E_\alpha) = (b - a)(1 - \alpha)$. The proof now follows from Theorem 5.21. □

Theorem 5.23 *For every interval $[a, b]$ there exists a function $f : [a, b] \to \mathbb{R}$ such that the derivative f' of f exists and is bounded in $[a, b]$ but f' is not Riemann integrable in $[a, b]$.*

The proof follows from Theorems 5.19 to 5.22.

Remark: From the time of the discovery of Calculus question arises regarding the relation between derivative and integration. In the primary stage integral was defined as the inverse process of differentiation and therefore integral was called anti derivative. When Riemann integration came various mathematicians began to study the relation between the derivative and integral and it was shown that if $f : [a, b] \to \mathbb{R}$ is differentiable then $\int_a^x f' dx = f(x) - f(a)$ for $x \in [a, b]$ provided the derivative f' of f is integrable in $[a, b]$. For Riemann integral of f', f' must be bounded. Now the question is whether f' is Riemann integrable when f' is bounded. Lebesgue-Vitali Theorem states that if a bounded function is Riemann integrable then it must be almost everywhere continuous. Voltera constructed a function f which has bounded derivative f' in $[0, 1]$ and f' is discontinuous on a set of positive measure in $[0, 1]$. Therefore integrability of f' must be assumed. Now the question is whether f' is Lebesgue integrable. If f' is bounded then the answer is yes (see the next theorem). But if f is unbounded then f' may not be Lebesgue integrable (See the last Section).

Theorem 5.24 *Let $f : [a, b] \to \mathbb{R}$ be differentiable. If the derivative f' is bounded in $[a, b]$ then f' is Lebesgue integrable in $[a, b]$ and*

$$\int_a^b f' dx = f(b) - f(a).$$

Proof Let $M > 0$ be such that $|f'(x)| \leq M$ for all $x \in [a, b]$. Extend f to $[a, b + 1]$ by defining $f(x) = f(b) + f'(b)(x - b)$ for $x \in [b, b + 1]$. So f is differentiable in

$[a, b+1]$ and $|f'(x)| \leq M$ for all $x \in [a, b+1]$. For each n, define $f_n : [a, b] \to \mathbb{R}$ by $f_n(x) = n[f(x + \frac{1}{n}) - f(x)]$. Since f is differentiable, f is continuous and so each f_n is continuous in $[a, b]$ and hence is measurable. Since $\lim_{n \to \infty} f_n = f'$, f' is measurable. Also for each $x \in [a, b]$ there is, by mean value theorem, a point $\xi \in (x, x + \frac{1}{n})$ such that $f_n(x) = f'(\xi)$ and so $|f_n(x)| \leq M$. Hence by bounded convergence theorem, i.e. by Theorem 5.17

$$\lim_{n \to \infty} \int_a^b f_n dx = \int_a^b f' dx.$$

Since f_n is continuous for all n the integral in the left side is Riemann integral and so by continuity of f

$$\lim_{n \to \infty} \int_a^b f_n dx = \lim_{n \to \infty} \left[n \int_a^b f(t + \frac{1}{n}) dt - n \int_a^b f(t) dt \right]$$
$$= \lim_{n \to \infty} n \int_a^{b+\frac{1}{n}} f dx - \lim_{n \to \infty} n \int_a^{a+\frac{1}{n}} f dx$$
$$= f(b) - f(a).$$

and so

$$\int_a^b f' dx = f(b) - f(a)$$

□

The boundedness of f' is assumed above to ensure that f' is integrable, since a bounded measurable function is Lebesgue integrable. After introducing an absolutely continuous function in Chap. 6 we prove a theorem where boundedness of f' will be removed but integrability condition of f' must be assumed.

5.5 Lebesgue Integral of Unbounded Function

Although for bounded functions the definition of Riemann integral and Lebesgue integral are analogous and comparable, for unbounded functions the definition of Riemann integral which is usually known as 'improper Riemann integral' is quite different from the definition of Lebesgue integral. The basic difference lies in a property of the Lebesgue integral which states that f is Lebesgue integrable if and only if $|f|$ is Lebesgue integrable. Riemann integral does not have this property. To get advantage of this property, Lebesgue integral of an unbounded function is defined as the sum of the integrals of its positive and negative parts. The improper Riemann integral and Lebesgue integral are not comparable in the sense that there are functions which are improperly Riemann integrable but not Lebesgue integrable.

As we have seen in Theorem 5.15 that a bounded function f is integrable if and only if f is measurable, we shall assume throughout that f is measurable.

Let E be a measurable set of finite measure and let f be a non-negative measurable function defined on E. For each n we define the function f_n on E such that $f_n(x) = f(x)$ if $f(x) \leq n$ and $f_n(x) = n$ if $f(x) > n$. The functions f_n are called truncated functions of f. Clearly $f_n \leq f_{n+1}$ for all n and $\lim_{n \to \infty} f_n = f$. Each f_n being bounded and measurable, it is integrable and $\int_E f_n dx \leq \int_E f_{n+1} dx$ and so $\lim_{n \to \infty} \int_E f_n dx$ exists which may be infinite. The truncated functions of f are also denoted by $[f]_n$ i.e $[f]_n(x) = f(x)$ if $f(x) \leq n$ and $[f]_n(x) = n$ if $f(x) > n$.

Definition 5.3 If $\{f_n\}$ is the sequence of truncated functions of a non-negative measurable function f on E then the integral of f on E is defined by $\int_E f dx = \lim_{n \to \infty} \int_E f_n dx$. If $\int_E f dx$ is finite then f is said to be integrable on E.

Definition 5.4 If f is a non-negative measurable function on E and if $S = \left\{ \int_E \phi dx : \phi \text{ is a measurable simple function such that } \phi \leq f \text{ on } E \right\}$ then the integral of f on E is defined by $\int_E f dx = \sup S$. If $\int_E f dx$ is finite then f is said to be integrable on E.

We show that the two definitions are equivalent. Before showing this note that Lebesgue integral of a bounded measurable function f is

$$\sup_{\tau} \{s_\tau(f) : s_\tau(f) \text{ is the lower sum corresponding to the partition } \tau \text{ of } E\}.$$

Also by the property of 'sup', there is a sequence $\{(s_\tau(f))_n\}$ from the set $\{(s_\tau(f))\}$ such that $\sup_{\tau}\{s_\tau(f)\} = \lim_{n}(s_\tau(f))_n$. So Definitions 5.3 and 5.4 are extension of Definitions 5.1 and 5.2 respectively. We have shown after proving Theorem 5.6 that Definitions 5.1 and 5.2 are equivalent. Now we show that Definitions 5.3 and 5.4 are equivalent. We are to show that $\sup S = \lim_{n \to \infty} \int_E f_n dx = T$ (say), where f_n are the truncated functions of f. Suppose $\sup S < T$. Choose λ_1, λ_2 such that $\sup S < \lambda_1 < \lambda_2 < T$. Then there is n such that $\lambda_2 < \int_E f_n dx \leq T$. Since f_n is bounded and Definitions 5.1 and 5.2 are equivalent, $\int_E f_n dx = \sup \left\{ \int_E \phi dx : \phi \leq f_n; \phi \text{ is a measurable simple function} \right\}$ and so there exists a measurable simple function ϕ_n such that $\phi_n \leq f_n$ and $\lambda_2 < \int_E \phi_n dx \leq \int_E f_n dx$. Since $\phi_n \leq f$, $\int_E \phi_n dx \in S$. Hence $\int_E \phi_n dx \leq \sup S < \lambda_1$, which is a contradiction.

Now suppose $\sup S > T$. Then there exists $\int_E \phi dx \in S$ such that $\int_E \phi dx > T$ and so $\int_E \phi dx > \int_E f_n dx$ for all n. But since f is unbounded there exist n_1, such that $\phi \leq f_{n_1} \leq f$ and so $\int_E \phi dx \leq \int_E f_{n_1} dx$ which is a contradiction.

Definitions 5.3 and 5.4 being equivalent we shall use either of them whenever convenient. But it may be noted that all the results which are proved here can also be proved by using only one of the Definitions 5.3 and 5.4.

To define Lebesgue integral of measurable functions of arbitrary sign we consider $f_+ = \max[f, 0]$ and $f_- = \max[-f, 0]$. Then f_+ and f_- are non-negative measurable functions and $f = f_+ - f_-$ and $|f| = f_+ + f_-$.

Definition 5.5 Let E be a measurable set and let f be a measurable function on E. The integral of f on E is defined by

$$\int_E f dx = \int_E f_+ dx - \int_E f_- dx,$$

provided at least one of f_+ and f_- is integrable on E (so that the right-hand side is not of the form $\infty - \infty$). If both f_+ and f_- are integrable then f is said to be integrable on E.

5.6 Properties of Lebesgue Integral

Theorem 5.25 Let E be a measurable set of finite measure.

(i) If E is of measure zero than $\int_E f dx = 0$ for every measurable function f on E.

(ii) If E_1 and E_2 are disjoint measurable subsets of E then f is integrable on $E_1 \cup E_2$ if and only if f is integrable on each of E_1 and E_2. In either case.

$$\int_{E_1} f dx + \int_{E_2} f dx = \int_{E_1 \cup E_2} f dx.$$

(iii) If f and g are integrable on E and $f \geq g$ a.e on E then

$$\int_E f dx \geq \int_E g dx.$$

(iv) If f and g are integrable on E then $f + g$ is also integrable on E and

$$\int_E (f + g) dx = \int_E f dx + \int_E g dx.$$

(v) If f is integrable on E and c is a constant then cf is also integrable on E and

$$\int_E (cf) dx = c \int_E f dx.$$

(vi) If f is measurable on E then f is integrable on E if and only if $|f|$ is integrable on E. In either case
$$\left| \int_E f\,dx \right| \leq \int_E |f|\,dx.$$

(vii) If f is integrable on E then for every $\epsilon > 0$ there is $\delta > 0$ such that $\left| \int_{E_1} f\,dx \right| < \epsilon$ for every measurable set $E_1 \subset E$ such that $\mu(E_1) < \delta$.

(viii) If f and g are integrable on E and $f = g$ a-e on E then
$$\int_E f\,dx = \int_E g\,dx.$$

(ix) If f is integrable on E then the set $\{x \in E : f(x) = \pm\infty\}$ is of measure zero.

(x) If f is integrable on E and $\int_{(-\infty,x]\cap E} f\,dx = 0$ for all $x \in E$ then $f = 0$ a.e. on E

(xi) If f is integrable on E and $f \geq 0$ on E and if $\int_E f\,dx = 0$ then $f = 0$ a.e. on E.

(xii) If $\{E_n\}$ in any countable collection of disjoint measurable subsets of E such that $E = \bigcup_{n=1}^{\infty} E_n$ and if f is integrable on E then f is integrable on each E_n and $\int_E f\,dx = \sum_{n=1}^{\infty} \int_{E_n} f\,dx$. The converse is not true.

Proof Definitions 5.3 and 5.4 being equivalent we shall use any of them which is suitable for simple proof. Each f_n denotes the truncated function of the non-negative measurable function f and each ϕ is a non-negative measurable simple function.

(i) If f is bounded then by Corollary 5.11 the proof follows. So if f is non-negative then $\int_E f_n\,dx = 0$ for each n and so the result is proved for f_+ and similarly for f_-. So the proof is complete by Definition 5.5.

(ii) Suppose that f is non-negative. Since by Theorem 5.5(i) the result is true for bounded function, for any n
$$\int_{E_1} f_n\,dx + \int_{E_2} f_n\,dx = \int_{E_1 \cup E_2} f_n\,dx$$

and so letting $n \to \infty$ the result is true for non-negative functions. For the general case, since f is integrable if and only if both f_+ and f_- are integrable and since f_+ and f_- are nonnegative, by the above
$$\int_{E_1} f_+\,dx + \int_{E_2} f_+\,dx = \int_{E_1 \cup E_2} f_+\,dx \quad \text{and}$$

5.6 Properties of Lebesgue Integral

$$\int_{E_1} f_- dx + \int_{E_2} f_- dx = \int_{E_1 \cup E_2} f_- dx$$

and hence

$$\int_{E_1} f dx + \int_{E_2} f dx = \int_{E_1} f_+ dx - \int_{E_1} f_- dx + \int_{E_2} f_+ dx - \int_{E_2} f_- dx$$

$$= \int_{E_1 \cup E_2} f_+ dx - \int_{E_1 \cup E_2} f_- dx$$

$$= \int_{E_1 \cup E_2} f dx$$

completing the proof.

(iii) First suppose that f and g are non-negative. Since $f \geq g$ a.e., $f_n \geq g_n$ a.e for each n.. So by Theorem 5.14(iii)

$$\int_E f_n dx \geq \int_E g_n dx \quad \text{for each } n.$$

Letting $n \to \infty$, the result is proved in this case.

For the general case, note that since $f \geq g$, $f_+ \geq g_+$ and $f_- \leq g_-$ a.e. on E and so by the above special case

$$\int_E f_+ dx \geq \int_E g_+ dx \quad \text{and} \quad \int_E f_- dx \leq \int_E g_- dx$$

and hence

$$\int_E f dx = \int_E f_+ dx - \int_E f_- dx \geq \int_E g_+ dx - \int_E g_- dx = \int_E g dx$$

completing the proof.

(iv) Suppose f and g are nonnegative. Let f and g be integrable and let $h = f + g$. Let $x \in E$ be arbitrary. Take any n. If $f(x) \leq n$ and $g(x) \leq n$ then $h_n(x) \leq h(x) = f(x) + g(x) = f_n(x) + g_n(x)$ and if one of $f(x)$ and $g(x)$ is greater than n then $h(x)$ is also greater than n and hence $h_n(x) = n \leq f_n(x) + g_n(n)$. So, in any case $h_n(x) \leq f_n(x) + g_n(x)$. Since x is any point of E, $h_n \leq f_n + g_n$. So by (iii) above and by Theorem 5.14(ii)

$$\int_E h_n dx \leq \int_E f_n dx + \int_E g_n dx$$

. Letting $n \to \infty$

$$\int_E h dx \leq \int_E f dx + \int_E g dx.$$

Since f and g are integrable, h is integrable. For the reverse inequality, since $f_n \le f$ and $g_n \le g$, $f_n + g_n \le f + g = h$ and so

$$\int_E f_n dx + \int_E g_n dx \le \int_E h dx$$

Letting $n \to \infty$ the proof is complete in this case.
For the general case, since $h_+ + h_- = |h| = |f + g| \le |f| + |g| = f_+ + f_- + g_+ + g_-$, we have

$$\int_E h_+ dx + \int_E h_- dx \le \int_E f_+ dx + \int_E f_- dx + \int_E g_+ dx + \int_E g_- dx.$$

So, if f and g are integrable, the right-hand side is finite and so the left-hand side is finite and hence h_+ and h_- are integrable and so h is integrable. Also $h_+ - h_- = h = f + g = f_+ - f_- + g_+ - g_-$ and hence $h_+ + f_- + g_- = f_+ + g_+ + h_-$ which gives by the above (since all are non-negative)

$$\int_E h_+ dx + \int_E f_- dx + \int_E g_- dx = \int_E f_+ dx + \int_E g_+ dx + \int_E h_- dx$$

and so

$$\int_E h dx = \int_E h_+ dx - \int_E h_- dx = \int_E f_+ dx - \int_E f_- dx + \int_E g_+ dx - \int_E g_- dx$$
$$= \int_E f dx + \int_E g dx$$

completing the proof.

(v) If $c = 0$ the proof is trivial. Let $c > 0$. We first suppose that f is non-negative. Noting that if ϕ is a measurable simple function then $\frac{\phi}{c}$ is also a measurable simple function. So putting $\frac{\phi}{c} = \psi$ we have using Theorem 5.14(i)

$$\int_E (cf) dx = \sup \left\{ \int_E \phi dx : \phi \le cf \right\}$$
$$= \sup \left\{ \int_E c\psi dx : c\psi \le cf \right\}$$
$$= \sup \left\{ c \int_E \psi dx : \psi \le f \right\}$$
$$= c \sup \left\{ \int_E \psi dx : \psi \le f \right\}$$
$$= c \int f dx.$$

5.6 Properties of Lebesgue Integral

So the result is true for non-negative functions. For the general case, since $c > 0$, $(cf)_+ = cf_+$ and $(cf_-) = cf_-$. Since f_+ and f_- are non-negative

$$\int_E (cf)_+ dx = \int_E (cf_+) dx = c \int_E f_+ dx$$

and

$$\int_E (cf)_- dx = \int_E (cf_-) dx = c \int_E f_- dx.$$

Since f is integrable, $\int_E f_+ dx$ and $\int_E f_- dx$ are finite and so $\int_E (cf)_+ dx$ and $\int_E (cf)_- dx$ are finite and hence cf is integrable and

$$\int_E (cf) dx = \int_E (cf)_+ dx - \int_E (cf)_- dx$$
$$= c \int_E f_+ dx - c \int_E f_- dx$$
$$= c \left[\int_E f_+ dx - \int_E f_- dx \right]$$
$$= c \int_E f dx.$$

Let $c < 0$ put $c = -k$. Then $k > 0$. So using (iv)

$$\int_E (cf) dx - c \int_E f dx = \int_E (cf) dx + k \int_E f dx$$
$$= \int_E (cf) dx + \int_E (kf) dx = \int_E (cf + kf) dx = 0$$

and so

$$\int_E (cf) dx = c \int_E f dx,$$

completing the proof.

(vi) Since $| f | = f_+ + f_-$ by (iv)

$$\int_E | f | dx = \int_E f_+ dx + \int_E f_- dx.$$

So, if f is integrable then by definition $\int_E f_+ dx$ and $\int_E f_- dx$ are finite and hence $| f |$ is integrable and if $| f |$ is integrable then by the above equality

$\int_E f_+ dx$ and $\int_E f_- dx$ are finite and hence f is intergable. Also

$$|\int_E f dx | = |\int_E f_+ dx - \int_E f_- dx | \leq \int_E f_+ dx + \int_E f_- dx = \int_E |f| dx.$$

(vii) Let f be integrable on E. Let $\epsilon > 0$ be arbitrary. By (vi) $|f|$ is integrable on E and so there is n such that

$$\int_E |f| dx - \int_E (|f|)_n dx < \frac{\epsilon}{2} \qquad (5.27)$$

where $(|f|)_n$ is a truncated function of $|f|$ defined by $(|f|)_n(x) = |f|(x)$ if $|f|(x) \leq n$ and otherwise its value is n. Let $\delta = \frac{\epsilon}{2n}$. Let $E_1 \subset E$ be measurable and $\mu(E_1) < \delta$. Then since $|f| - (|f|)_n \geq 0$ and $E_1 \subset E$ we have by (5.27)

$$\int_{E_1} [|f| - (|f|)_n] dx \leq \int_E [|f| - (|f|)_n] dx < \frac{\epsilon}{2} \qquad (5.28)$$

Since $(|f|)_n \leq n$ we have from (5.28)

$$\int_{E_1} |f| dx < \frac{\epsilon}{2} + \int_{E_1} (|f|)_n dx \leq \frac{\epsilon}{2} + n\mu(E_1) < \frac{\epsilon}{2} + n\delta < \frac{\epsilon}{2} + \frac{\epsilon}{2} = \epsilon$$

Hence by (vi)

$$|\int_{E_1} f dx | \leq \int_{E_1} |f| dx < \epsilon.$$

(viii) Let $E_1 = \{x \in E : f(x) \neq g(x)\}$. Then $\mu(E_1) = 0$. So by (i)

$$\int_E f dx = \int_{E_1} f dx + \int_{E \sim E_1} f dx = \int_{E \sim E_1} g dx = \int_{E_1} g dx + \int_{E \sim E_1} g dx = \int_E g dx.$$

(ix) Let $E_1 = \{x \in E : |f| = \infty\}$. Since f is integrable on E and E_1 is a measurable subset of E, the function f is integrable on E_1 by (ii) and so $|f|$ is integrable on E_1 by (vi) and hence $\int_{E_1} |f| dx$ is finite. Since for any positive integer n, $|f(x)| \geq n$ for $x \in E_1$, $\int_{E_1} |f| dx \geq n\mu(E_1)$ and so $\mu(E_1) = 0$, for otherwise, letting $n \to \infty$, $\int_{E_1} |f| dx = \infty$ which is a contradiction.

(x) We first suppose that E is bounded. Let $E = [a, b]$. Then by the given condition $\int_a^x f dx = 0$ for all x. Let $P = \{x \in E : f(x) > 0\}$. If possible, suppose $\mu(P) > 0$. Then for some positive integer k the set $E_k = \{x \in E : f(x) \geq \frac{1}{k}\}$ is of positive measure. So, by Theorem 2.14(ii) there is a closed set $F \subset E_k$ such that $\mu(F) > 0$. Let $G = [a, b] \sim F$, Then since $\int_F f dx \geq \frac{1}{k}\mu(F) > 0$ and

5.6 Properties of Lebesgue Integral

$\int_F f dx + \int_G f dx = \int_a^b f dx = 0$, the integral $\int_G f dx \neq 0$. Since F is closed. G is the union of a countable collection of disjoint sub-intervals $\{(a_n, b_n)\}$ of $[a, b]$. So $\int_{a_n}^{b_n} f dx \neq 0$ for at least one (a_n, b_n). But

$$\int_{a_n}^{b_n} f dx = \int_a^{b_n} f dx - \int_a^{a_n} f dx = 0.$$

which is a contradiction. Hence $\mu(P) = 0$. Similarly $\mu(Q) = 0$ where $Q = \{x \in E : f(x) < 0\}$. So, $f = 0$ a.e on $E = [a, b]$.

Let E be a bounded measurable set. Let $a = \inf E$ and $b = \sup E$ and $G = [a, b] \sim E$. Let $\phi(x) = f(x)$ if $x \in E$ and $\phi(x) = 0$ if $x \in G$. Then ϕ is integrable on $[a, b]$ and

$$\int_a^x \phi dx = \int_{[a,x] \cap E} \phi dx + \int_{[a,x] \cap G} \phi dx = \int_{[a,x] \cap E} f dx$$

for all $x \in E$. So by the above $\phi = 0$ a.e. on $[a, b]$ and so $f = 0$ a.e on E, completing the proof when E is bounded.

For the general case let n be any positive integer and let $E_n = [-n, n] \cap E$. Then E_n is bounded measurable set and f is integrable on E_n and $\int_{(-\infty,x] \cap E_n} f dx = 0$ for all $x \in E_n$. So by the above $f = 0$ a,e on E_n. Since n is arbitrary $f = 0$ a.e. on E_n for all n. Since $E = \bigcup_{n=1}^{\infty} E_n$, $f = 0$ a.e. on E.

(xi) Let $P = \{x \in E : f(x) > 0\}$ and let $E_k = \{x \in E : f(x) \geq \frac{1}{k}\}$ where k is a positive integer. Then $P = \bigcup_{k=1}^{\infty} E_k$. We have since $E_K \subset E$

$$0 = \int_E f dx \geq \int_{E_k} f dx \geq \frac{1}{k} \mu(E_k) \text{ and so } \mu(E_k) = 0$$

Since k is arbitrary $\mu(E_k) = 0$ for all k and so $\mu(P) = 0$ completing the proof.

(xii) Let f be integrable on E. Take n arbitrary and let $F_n = E \sim E_n$. Then $E = E_n \cup F_n$ and $E_n \cap F_n = \phi$. So by (ii) f is integrable on E_n. Since n is arbitrary, f is integrable on each E_n

Let $\epsilon > 0$ be arbitrary. Since f is integrable on E, by (vii) there is $\delta > 0$ such that

$$\left| \int_A f dx \right| < \epsilon \text{ for every measurable set } A \subset E \text{ with } \mu(A) < \delta \quad (5.29)$$

Since $E = \bigcup_{k=1}^{\infty} E_k$ and the sets E_k one measurable and disjoint, $\mu(E) = \sum_{k=1}^{\infty} \mu(E_k)$ and so the series $\sum_{k=1}^{\infty} \mu(E_k)$ is convergent. Hence there is N such that $\sum_{k=n+1}^{\infty} E_k < \delta$ whenever $n \geq N$. Let $S_n = \bigcup_{k=1}^{n} E_k$ and $R_n = \bigcup_{k=n+1}^{\infty} E_k$. Then $\mu(R_n) = \sum_{k=n+1}^{\infty} \mu(E_k) < \delta$ whenever $n \geq N$. So by, (5.29)

$$\left| \int_{R_n} f\,dx \right| < \epsilon \text{ whenever } n \geq N. \tag{5.30}$$

By (ii)

$$\int_E f\,dx = \int_{S_n} f\,dx + \int_{R_n} f\,dx = \sum_{K=1}^{n} \int_{E_k} f\,dx + \int_{R_n} f\,dx$$

and so by (5.30)

$$\left| \int_E f\,dx - \sum_{k=1}^{n} \int_{E_k} f\,dx \right| = \left| \int_{R_n} f\,dx \right| < \epsilon \text{ whenever } n \geq N$$

Hence

$$\int_E f\,dx = \lim_{n \to \infty} \sum_{k=1}^{n} \int_{E_k} f\,dx = \sum_{k=1}^{\infty} \int_{E_k} f\,dx$$

For the converse part ler $E = (0, 1]$ and $E_n = \left(\frac{1}{n+1}, \frac{1}{n}\right]$ for each n and let $f(x) = n$ for $x \in E_n$. Clearly f is integrable on each E_n and $\int_{E_n} f\,dx = n\left(\frac{1}{n} - \frac{1}{n+1}\right)$. So, $\sum_{k=1}^{n} \int_{E_k} f\,dx = \sum_{n=1}^{\infty} \frac{1}{n+1} = \infty$. Since $E = \bigcup_{n=1}^{\infty} E_n$, the function f is not integrable on E.

Note The converse is true if f is bounded and integrable (See Theorem 5.5(ii)).

□

Example 5.1 Counter examples of :
(iv) If f and g are not integrable on a measurable set E then $f + g$ may be integrable on E.

Since f and g are not integrable on E, the set E is of positive measure. Then E contains a nonmeasurable set $E_1 \subset E$. Define $f(x) = 1$ for $x \in E_1$ and $= 0$ for $x \in E \sim E_1$ and $g(x) = 0$ for $x \in E_1$ and $= 1$ for $x \in E \sim E_1$. Then $f(x) + g(x) = 1$ for all $x \in E$. So $f + g$ is integrable on E but f and g are not integrable on E.

5.6 Properties of Lebesgue Integral

(vi) Measurability of f is necessary. For if E is any measurable set of positive measure then E contains a nonmeasurable set $E_1 \subset E$. Define $f(x) = 1$ for $x \in E_1$ and $= -1$ for $x \in E \sim E_1$. Then $|f(x)| = 1$ for all $x \in E$ and so $|f|$ is integrable but f is not measurable.

(viii) If $\int_E f = \int_E g$ then $f = g$ a.e. may not hold.
  For consider $E = [0, 2]$. Let $f(x) = 1$ on $[0, 1]$ and $= 0$ on $(1, 2]$ and $g(x) = 0$ on $[0, 1]$ and $= 1$ on $(1, 2]$. Then $\int_E f = 1$ and $\int_E g = 1$ but $f = g$ a.e. does not hold.

(xi) The condition $f \geq 0$ is necessary. For consider the function f on $[0, 1]$ such that $f(x) = 1$ on $[0, \frac{1}{2}]$ and $f(x) = -1$ on $(\frac{1}{2}, 1]$. Then $\int_{[0,1]} f = 0$ but $f = 0$ a.e. does not hold.

Theorem 5.26 *If f is integrable on $[a, b]$ then for every $\epsilon > 0$ there is a continuous function g_ϵ on $[a, b]$ such that $\int_a^b | f - g_\epsilon | < \epsilon$.*

Proof We first suppose that f is bounded and let $|f| < M$ on $[a, b]$. Let $\epsilon > 0$ be arbitrary. Then by Lusin's Theorem (Theorem 3.24) there is a continues function g_ϵ on $[a, b]$ and a measurable set $E \subset [a, b]$ such that $g_\epsilon = f$ on $[a, b] \sim E$, $|g_\epsilon| < M$ and $\mu(E) < \frac{\epsilon}{2M}$. So,

$$\int_a^b |f - g_\epsilon| = \int_E |f - g_\epsilon| \leq 2M\mu(E) < \epsilon. \tag{5.31}$$

So, the theorem is proved when f is bounded. For the general case let $\epsilon > 0$ be arbitrary. Then by Theorem 5.25(vi) there is $\delta > 0$ such that

$$\int_E |f| < \frac{\epsilon}{2} \text{ for every measurable set } E \subset [a, b] \text{ such that } \mu(E) < \delta \tag{5.32}$$

Let $E_n = \{x \in [a, b] : |f(x)| > n\}$. Since $E_{n+1} \subset E_n$ for all n and $\bigcap_{r=1}^{\infty} E_n$ is of measure zero by Theorem 5.25(ix) we have by Theorem 2.9, $\lim_{n \to \infty} \mu(E_n) = 0$ and so there is a positive integer N such that $\mu(E_n) < \delta$ for $n \geq N$. So, by (5.32)

$$\int_{E_n} |f| < \frac{\epsilon}{2} \tag{5.33}$$

Let $A = [a, b] \sim E_N$. Then $|f| \leq N$ on A. So, the function $f\chi_A$ is integrable on $[a, b]$ and $|f\chi_A| \leq N$ on $[a, b]$, where χ_A is the characteristic function of A. Since $f\chi_A$ is bounded, by the above special case there is a continuous function g_ϵ on $[a, b]$ such that

$$\int_a^b |f\chi_A - g_\epsilon| < \frac{\epsilon}{2}. \tag{5.34}$$

So by (5.33) and (5.34)

$$\int_a^b |f - g_\epsilon| \le \int_a^b |f - f\chi_A| + \int_a^b |f\chi_A - g_\epsilon| < \int_{E_N} |f| + \frac{\epsilon}{2} < \epsilon$$

completing the proof. □

5.7 Convergence Theorems for Lebesgue Integral

Theorem 5.27 (Fatou's Lemma) *If $\{f_n\}$ is a sequence of non-negative measurable functions defined on a measurable set E of finite measure. Then*

$$\int_E (\liminf_{n \to \infty} f_n) \le \liminf_{n \to \infty} \int_E f_n$$

Proof Let $f = \liminf\limits_{n \to \infty} f_n$. Then f is non-negative and measurable by Theorem 3.9. To prove the theorem we first show that for every bounded measurable function g such that $0 \le g \le f$ on E,

$$\int_E g \le \liminf_{n \to \infty} \int_E f_n. \tag{5.35}$$

Let g be fixed. Let $g_n = min[g, f_n]$. Then by Theorem 3.9 $\{g_n\}$ is a sequence of measurable functions. Let $x \in E$ and let $\epsilon > 0$. Since $g(x) \le f(x) = \liminf\limits_{n \to \infty} f_n(x)$ there is N such that $g(x) - \epsilon < f_n(x)$ for all $n \ge N$. Hence $g(x) - \epsilon = min[g(x), g(x) - \epsilon] \le min[g(x), f_n(x)] = g_n(x) \le g(x)$ for $n \ge N$ and so $\lim\limits_{n \to \infty} g_n(x) = g(x)$ for all $x \in E$. Since g is bounded, by Theorem 5.17

$$\lim_{n \to \infty} \int_E g_n = \int_E g \tag{5.36}$$

Since $g_n \le f_n$ for all n, $\int_E g_n \le \int_E f_n$ for all n and hence

$$\lim_{n \to \infty} \int_E g_n \le \liminf_{n \to \infty} \int_E f_n \tag{5.37}$$

So, (5.35) is proved by (5.36) and (5.37).

Now we come to the proof of the theorem. Consider the truncated function $[f]_k$ defined by $[f]_k(x) = f(x)$ if $f(x) \le k$ and $[f]_k(x) = k$ if $f(x) > k$. Then $[f]_k$ is a bounded measurable function such that $0 \le [f]_k \le f$. So by (5.35)

$$\int_E [f]_k \le \liminf_{n \to \infty} \int_E f_n.$$

5.7 Convergence Theorems for Lebesgue Integral

Letting $k \to \infty$, we have by definition

$$\int_E f \leq \liminf_{n \to \infty} \int_E f_n.$$

completing the proof. □

Remark In the above theorem strict inequality may occur. For, let

$$f_n(x) = n \quad \text{for } 0 < x < \frac{1}{n}$$
$$= 0 \quad \text{otherwise.}$$

Then $\lim_{n \to \infty} f_n(x) = 0$ for all x and

$$\lim_{n \to \infty} \int_0^1 f_n = \lim_{n \to \infty} \int_0^{\frac{1}{n}} n = 1. \text{ So } \int_0^1 \lim_{n \to \infty} f_n(x) = 0 < 1 = \lim_{n \to \infty} \int_0^1 f_n.$$

Also both sides may be infinite. For, let

$$f_n(x) = \frac{1}{x} \quad \text{for } \frac{1}{n} \leq x \leq 1$$
$$= 0 \quad \text{otherwise}$$

Then

$$\lim_{n \to \infty} f_n(x) = \frac{1}{x} \quad \text{for } 0 < x \leq 1$$
$$= 0 \quad at\ x = 0$$

So if $f = \lim_{n \to \infty} f_n$ then $f(x) = \frac{1}{x}$ for $0 < x \leq 1$ and $f(0) = 0$. Hence the truncated function $[f]_n$ is given by

$$[f]_n(x) = \frac{1}{x} \quad \text{for } \frac{1}{n} \leq x \leq 1$$
$$= n \quad \text{for } 0 \leq x < \frac{1}{n}.$$

Hence

$$\int_0^1 [f]_n = \int_{\frac{1}{n}}^1 \frac{1}{x} + \int_0^{\frac{1}{n}} n = \log x \Big|_{1/n}^1 + 1 = \log n + 1.$$

So, $\lim_{n\to\infty} \int_0^1 [f]_n = \infty$ and hence $\int_0^1 f = \infty$. (Note that f is not integrable), i.e. $\int_0^1 \lim_{n\to\infty} f_n = \infty$. Also

$$\lim_{n\to\infty} \int_0^1 f_n = \lim_{n\to\infty} \int_{\frac{1}{n}}^1 \frac{1}{x} = \lim_{n\to\infty} \log n = \infty.$$

Theorem 5.28 (Monotone Convergence Theorem) *Let $\{f_n\}$ be a non-decreasing sequence of non-negative measurable functions defined on a measurable set E of finite measure. Then*

$$\int_E \lim_{n\to\infty} f_n = \lim_{n\to\infty} \int_E f_n$$

Proof Note that since $f_n(x) \leq f_{n+1}(x)$ for all n and $x \in E$ the left-hand limit exists (may be infinite) and since $\int_E f_n \leq \int_E f_{n+1}$ for all n the right-hand limit exists. Since $f_n(x) \leq \lim_{n\to\infty} f_n(x)$ for all n and $x \in E$, $\int_E f_n \leq \int_E \lim_{n\to\infty} f_n$ and so applying Fatou's lemma,

$$\lim_{n\to\infty} \int_E f_n \leq \int_E \lim_{n\to\infty} f_n \leq \lim_{n\to\infty} \int_E f_n$$

and so

$$\int_E \lim_{n\to\infty} f_n = \lim_{n\to\infty} \int_E f_n.$$

□

Corollary 5.29 (Monotone Convergence Theorem) *Let $\{f_n\}$ be a non-increasing sequence of non-positive measurable functions defined on a measurable set E of finite measure. Then*

$$\int_E \lim_{n\to\infty} f_n = \lim_{n\to\infty} \int_E f_n$$

The proof follows by putting $g_n = -f_n$ and applying the above theorem.

Corollary 5.30 *Let $\{f_n\}$ be a monotone sequence of integrable functions defined on a measurable set E of finite measure and let $\{f_n(x)\}$ converge to $f(x)$ for $x \in E$. If $\lim_{n\to\infty} \int_E f_n$ is finite then f is integrable on E and*

$$\int_E f = \lim_{n\to\infty} \int_E f_n.$$

5.7 Convergence Theorems for Lebesgue Integral

Proof We first prove the result when $\{f_n\}$ is non-increasing. In this case the sequence $\{f_1 - f_n\}$ is non-decreasing and non-negative and $\{f_1(x) - f_n(x)\}$ converges to $f_1(x) - f(x)$ for $x \in E$.
So, by Theorem 5.28

$$\int_E (f_1 - f) = \lim_{n \to \infty} \int_E (f_1 - f_n) = \int_E f_1 - \lim_{n \to \infty} \int_E f_n.$$

Since the right-hand side is finite, $f_1 - f$ is integrable on E and since f_1 is integrable, f is integrable and

$$\int_E f = \lim_{n \to \infty} \int_E f_n.$$

Suppose that $\{f_n\}$ is non-decreasing. Put $g_n = -f_n$. Then $\{g_n\}$ is non-increasing and $\{g_n(x)\}$ converges to $g(x)$ for $x \in E$ where $g = -\lim_{n \to \infty} f_n = -f$. So by the first part $\int_E g = \lim_{n \to \infty} \int_E g_n$. So $\int_E f = \lim_{n \to \infty} \int_E f_n$. □

Corollary 5.31 *If $\{f_n\}$ is a sequence of non-negative measurable functions defined on a measurable set E of finite measure. Then*

$$\int_E (\sum f_n) = \sum \int_E f_n$$

Proof Putting $S_n = \sum_{i=1}^{n} f_i$ and applying Theorem 5.28 and then Theorem 5.24 we have

$$\int_E (\sum_{n=1}^{\infty} f_n) = \int_E (\lim_{n \to \infty} S_n) = \lim_{n \to \infty} \int_E S_n = \lim_{n \to \infty} \sum_{i=1}^{n} \int_E f_i = \sum_{n=1}^{\infty} \int_E f_n$$

□

Corollary 5.32 *Let E be a measurable set of finite measure and $\{E_n\}$ be a sequence of disjoint measurable sets such that $\bigcup_{n=1}^{\infty} E_n = E$. If f is integrable on E then f is integrable on each E_n and*

$$\int_E f = \sum_{n=1}^{\infty} \int_{E_n} f$$

Proof We first suppose that $f \geq 0$ on E. Let $f_n = f \chi_{E_n}$ where χ_{E_n} is the characteristic function of E_n. Then $\{f_n\}$ is a sequence of non-negative integrable function on E such that $f = \sum_{n=1}^{\infty} f \chi_{E_n}$. So by Corollary 5.31

$$\int_E f = \sum_{n=1}^{\infty} \int_E f \chi_{E_n} = \sum_{n=1}^{\infty} \int_{E_n} f$$

So the result is proved for non-negative function f. For the general case, since f is integrable, its positive part f_+ and the negative part f_- are integrable on E and since f_+ and f_- one nonnegative, by the above special case

$$\int_E f_+ = \sum_{n=1}^{\infty} \int_{E_n} f_+ \quad \text{and} \quad \int_E f_- = \sum_{n=1}^{\infty} \int_{E_n} f_-$$

and so

$$\int_E f = \int_E f_+ - \int_E f_- = \sum_{n=1}^{\infty} \int_{E_n} f_+ - \sum_{n=1}^{\infty} \int_{E_n} f_-$$
$$= \sum_{n=1}^{\infty} \left(\int_{E_n} f_+ - \int_{E_n} f_- \right)$$
$$= \sum_{n=1}^{\infty} \int_{E_n} f.$$

Remark The monotone convergence theorem (Theorem 5.28) is valid if the condition that each f_n is non-negative is replaced by $f_n \geq g$ for all n where g is a Lebesgue integrable function. In this consider the sequence $\{f_n - g\}$. A similar remark holds for Corollary 5.29.

We now come to prove another important theorem.

Theorem 5.33 (Lebesgue Dominated Convergence Theorem) *Let $\{f_n\}$ be a sequence of measurable function defined on a measurable set E of finite measure and let $\lim_{n \to \infty} f_n(x) = f(x)$ for $x \in E$. If there exists an integrable function g on E such that $\mid f_n(x) \mid \leq g(x)$ for all n and all $x \in E$ then f is integrable on E and*

$$\int_E f = \lim_{n \to \infty} \int_E f_n$$

Proof The sequences $\{g + f_n\}$ and $\{g - f_n\}$ are non-negative which converge on E to $g + f$ and $g - f$ respectively. So by Fatou's lemma, i.e. by Theorem 5.27

$$\int_E (g+f) \leq \liminf_{n \to \infty} \int_E (g+f_n), \quad \text{and} \quad \int_E (g-f) \leq \liminf_{n \to \infty} \int_E (g-f_n)$$

which give

$$\int_E f \leq \liminf_{n \to \infty} \int_E f_n \quad \text{and} \quad -\int_E f \leq \liminf_{n \to \infty} \left(-\int_E f_n \right) = -\limsup_{n \to \infty} \int_E f_n$$

5.7 Convergence Theorems for Lebesgue Integral

and so

$$\liminf_{n\to\infty} \int_E f_n \geq \int_E f \geq \limsup_{n\to\infty} \int_E f_n$$

and hence $\int_E f = \lim_{n\to\infty} \int_E f_n$. □

Remark The above Lebesgue Dominated Convergence Theorem is a generalization of Theorem 5.22 (Bounded Convergence Theorem). The last three theorems, Fatou's Lemma, Monotone Convergence Theorem, and Dominated convergence Theorem are very useful in application. In the following, we generalize Theorem 5.33.

Theorem 5.34 (Dominated Convergence Theorem) *Let $\{f_n\}$ be a sequence of measurable functions defined on a measurable set E of finite measure which converges in measure to a measurable function f on E. If there exists an integrable function g on E such that $|f_n(x)| \leq g(x)$ for all n and all $x \in E$ then f is integrable on E and*

$$\int_E f = \lim_{n\to\infty} \int_E f_n$$

Proof We may suppose that $\mu(E) > 0$. Since g is integrable, the function f_n is also integrable on E for each n. Also since $\{f_n\}$ converges in measure to f, there is, by Theorem 3.31, a subsequence $\{f_{n_k}\}$ of $\{f_n\}$ which converges to f a.e. on E. Hence $|f| \leq g$ a.e on E, and so f is integrable on E. Let $\epsilon > 0$ be arbitrary. Then by Theorem 5.25(vi) there is $\delta > 0$ such that

$$\left|\int_{E_1} g\right| < \frac{\epsilon}{4} \text{ for every measurable set } E_1 \subset E \text{ such that } \mu(E_1) < \delta \quad (5.38)$$

Choose $0 < \sigma < \frac{\epsilon}{2\mu(E)}$ and set

$$A_n(\sigma) = \{x \in E : |f_n(x) - f(x)| \geq \sigma\}, \quad B_n(\sigma) = E \sim A_n(\sigma).$$

Then $\lim_{n\to\infty} \mu(A_n(\sigma)) = 0$. Hence there is a positive integer N such that

$$\mu(A_n(\sigma)) < \delta \text{ for } n \geq N \quad (5.39)$$

Since $|f_n(x) - f(x)| \leq 2g$ for almost all $x \in E$ and for all n, from (5.38) and (5.39)

$$\int_{A_n(\sigma)} |f_n - f| \leq 2 \int_{A_n(\sigma)} g < \frac{\epsilon}{2} \text{ for } n \geq N \quad (5.40)$$

Also for $x \in B_n(\sigma)$, $|f_n(x) - f(x)| < \sigma$ and hence

$$\int_{B_n(\sigma)} |f_n - f| \leq \sigma\mu(B_n(\sigma)) \leq \sigma\mu(E) < \frac{\epsilon}{2} \text{ for all } n \quad (5.41)$$

Hence from (5.40) and (5.41)

$$\left|\int_E (f_n - f)\right| \leq \int_E |f_n - f| \leq \int_{A_n(\sigma)} |f_n - f| + \int_{B_n(\sigma)} |f_n - f| < \epsilon \text{ for } n \geq N.$$

Since ϵ is arbitrary,

$$\lim_{n\to\infty} \int_E (f_n - f) = 0 \text{ i.e. } \lim_{n\to\infty} \int_E f_n = \int_E f.$$

$\square$

Theorem 5.35 (Vitali Convergence Theorem) *Let $\{f_n\}$ be a sequence of functions which are integrable on a bounded measurable set E and let $\{f_n\}$ converges in measure to a measurable function f on E. If for every $\epsilon > 0$ there is a $\delta > 0$ such that $\left|\int_A f_n\right| < \epsilon$ for every set $A \subset E$ with $\mu(A) < \delta$ and for all n. Then f is integrable on E and*

$$\int_E f = \lim_n \int_E f_n.$$

Proof We first show that f is integrable on E. let $\epsilon > 0$ be arbitrary. By the given condition there is $\delta > 0$ such that
$\left|\int_A f_n\right| < \frac{\epsilon}{2}$ for every $A \subset E$ with $\mu(A) < \delta$ and for all n.

Let A be chosen and fixed. Let $A_+ = \{x \in A : f_n(x) \geq 0\}$, $A_- = A \sim A_+$. Since $A_+, A_- \subset E$ and $\mu(A_+) < \delta$, $\mu(A_-) < \delta$,

$$\int_{A_+} |f_n| = \left|\int_{A_+} f_n\right| < \frac{\epsilon}{2} \text{ and } \int_{A_-} |f_n| = \left|\int_{A_-} f_n\right| < \frac{\epsilon}{2} \text{ for all n}$$

and hence

$$\int_A |f_n| = \int_{A_+} |f_n| + \int_{A_-} |f_n| < \epsilon \text{ for all } n \tag{5.42}$$

Since $\{f_n\}$ converges to f in measure, by Theorem 3.32 there is a subsequence $\{f_{n_k}\}$ of $\{f_n\}$ which converges to f pointwise on a set $E_1 \subset E$ where $\mu(E_1) = \mu(E)$. Letting $A_1 = A \cap E_1$, the sequence $\{f_{n_k}\}$ converges pointwise to f on A_1. Hence the sequence $\{|f_{n_k}|\}$ converges pointwise to $|f|$ on A_1. So by Theorem 5.27

$$\int_{A_1} |f| \leq \liminf_{k\to\infty} \int_{A_1} |f_{n_k}| \text{ that is } \int_A |f| \leq \liminf_{k\to\infty} \int_A |f_{n_k}| \tag{5.43}$$

Since (5.42) is true for all n, it is also true for all k if n is replaced by n_k in (5.42) and hence from (5.42) and (5.43) $\int_A |f| \leq \epsilon$. So $|f|$ is integrable on A. Since A is arbitrary, $|f|$ is integrable on every measurable subset of E of measure $< \delta$. Since

5.7 Convergence Theorems for Lebesgue Integral

E is bounded we can find measurable sets $E_1, E_2, \ldots, E_r$ such that $\mu(E_i) < \delta$ for all i, $1 \leq i \leq r$ and $E = \bigcup_{i=1}^{r} E_i$. Therefore, since $|f|$ is integrable on each E_i, $|f|$ is integrable on E by Theorem 5.25(v). This proves the first part of the theorem.

For the second part, let $\epsilon > 0$ be arbitrary, Since $|f|$ is integrable on E by Theorem 5.25(vi) there is $\delta_1 > 0$ such that

$$\int_A |f| < \frac{\epsilon}{3} \text{ whenever } A \subset E \text{ with } \mu(A) < \delta_1. \tag{5.44}$$

Also applying arguments which are used to prove (5.42) it can be shown that there exist $\delta_2 > 0$ such that

$$\int_A |f_n| < \frac{\epsilon}{3} \text{ whenever } A \subset E \text{ with } \mu(A) < \delta_2 \text{ and for all } n. \tag{5.45}$$

Let $\delta = min[\delta_1, \delta_2]$. Choose σ such that $0 < \sigma < \frac{\epsilon}{3\mu(E)}$. Let $A_n(\sigma) = \{x \in E : |f_n(x) - f(x)| \geq \sigma\}$ and $B_n(\sigma) = E \sim A_n(\sigma)$. Since $\{f_n\}$ converges to f in measure on E,

$$\lim_{n \to \infty} \mu(A_n(\sigma)) = 0$$

and so there is N such that $\mu(A_n(\sigma)) < \delta$ for all $n \geq N$. Since $\delta \leq \delta_2$, $\mu(A_n(\sigma)) < \delta_2$ for all $n \geq N$ and hence by (5.45)

$$\int_{A_n(\sigma)} |f_n| < \frac{\epsilon}{3} \text{ for all } n \geq N. \tag{5.46}$$

Also since $\delta \leq \delta_1$ by (5.44)

$$\int_{A_n(\sigma)} |f| < \frac{\epsilon}{3} \text{ for all } n \geq N. \tag{5.47}$$

Finally by our choice of σ.

$$\int_{B_n(\sigma)} |f_n - f| \leq \sigma \mu(B_n(\sigma)) \leq \sigma \mu(E) < \frac{\epsilon}{3} \text{ for all } n. \tag{5.48}$$

So, by (5.46),(5.47) and (5.48).

$$\left| \int_E f_n - \int_E f \right| \leq \int_E |f_n - f|$$
$$= \int_{A_n(\sigma)} |f_n - f| + \int_{B_n(\sigma)} |f_n - f|$$
$$\leq \int_{A_n(\sigma)} |f_n| + \int_{A_n(\sigma)} |f| + \int_{B_n(\sigma)} |f_n - f| < \epsilon \text{ for } n \geq N.$$

Hence
$$\lim_{n\to\infty}\int_E f_n = \int_E f.$$

□

Corollary 5.36 (Bounded Convergence Theorem) *Let $\{f_n\}$ be a sequence of measurable functions defined on a bounded measurable set E and let $|f_n(x)| \leq M$ for all n and for all $x \in E$ where M is a constant. If $\{f_n\}$ converges in measure to a measurable function f on E then f is integrable on E and*

$$\int_E f = \lim_{n\to\infty}\int_E f_n.$$

Proof For $\epsilon > 0$, taking $\delta = \frac{\epsilon}{M}$ we have

$$\left|\int_A f_n\right| \leq \int_A |f_n| \leq \int_A M = M\mu(A) < \epsilon$$

for all n and for all $A \subset E$ with $\mu(A) < \delta$, and so the result follows by the above theorem. □

Corollary 5.37 (Dominated Convergence Theorem) *Let $\{f_n\}$ be a sequence of measurable functions defined on a bounded measurable set E and let $\{f_n\}$ converges in measure to a measurable function f on E. If there exist an integrable function g on E such that $|f_n(x)| \leq g(x)$ for all n and all $x \in E$ then f is integrable on E and*

$$\int_E f = \lim_{n\to\infty}\int_E f_n$$

Proof Let $\epsilon > 0$ be arbitrary. Since g is integrable and non-negative on E by Theorem 5.25 (vii) there is $\delta > 0$ such that

$$\int_A g < \epsilon \text{ for every } A \subset E \text{ with } \mu(A) < \delta.$$

Since $|f_n(x)| \leq g(x)$ for all n and all $x \in E$,

$$\left|\int_A f_n\right| \leq \int_A |f_n| \leq \int_A g < \epsilon \text{ for all } n \text{ and for every } A \subset E \text{ with } \mu(A) < \delta,$$

and so the result follows by the above theorem. □

The (ϵ, δ) condition imposed on the sequence $\{f_n\}$ in Vitali Convergence Theorem above is sufficient but not necessary. For, consider

$$f_n(x) = n \text{ if } \frac{1}{n} < x < \frac{2}{n}$$
$$= -n \text{ if } \frac{2}{n} < x < \frac{3}{n}$$
$$= 0 \text{ otherwise in } [0, 1]$$

Then $\lim_{n \to \infty} f_n(x) = 0$ for all $x \in [0, 1]$. Also $\lim_{n \to \infty} \int_0^1 f_n = 0$. So the result is true, but $\{f_n\}$ does not satisfy the (ϵ, δ) condition.

5.8 Lebesgue Integral of Functions on Sets of Infinite Measure

While a bounded measurable set must be of finite measure, the measure of an unbounded measurable set may be finite or infinite. For, consider $E = \bigcup_{n=1}^{\infty} E_n$ and $F = \bigcup_{n=1}^{\infty} F_n$ where $E_n = [n, n + \frac{1}{n^2})$ and $F_n = [n, n + \frac{1}{n})$. The Lebesgue integral of function on sets of finite measure is defined. We now define it on sets of infinite measure which are clearly unbounded.

Definition 5.6 Let E be a measurable unbounded set and let f be a measurable non-negative function on E. The integral of f on E is defined by

$$\int_E f dx = \lim_{n \to \infty} \int_{[-n,n] \cap E} f dx.$$

Note Since f is non-negative, the integral in the right exists, finitely or infinitely for each n and it increases with n and so the limit exists. If this limit is finite then f is said to be integrable on E.

Definition 5.7 Let E be a measurable unbounded set and let f be a measurable function on E. The integral of f on E is defined by

$$\int_E f dx = \int_E f_+ dx - \int_E f_- dx$$

provided at least one of f_+ and f_- is integrable on E. If both are integrable on E so that the integral in the left is finite then f is said to be integrable on E.

The integral on sets of infinite measure has unusual properties such as very good functions may not be integrable where as very bad functions may be integrable. For, consider (i) $f(x) = 1$ for all $x \in E = [1, \infty)$

(ii) $f(x) = \frac{\sin \frac{1}{x} + \cos 2x}{2x^2}$ for all $x \in E = [1, \infty)$.

Clearly the function in (i) is not integrable on E. For the function in (ii) consider $g(x) = \frac{1}{x^2}$ for all $x \in E$. Then g in Riemann integrable on $[1, n]$ for all n and $(R) \int_1^n g \, dx = 1 - \frac{1}{n}$ and so $(L) \int_{[0,n] \cap E} g \, dx = 1 - \frac{1}{n}$ and hence $\int_E g \, dx = 1$. Since $|f| \leq g$ on E, $\int_E |f| \, dx \leq \int_E g \, dx = 1$ and so $|f|$ is integrable on E and hence f_+ and f_- are both integrable on E and therefore f is integrable on E.

Since a set of finite measure may be unbounded, we must show that Definition 5.5 is consistent with Definition 5.7. Let E be any set of finite measure and suppose that f is integrable on E. Then f_+ and f_- are integrable on E. Let $E_1 = [-1, 1] \cap E$ and $E_r = ([-r, r] \cap E) \sim ([-r+1, r-1] \cap E)$ for $r \geq 2$. Then $E_r, r = 1, 2, \ldots$ are disjoint measurable subsets of E and $E = \bigcup_{r=1}^{\infty} E_r$. So, by Theorem 5.25(xii), f_+ and f_- are integrable on each E_r and

$$\int_E f_+ dx = \sum_{r=1}^{\infty} \int_{E_r} f_+ dx = \lim_{n \to \infty} \sum_{r=1}^{n} \int_{E_r} f_+ dx$$

$$= \lim_{n \to \infty} \int_{E_n} f_+ dx = \lim_{n \to \infty} \int_{[-n,n] \cap E} f_+ dx$$

with a similar relation for f_-. So using Definition 5.6 it now follows that f is integrable in the sense of Definition 5.7.

5.9 Improper Riemann Integral and Lebesgue Integral

Riemann integration requires that

(i) the interval in which the function f is defined should be finite, and
(ii) the function f should be bounded there.

If any one of these two conditions are not satisfied then it is not possible to apply the definition of Riemann integration. But even when these two conditions are not satisfied, if f satisfies some other conditions then f may be integrable in a sense which is called improper Riemann integral which is obtained as a limit of Riemann integral. Although every Riemann integrable function is Lebesgue integrable improper Riemann integral and Lebesgue integral are not comparable in the sense that there are functions which are improper Riemann integrable but not Lebesgue integrable and there are functions which are Lebesgue integrable but not improper Riemann integrable.

Regarding (i) above suppose that f is defined in $[a, \infty)$ and is Riemann integrable in $[a, X]$ for every $X, a < X < \infty$. If $(R) \int_a^X f$ tends to a finite limit l as $X \to \infty$ then f is improper Riemann integrable in $[a, \infty)$ and $(R_i) \int_a^{\infty} f = l$ where R_i denotes improper Riemann integral. Hence $(L) \int_{[a,x]} f \to l$ as $X \to \infty$ and so it is natural

5.9 Improper Riemann Integral and Lebesgue Integral

to assume that f should be Lebesgue integrable on $[\alpha, \infty)$ and the integral is l. But this is not true. Consider.

Example 5.2 Let $f(x) = \frac{\sin x}{x}$ for $x \in [1, \infty)$. Then f improper Riemann integrable but not Lebeague integrable in $[1, \infty)$.

By Dirichlet test for improper Riemann integral, the integral $\int_1^\infty \frac{\sin x}{x}$ is convergent and so by definition f is improper Riemann integrable in $[1, \infty)$. We show that $|f|$ is not Lebesgue integrable in $[1, \infty)$. This will prove that f is not Lebesgue integrable in $[1, \infty)$. We have

$$\int_0^{n\pi} \frac{|\sin x|}{x} dx = \sum_{r=1}^n \int_{(r-1)\pi}^{r\pi} \frac{|\sin x|}{x} dx = \sum_{r=1}^n \int_0^\pi \frac{\sin t}{(r-1)\pi + t} dt$$

$$\geq \sum_{r=1}^n \frac{1}{r\pi} \int_0^\pi \sin t \, dt = \frac{2}{\pi} \sum_{r=1}^n \frac{1}{r}$$

and therefore, since $\int_0^1 \frac{\sin x}{x} dx$ is finite

$$\int_1^{n\pi} \frac{|\sin x|}{x} dx \geq \frac{2}{\pi} \sum_{r=1}^n \frac{1}{r} - \int_0^1 \frac{\sin x}{x} dx \quad (5.49)$$

Let $k > 4$. Let p be the greatest integer not greater than $\frac{k}{\pi}$. Then $p \to \infty$ as $k \to \infty$. Taking $E = [1, \infty)$ in the definition of the Lebesgue integral $(L) \int_E |f|$ we have, since $|f| \leq 1$ in $[1, \infty]$ and $k \geq p\pi$ by (5.49)

$$(L) \int_E |f| \, dx = \lim_{k \to \infty} (L) \int_{E \cap [-k,k]} |f| \, dx$$

$$= \lim_{k \to \infty} (L) \int_{[1,k]} |f| \, dx$$

$$= \lim_{k \to \infty} (R) \int_1^k |f| \, dx$$

$$\geq \lim_{p \to \infty} (R) \int_1^{p\pi} |f| \, dx$$

$$\geq \lim_{p \to \infty} \left[\frac{2}{\pi} \sum_{r=1}^p \frac{1}{r} - \int_0^1 \frac{\sin x}{x} dx \right]$$

$$= \frac{2}{\pi} \sum_{r=1}^\infty \frac{1}{r} - \int_0^1 \frac{\sin x}{x} dx$$

Therefore $|f|$ is not Lebesgue integrable in $E = [1, \infty)$ and so f is not Lebesgue integrable in $[1, \infty)$.

Regarding (ii) above, suppose that f is defined in $[a, b]$ and has infinite discontinuity at c, $a < c < b$. If f is bounded and Riemann integrable in $[a, c - \epsilon_1]$ and in $[c + \epsilon_2, b]$ for every ϵ_1, $0 < \epsilon_1 < c - a$ and every ϵ_2, $0 < \epsilon_2 < b - c$ and if $(R) \int_a^{c-\epsilon_1} f$ and $(R) \int_{c+\epsilon_2}^b f$ tend to finite limits l_1 and l_2 respectively as $\epsilon_1 \to 0$ and $\epsilon_2 \to 0$ then f is improper Riemann integrable in $[a, b]$ and $(R_i) \int_a^b f = l_1 + l_2$. Therefore

$$(L) \int_{[a, c - \epsilon_1]} f + (L) \int_{[c + \epsilon_2, b]} f \to l_1 + l_2 \text{ as } \epsilon_1 \to 0 \text{ and } \epsilon_2 \to 0.$$

Since $[a, c - \epsilon_1] \cup [c + \epsilon_2, b]$ tends to $[a, c] \cup [c, b]$ as $\epsilon_1 \to 0$ and $\epsilon_2 \to 0$ it is expected that f should be Lebesgue integrable in $[a, b]$. But this is not true. Consider

Example 5.3

$$\text{Let } f(x) = (-1)^{r+1}(r + 1) \text{ for } \frac{1}{r + 1} < x \leq \frac{1}{r}, r = 1, 2, \ldots$$
$$= 0 \text{ for } x = 0$$

Then f is improper Riemann integrable but not Lebesgue integrable in $[0, 1]$. Since f is unbounded in every neighborhood of the point $x = 0$, f is not Riemann integrable in $[0, 1]$. But for every ϵ, $0 < \epsilon < 1$, f is bounded and has only finite member of points of discontinuity in $[\epsilon, 1]$ and so f is Riemann integrable in $[\epsilon, 1]$. Also for any positive integer m

$$(R) \int_{\frac{1}{m+1}}^1 f dx = \sum_{r=1}^m \frac{(-1)^{r+1}(r+1)}{r(r+1)} = \sum_{r=1}^m \frac{(-1)^{r+1}}{r}$$

which tends to a finite limit as $m \to \infty$. Since for any ϵ, $0 < \epsilon < 1$, there is m such that $(R) \int_\epsilon^1 f dx$ lies between $(R) \int_{\frac{1}{m+1}}^1 f dx$ and $(R) \int_{\frac{1}{m}}^1 f dx$, the integral $(R) \int_\epsilon^1 f dx$ tends to a finite limit on $\epsilon \to 0$ and so f is improper Riemann integrable in $[0, 1]$. We show that $|f|$ is not Lebesgue integrable in $[0, 1]$. Since

$$|f(x)| = r + 1 \text{ for } \frac{1}{r+1} < x \leq \frac{1}{r}, r = 1, 2, \ldots.$$

the sequence of truncated function $\{|f|_n\}$ is given by

5.9 Improper Riemann Integral and Lebesgue Integral

$$[f]_n(x) = r + 1 \text{ for } \frac{1}{r+1} < x \leq \frac{1}{r}, r = 1, 2, \ldots, n-1$$
$$= n \text{ for } 0 < x \leq \frac{1}{n}$$
$$= 0 \text{ for } x = 0$$

Therefore

$$(L)\int_{[0,1]} |f|_n dx \geq (L)\int_{\frac{1}{n}}^{1} |f| dx = \sum_{r=1}^{n-1} \int_{\frac{1}{r+1}}^{\frac{1}{r}} |f| dx = \sum_{r=1}^{n-1} \frac{1}{r}.$$

Hence

$$\lim_{n \to \infty} (L) \int_{[0,1]} |f|_n dx \geq \sum_{r=1}^{\infty} \frac{1}{r}.$$

So $|f|$ is not Lebesgue integrable in $[0, 1]$. Hence f is not Lebesgue integrable in $[0, 1]$.

In both of the above examples the function f are such that $|f|$ are not improper Riemann integrable. If however $|f|$ is improper Riemann integrable then f is Lebesgue integrable. We have

Theorem 5.38 *Let $f : [a, \infty) \to \mathbb{R}$ where $-\infty < a < \infty$ and let f be Riemann integrable in $[a, b]$ for every b, $a < b < \infty$. Then $|f|$ is improper Riemann integrable in $[a, \infty)$ if and only if f is Lebesgue integrable in $[a, \infty)$. In either case the integrals are equal.*

Proof Suppose that $|f|$ is improper Riemann integrable in $[a, \infty)$. Then $|f|$ is Riemann integrable in $[a, n]$ for every $n > a$ and so $|f|$ is Lebesgue integrable in $[a, n]$ for every $n > a$ and

$$(L)\int_{[a,n]} |f| dx = (R)\int_a^n |f| dx.$$

Since $|f|$ is improper Riemann integrable in $[a, \infty)$, $\lim_{n \to \infty} (R) \int_a^n |f| dx$ is finite and so $\lim_{n \to \infty} (L) \int_{[a,n]} |f| dx$ is finite. Hence $\lim_{n \to \infty} (L) \int_{[a,n]} f_+ dx$ and $\lim_{n \to \infty} (L) \int_{[a,n]} f_- dx$ are finite and hence f is Lebesgue integrable in $[a, \infty)$.

Conversely, suppose that f is Lebesgue integrable in $[a, \infty)$. Then $|f|$ is Lebesgue integrable in $[a, \infty)$. Let $n > a$ be any positive integer. Since f is Riemann integrable in $[a, b]$ for every b, $a < b < \infty$, f is Riemann integrable in $[a, n]$ and so $|f|$ is Riemann integrable in $[a, n]$ and hence $(R) \int_a^n |f| dx = (L) \int_{[a,n]} |f| dx$. Since n is arbitrary, this is true for all n. Since $|f|$ is Lebesgue integrable in $[a, \infty)$,

$$\lim_{n\to\infty} (L) \int_{[a,n]} |f|\, dx = (L) \int_{[a,\infty]} |f|\, dx$$

and hence

$$\lim_{n\to\infty} (R) \int_a^n |f|\, dx = (L) \int_{[a,\infty]} |f|\, dx$$

and so $|f|$ is improper Riemann integrable in $[a, \infty)$. The rest is clear. □

Theorem 5.39 *Let* $f : [a, b] \to \mathbb{R}$ *and let* f *have an infinite discontinuity at* c, $a < c < b$. *If* f *is Riemann integrable in* $[a, c - \epsilon_1]$ *and in* $[c + \epsilon_2, b]$ *for every* ϵ_1, $0 < \epsilon_1 < c - a$ *and for every* ϵ_2, $0 < \epsilon_2 < b - c$, *then* $|f|$ *is improper Riemann integrable in* $[a, b]$ *if and only if* f *is Lebesgue integrable in* $[a, b]$. *In either case the integrals are equal.*

Proof Let f be Lebesgue integrable in $[a, b]$. For each positive integer $n \geq \frac{1}{c-a}$ let

$$f_n(x) = f(x) \text{ for } a \leq x \leq c - \frac{1}{n}$$
$$= 0 \text{ for } c - \frac{1}{n} < x \leq c.$$

Then $|f_n(x)| \to |f(x)|$ as $n \to \infty$ for all $x \in [a, c)$. Also f_n is Lebesgue integrable in $[a, c)$ for all n. Since $|f_n(x)| \leq |f(x)|$ for all $n \geq \frac{1}{c-a}$ and for all $x \in [a, c]$, by Dominated convergence theorem applied on $\{|f_n|\}$,

$$\lim_{n\to\infty} (L) \int_{[a,c]} |f_n|\, dx = (L) \int_{[a,c]} |f|\, dx$$

i.e.

$$\lim_{n\to\infty} (R) \int_a^{c-\frac{1}{n}} |f|\, dx = (L) \int_{[a,c]} |f|\, dx.$$

Hence $|f|$ is improper Riemann integrable in $[a, c]$. Similarly $|f|$ is improper Riemann integrable in $[c, b]$. So $|f|$ is improper Riemann integrable in $[a, b]$.

Conversely, suppose that $|f|$ is improper Riemann integrable in $[a, b]$. Then $|f|$ is improper Riemann integrable in $[a, c]$ and in $[c, b]$. Since f is Riemann integrable in $[a, c - \epsilon_1]$ for every $\epsilon_1, 0 < \epsilon_1 < c - a$, f is measurable in $[a, c]$. Let

$$f_n(x) = |f(x)| \text{ for } a \leq x \leq c - \frac{1}{n}$$
$$= 0 \text{ for } c - \frac{1}{n} < x \leq c.$$

Then $\{f_n\}$ is a non-decreasing sequence of non-negative measurable functions in $[a, c]$ which converges to $|f|$ in $[a, c)$. So by Monotone convergence theorem,

$$\lim_{n\to\infty} (L) \int_{[a,c]} f_n dx = (L) \int_{[a,c]} |f|\, dx$$

i.e

$$\lim_{n\to\infty} (R) \int_a^{c-\frac{1}{n}} |f|\, dx = (L) \int_{[a,c]} |f|\, dx.$$

Since $|f|$ is improper Riemann integrable in $[a, c]$, the left-hand side is finite and so $|f|$ is Lebesgue integrable in $[a, c]$. Similarly $|f|$ is Lebesgue integrable in $[c, b]$ and so $|f|$ is Lebesgue integrable in $[a, b]$. Hence f is Lebesgue integrable in $[a, b]$. The rest is clear. □

Theorem 5.40 *Let $f : [a, \infty] \to \mathbb{R}$ where $-\infty < a < \infty$. If f is improper Riemann integrable as well as Lebesgue integrable in $[a, \infty]$. then the integrals are equal. This result also true for function $f : [a, b] \to \mathbb{R}$ when f has infinite discontinuity in $[a, b]$*

Proof Let f be improper Riemann integrable as well as Lebesgue integrable in $[a, \infty]$. As in Theorem 5.38 define $f_n = f \chi_{[a,n]}$. Since $f_n(x) \to f(x)$ as $n \to \infty$ for all x, $a \leq x < \infty$ and $|f_n(x)| \leq |f(x)|$ for all $n > a$ and all $x \in [a, \infty]$, by Dominated convergence theorem applied on $\{f_n\}$ we have

$$\lim_{n\to\infty} (L) \int_{[a,\infty)} f_n dx = \lim_{n\to\infty} (L) \int_{[a,n]} f_n dx = (L) \int_{[a,\infty)} f\, dx$$

Since f is Riemann integrable in $[a, n]$, from the above we have

$$(R_i) \int_a^\infty f\, dx = \lim_{n\to\infty} (R) \int_a^n f\, dx = \lim_{n\to\infty} (L) \int_{[a,n]} f_n dx = (L) \int_{[a,\infty)} f\, dx$$

where R_i denotes improper Riemann integral proving the first part.

The proof of the second part is similarly obtained by applying Theorem 5.39 instead of Theorem 5.38 □

Theorem 5.41 *Let $f : [a, \infty) \to \mathbb{R}$ where $-\infty < a < \infty$. if f and $|f|$ are both improper Riemann integrable then f is Lebesgue integrable in $[a, \infty)$, and*

$$(R_i) \int_a^\infty f\, dx = (L) \int_{[a,\infty)} f\, dx.$$

This result is also true for function $f : [a, b] \to \mathbb{R}$ when f has infinite discontinuity in $[a, b]$.

Proof By Theorem 5.38, f is Lebesgue integrable in $[a, \infty)$ and so by Theorem 5.40 the result follows. For $f : [a, b] \to \mathbb{R}$ apply Theorem 5.39. □

5.10 Newton Integral and Lebesgue Integral

Definition 5.8 A function $f : [a, b] \to \mathbb{R}$ is said to be Newton integrable if there exists $F : [a, b] \to \mathbb{R}$ such that $F'(x) = f(x)$ for all $x \in [a, b]$ and the Newton integral of f is defined by

$$(N) \int_a^b f dx = F(b) - F(a).$$

In other words a function f is Newton integrable if and only if f is a derivative of some function. The function F is also called anti derivative of f. Newton integrable functions may not be Lebesgue integrable. Consider

Example 5.4 Let

$$F(x) = x^2 \cos \frac{1}{x^2} \quad \text{for } x \neq 0$$
$$= 0 \quad \text{for } x = 0.$$

Then the derivative F' of F exists finitely everywhere but is not Lebesgue integrable in any interval containing the point zero. To see this we have

$$F'(x) = 2x \cos \frac{1}{x^2} + \frac{2}{x} \sin \frac{1}{x^2} \quad \text{for } x \neq 0$$
$$= 0 \quad \text{for } x = 0.$$

So F' is bounded and continuous in every interval $[\alpha, \beta], 0 < \alpha < \beta < 1$, and hence by the Fundamental Theorem of Integral Calculus

$$(R) \int_\alpha^\beta F'(x) dx = F(\beta) - F(\alpha) = \beta^2 \cos \frac{1}{\beta^2} - \alpha^2 \cos \frac{1}{\alpha^2}$$

Putting $\alpha_n = \sqrt{\frac{2}{(4n+1)\pi}}$ and $\beta_n = \sqrt{\frac{1}{2n\pi}}$ We have

$$(R) \int_{\alpha_n}^{\beta_n} F'(x) dx = \frac{1}{2n\pi}$$

Since Riemann integrability implies Lebesgue integrability and the integrals being equal,

$$(L) \int_{[0,1]} | F'(x) | dx \geq \sum_{n=1}^{\infty} (L) \int_{[\alpha_n, \beta_n]} | F'(x) | dx$$

5.10 Newton Integral and Lebesgue Integral

$$\geq \sum_{n=1}^{\infty}(L)\int_{[\alpha_n,\beta_n]} F'(x)dx = \sum_{n=1}^{\infty}(R)\int_{\alpha_n}^{\beta_n} F'(x)dx = \frac{1}{2\pi}\sum_{n=1}^{\infty}\frac{1}{n}$$

and so $|F'|$ is not Lebesgue integrable in $[0, 1]$ and hence F' is not Lebesgue integrable in $[0, 1]$.

Example 5.5 Let

$$f(x) = \frac{1}{x}\sin\frac{1}{x^2} \quad \text{for} \quad x \neq 0$$
$$= 0 \quad \text{for} \quad x = 0.$$

Then f is Newton integrable but not Lebesgue integrable in any interval containing the point $x = 0$. Let

$$F(x) = x^2 \cos\frac{1}{x^2} \quad \text{for} \quad x \neq 0$$
$$= 0 \quad \text{for} \quad x = 0.$$

and

$$g(x) = x \cos\frac{1}{x^2} \quad \text{for} \quad x \neq 0$$
$$= 0 \quad \text{for} \quad x = 0.$$

Since g is continuous it is Riemann integrable. Also

$$\frac{1}{2}\frac{d}{dx}\left[F(x) - 2(R)\int_0^x g(t)dt\right] = f(x).$$

So, f is the derivative of the function $\frac{1}{2}F - (R)\int_a^x g(t)dt$ and hence f is Newton integrable. But f is not Lebesgue integrable in $[0, 1]$. For, by Example 5.4 above F' is not Lebesgue integrable in $[0, 1]$. Therefore g being continuous the function $\frac{1}{2}F' - g$ is not Lebesgue integrable in $[0, 1]$ and so f is not Lebesgue integrable in $[0, 1]$.

Example 5.6 Let $p > 0, q \geq p + 1$. Define

$$f(x) = \frac{1}{x^{q-p}}\cos\frac{1}{x^q} \quad \text{for} \quad x \neq 0$$
$$= 0 \quad \text{for} \quad x = 0.$$

Then f is Newton integrable but not Lebesgue integrable in $[0, 1]$. Let

$$F(x) = x^{p+1} \sin \frac{1}{x^q} \quad \text{for} \quad x \neq 0$$
$$= 0 \quad \text{for} \quad x = 0.$$

$$G(x) = x^p \sin \frac{1}{x^q} \quad \text{for} \quad x \neq 0$$
$$= 0 \quad \text{for} \quad x = 0.$$

Then $F'(0) = 0$ and $F'(x) = (p+1)x^p \sin \frac{1}{x^q} - \frac{q}{x^{q-p}} \cos \frac{1}{x^q}$ for $x \neq 0$. So $F' = (p+1)G - qf$ i.e. $f = \frac{1}{q}[(p+1)G - F']$. Since G is continuous, G is a derivative function. Also F' is a derivative function. Hence f is a derivative function and so it is Newton integrable. Since

$$f(x) = \frac{1}{q}[(p+1)G(x) - F'(x)] = \frac{1}{q}\frac{d}{dx}\left\{(p+1)(R)\int_0^x G dx - F(x)\right\},$$

we get

$$(N)\int_0^1 f dx = \frac{p+1}{q}\left\{(R)\int_0^1 G dx - F(1)\right\}.$$

But f is not Lebesgue integrable in $[0, 1]$. For suppose that f is Lebesgue integrable in $[0, 1]$. Then since G is continuous F' is Lebesgue integrable in $[0, 1]$. Since F' is continuous in every closed sub-interval $[\alpha, \beta] \subset (0, 1)$,

$$(R)\int_\alpha^\beta F' dx = F(\beta) - F(\alpha) = \beta^{p+1} \sin \frac{1}{\beta^q} - \alpha^{p+1} \sin \frac{1}{\alpha^q}$$

So putting $\alpha_n = \left(\frac{1}{(2n+1)\pi}\right)^{\frac{1}{q}}$, $\beta_n = \left(\frac{2}{(4n+1)\pi}\right)^{\frac{1}{q}}$,

$$(R)\int_{\alpha_n}^{\beta_n} F' dx = \left(\frac{2}{(4n+1)\pi}\right)^{\frac{p+1}{q}}$$

Therefore

$$(L)\int_{[0,1]} |F'(x)| dx \geq \sum_{n=1}^\infty (L)\int_{[\alpha_n, \beta_n]} |F'(x)| dx \geq \sum_{n=1}^\infty (L)\int_{[\alpha_n, \beta_n]} F'(x) dx$$

$$= \sum_{n=1}^\infty (R)\int_{\alpha_n}^{\beta_n} F'(x) dx = \left(\frac{2}{\pi}\right)^{\frac{p+1}{q}} \sum_{n=1}^\infty \left(\frac{1}{4n+1}\right)^{\frac{p+1}{q}}$$

Since $p + 1 \leq q$, the last series is divergent and so F' is not Lebesgue integrable in $[0, 1]$.

From the above examples it is clear that Lebesgue integral does not include Newton integral. Also Newton integral does not include Lebesgue integral. For the function

$$f(x) = 1 \text{ for } x \text{ } rational$$
$$= 0 \text{ for } x \text{ } irrational.$$

is Lebesgue integrable but it cannot be derivative of any function. However the following theorem holds.

Theorem 5.42 *Let $f : [a, b] \to \mathbb{R}$. Then*

(i) if f is Newton integrable then f is Lebesgue integrable if and only if F is absolutely continuous in $[a, b]$ where $F(x) = (N) \int_a^x f dx, \quad x \in [a, b]$

(ii) if f is Lebesgue integrable then f is Newton integrable if and only if $F' = f$ in $[a, b]$ where $F(x) = (L) \int_a^x f dx, \quad x \in [a, b]$.

In either case the integrals are equal.

The proof of this theorem is a consequence of a theorem which gives a characterization of indefinite Lebesgue integral in Chap. 6 (Theorem 6.41).

5.11 Conclusion

From the previous two sections, it follows that Lebesgue integral and improper Riemann integral are mutually exclusive and that Lebesgue integral and Newton integral are mutually exclusive. Also improper Riemann integral and Newton integral are mutually exclusive. To see this note that there are bounded derivatives which are not Riemann integrable (see Theorem 5.23) and hence not improper Riemann integrable. This shows that there are Newton integrable functions which are not improper Riemann integrable. On the other hand consider the function $f(x) = 0$ for $x \neq 0$ and $f(0) = 1$. This function is Riemann integrable in $[-1, 1]$ and hence is improper Riemann integrable in $[-1, 1]$ but it is not Newton integrable in $[-1, 1]$, because the type of discontinuity of f at $x = 0$ shows that f cannot be a derivative of any function. Therefore, Lebesgue integral, Newton integral, and improper Riemann integral are mutually exclusive. So the natural question is that whether there is any integral which includes all these integrals. The answer is yes and it is the Denjoy-Perron integral which is equivalent to the Henstock-Kurzweil integral. We shall not discuss these integrals. These integrals can be found in [Saks, Meleod, Henstock, Gordon, Lee].

5.12 Exercises

1. Let $x_1, x_2, \ldots$ be an enumeration of the set of rational numbers in [0.1]. Let

   $$f_n(x) = 1 \text{ if } x \in \{x_1, x_2, \ldots, x_n\}$$
   $$= 0 \text{ otherwise}$$

   Show that $\lim f_n$ is not Riemann integrable but is Lebesgue integrable.

2. Let f be Lebesgue integrable in $[a, b]$. Show that for every $\epsilon > 0$ there is a continuous function g on $[a, b]$ such that

   $$\int_a^b |f - g| < \epsilon$$

3. Let f be Lebesgue integrable in $[a, b]$. Show that for every $\epsilon > 0$ there is a Riemann integrable function g on $[a, b]$ such that

   $$\left| (L) \int_a^b f - (R) \int_a^b g \right| < \epsilon$$

4. Let $0 < x \leq 1$ and let C_α be the Cantor set in $[0, 1]$ corresponding to α. Define

   $$f(x) = x \text{ if } x \in [0, 1] \sim C_\alpha$$
   $$= 0 \text{ for } x \in C_\alpha$$

   Examine the L-integrability and R-integrability of f in $[0, 1]$ when (i) $\alpha = 1$ and when (ii) $\alpha < 1$. Find the integrals whenever integrable.

5. If f and g are Lebesgue integrable on a measurable set E then show that $max[f, g]$ and $min[f, g]$ are also integrable on E and

   $$\int_E max[f, g] = \frac{1}{2} \left[\int_E (f + g) + \int_E |f - g| \right]$$

   and

   $$\int_E min[f, g] = \int_E (f + g) - \int_E max[f, g]$$

6. If f and g are bounded and measurable on E then show that fg is L-integrable on E.

7. Let $f : [a, b] \to \mathbb{R}$ be Lebesgue integrable and let $f \geq 0$ almost everywhere in $[a, b]$. If $\int_a^b f dx = 0$ prove that $f = 0$ almost everywhere in $[a, b]$.

8. Let f, g, h be measurable functions on $[a, b]$ such that $g \leq f \leq h$ almost everywhere in $[a, b]$. If g and h are Lebesgue integrable then prove that f is also Lebesgue integrable in $[a, b]$.

9. Give an example of a function $f : [a, b] \to \mathbb{R}$ such that f is Lebesgue integrable in every interval $[a + \epsilon, b], 0 < \epsilon < b - a$ but not Lebesgue integrable on $[a, b]$.
10. Give an example of a function $f : [a, b] \to \mathbb{R}$ such that f is unbounded in every sub-interval of $[a, b]$ but is Lebesgue integrable in $[a, b]$.
11. Let $f : [a, b] \to \mathbb{R}$ be Lebesgue integrable. If $E_n = \{x \in [a, b] :| f(x) |> n\}$ then prove that $\lim_{n \to \infty} (L) \int_{E_n} f dx = 0$.
12. Let $f : E \to \mathbb{R}$ be Lebesgue integrable. Then show that f is finite almost everywhere in E.
13. Let $f : [a, b] \to \mathbb{R}$ be Lebesgue integrable in $[a, b]$ and let $F(x) = \int_{[a,x]} f dx$ prove that F is a continues function on $[a, b]$.
14. Let $\{f_n\}$ be a sequence if integrable function on a measurable set E. If $\int_E |f_n| \to 0$ as $n \to \infty$ show that $\{f_n\}$ converges to 0 in measure
15. Let $0 < \alpha \leq 1$. and let E_α be the Cantor set corresponding to α. Show that the characteristic function of E_α is Riemann integrable if $\alpha = 1$ and that it is Lebesgue integrable but not Riemann integrable if $0 < \alpha < 1$.
16. Let $\{f_n\}$ be a sequence of measurable functions defined on a measurable set E and let g be a function integrable on E. If $|f_n(x)| \leq g(x)$ for all n and all $x \in E$ show that
$$\int_E (\liminf f_n) dx \leq \liminf \int_E f_n dx \leq \limsup \int_E f_n dx \leq \int_E (\limsup f_n) dx.$$
17. If f and g are integrable functions on a measurable set E then show that
$$\min. \left[\int_E f dx, \int_E g dx \right] \geq \int_E \min.[f, g] dx$$
what conclusion you can draw about the relation between f and g if the equality holds?
18. If f is integrable on a measurable set E of finite measure then show that for every $\epsilon > 0$ there is a measurable set $E_\epsilon \subset E$ such that $\left| \int_E f dx - \int_{E_\epsilon} f dx \right| < \epsilon$.
19. If f is measurable and g is integrable on a measurable set E and if $\alpha \leq f(x) \leq \beta$ almost everywhere on E. Then show that there is γ such that $\alpha \leq \gamma \leq \beta$ and
$$\int_E f|g| dx = \gamma \int_E |g| dx$$
Show by an example that $|g|$ cannot be replaced by g.
20. Let f be a non-negative measurable function on a measurable set E. Show that
$$\int_E f dx = \sup \sum_{k=1}^{n} \mu(E_k) \inf\{f(x) : x \in E_k\}$$
where the supremum is taken over the collection of all finite classes $\{E_1, E_2, \ldots, E_n\}$ of disjoint measurable sets such that $E = \bigcup_{k=1}^{n} E_k$.
21. Let f a bounded non-negative measurable function on a measurable set E show that
$$\int_E f dx = \lim_{n \to \infty} \sum_{k=0}^{nM} \frac{k}{n} \mu\left(E \cap \{x \in E : \frac{k}{n} \leq f(x) < \frac{k+1}{n}\}\right)$$

where M is an upper bound of f.

22. Let f and g be measurable functions on a measurable set E. If f and g are integrable then show that fg may not be integrable. If f^2 and g^2 are integrable then show that fg is integrable.

23. If f is bounded and measurable and g is integrable on a measurable set E then prove that fg is integrable on E.

24. Let f be continuous on $[a, b]$ and let $F(x) = \int_{[a,x]} f \, dx$ for $x \in [a, b]$. Prove that $F'(x) = f(x)$ for each $x \in [a, b]$.

25. Let $\{f_n\}$ be a sequence of non-negative integrable functions on a measurable set E and let $\sum_{n=1}^{\infty} f_n(x) = f_0(x)$ for almost all $x \in E$. Show that f_0 is integrable on E if and only if $\sum_{n=1}^{\infty} \int_E f_n(x)$ is convergent and

$$\int_E f_0 \, dx = \sum_{n=1}^{\infty} \int_E f_n(x).$$

[Hint: Use Monotone Convergence theorem]

26. Show that the function

$$f(x) = \frac{2}{x^3} \sin \frac{1}{x^2} \text{ for } x \neq 0$$
$$= 0 \text{ for } x = 0.$$

is not Lebesgue integrable in $[0, 1]$.

27. If E is a measurable set and f and g are integrable on E and if $f > g$ a.e on E then prove that $\int_E f > \int_E g$.

Chapter 6
Differentiation of Functions

Integration is associated with differentiation and so it is essential to see how they are related. In Riemann integration if a function f is Riemann integrable then there is a function F such that the derivative F' exists $a.e.$ and $F' = f$. In Lebesgue integration a function f is Lebesgue integrable if and only if there is a function F such that F is absolutely continuous and the derivative $F' = f$ $a.e.$ Before proving this result we need to study various properties of derivatives and related topics.

6.1 Limits of a Function and Their Properties

Let f be defined in some right neighborhood $(a, a+k)$ of a and let for δ, $0 < \delta < k$,

$$u(\delta) = \sup_{a<x<a+\delta} f(x) \text{ and } l(\delta) = \inf_{a<x<a+\delta} f(x).$$

Then as δ decreases $u(\delta)$ decreases and $l(\delta)$ increases. So, $u(\delta)$ and $l(\delta)$ tend to definite limits as $\delta \to 0$ and the limits are $\inf_{\delta>0} u(\delta)$ and $\sup_{\delta>0} l(\delta)$ respectively. The right-hand upper and lower limits of f at a are defined by

$$\limsup_{x \to a+} f(x) = \overline{f}(a+0) = \lim_{\delta \to 0} u(\delta) = \inf_{\delta>0} \sup_{a<x<a+\delta} f(x), \text{ and}$$

$$\liminf_{x \to a+} f(x) = \underline{f}(a+0) = \lim_{\delta \to 0} l(\delta) = \sup_{\delta>0} \inf_{a<x<a+\delta} f(x), \text{ respectively.}$$

Similarly, if f is defined in some left neighborhood of a then the left-hand upper and lower limits of f at a are defined by

$$\limsup_{x \to a-} f(x) = \overline{f}(a-0) = \inf_{\delta>0} \sup_{a-\delta<x<a} f(x) \text{ and}$$

$$\liminf_{x \to a-} f(x) = \underline{f}(a-0) = \sup_{\delta>0} \inf_{a-\delta<x<a} f(x), \text{ respectively.}$$

It is clear that $\underline{f}(a+0) \leq \overline{f}(a+0)$ and $\underline{f}(a-0) \leq \overline{f}(a-0)$ and that if the limit of f at a exists then all these four limits are equal to $\lim_{x \to a} f(x)$ and conversely if all these four limits are equal then the limit of f at a exists which is equal to the common value.

Theorem 6.1 *If $f : (a, a+k) \to \mathbb{R}$ then $\overline{f}(a+0) = A$ if and only if*

(i) $\limsup_{n \to \infty} f(x_n) \leq A$ *for every sequence $\{x_n\}$ such that $x_n > a$ for all n and $\lim_{n \to \infty} x_n = a$, and*

(ii) *there exists a sequence $\{x_n\}$ such that $x_n > a$ for all n and $\lim_{n \to \infty} x_n = a$ for which $\lim_{n \to \infty} f(x_n) = A$.*

Proof Suppose $\overline{f}(a+0) = A$. Let $\{x_n\}$ be any sequence such that $x_n > a$ for all n and $\lim_{n \to \infty} x_n = a$. If possible, suppose $\limsup_{n \to \infty} f(x_n) > A$. Choose $\epsilon > 0$ such that $\limsup_{n \to \infty} f(x_n) > A + \epsilon$. Let $\delta > 0$ be arbitrary. Then there is $x_n \in (a, a+\delta)$ such that $f(x_n) > A + \epsilon$. Hence $\sup_{a < x < a+\delta} f(x) > A + \epsilon$. So, taking the limit as $\delta \to 0$, $\overline{f}(a+0) \geq A + \epsilon > A$ which is a contradiction. This proves (i). To prove (ii) let n be any positive integer. Since $\overline{f}(a+0) = A$, $\lim_{\delta \to 0} \sup_{a < x < a+\delta} f(x) > A - \frac{1}{n}$ and so there is δ_n, $0 < \delta_n < \frac{1}{n}$, such that $\sup_{a < x < a+\delta_n} f(x) > A - \frac{1}{n}$ and so there is x_n, $a < x_n < a + \delta_n$, such that $f(x_n) > A - \frac{1}{n}$. So, we get the sequence $\{x_n\}$ such that $x_n > a$ for all n and since $a < x_n < a + \frac{1}{n}$, $\lim_{n \to \infty} x_n = a$. Since $f(x_n) > A - \frac{1}{n}$ for all n, $\liminf_{n \to \infty} f(x_n) \geq A$. Also by (i) $\limsup_{n \to \infty} f(x_n) \leq A$ and so (ii) is proved.

Next suppose that (i) and (ii) hold. Let $\epsilon > 0$ be arbitrary. Then by (i) there is a right neighborhood $(a, a+k)$ such that $f(x) \leq A + \epsilon$ for all $x \in (a, a+k)$. Hence $\sup_{a < x < a+\delta} f(x) \leq A + \epsilon$ for all δ, $0 < \delta < k$ and so $\overline{f}(a+0) \leq A + \epsilon$. Also by (ii), there are points in every right neighborhood of a for which $f(x) > A - \epsilon$. Hence $\sup_{a < x < a+\delta} f(x) > A - \epsilon$ for every $\delta > 0$ and so $\overline{f}(a+0) \geq A - \epsilon$. Since ϵ is arbitrary $\overline{f}(a+0) = A$, completing the proof. □

Analogous theorems hold for the other three limits, $\underline{f}(a+0), \overline{f}(a-0)$ and $\underline{f}(a-0)$. The both sided upper and lower limits of f at a are defined by

$$\limsup_{x \to a} f(x) = \max[\overline{f}(a+0), \overline{f}(a-0)] \text{ and } \liminf_{x \to a} f(x) = \min[\underline{f}(a+0), \underline{f}(a-0)]$$

and when they are equal, the equal value is the $\lim_{x \to a} f(x)$, i.e

$$\lim_{x \to a} f(x) = \limsup_{n \to a} f(x) = \liminf_{x \to a} f(x) = \overline{f}(a+0) = \underline{f}(a+0) = \overline{f}(a-0) = \underline{f}(a-0).$$

From the above theorem we thus get.

6.1 Limits of a Function and Their Properties

Corollary 6.2 *Let $f : (a - \delta, a) \cup (a, a + \delta) \to \mathbb{R}$. Then $\lim_{x \to a} f(x) = l$ (l may be infinite) if and only if $\lim_{n \to \infty} f(x_n) = l$ for every sequence $\{x_n\}$ such that $\lim_{n \to \infty} x_n = a$.*

Theorem 6.3 *Let $E \subset \mathbb{R}$ be any set. Then the set of points of E which are isolated at least on one side is countable.*

Proof Let A be the set of all points of E which are isolated points of E on the right. For each positive integer n let A_n be the set of all points x of A such that $[x, x + \frac{1}{n}]$ contains no points of E other than x. Then it is clear that for each integer k the interval $[\frac{k}{n}, \frac{k+1}{n}]$ can have at most one point in common with A_n. Hence each set A_n is countable. Since $A = \bigcup_{n=1}^{\infty} A_n$, A is countable. Similarly, if B is the set of all points of E which are isolated points of on the left then B is countable. □

Theorem 6.4 *For any real-valued function f each of the following sets is countable.*

(i) the set of points where f is strictly maximum or strictly minimum,
(ii) the set of points x where

$$\overline{f}(x + 0) \neq \overline{f}(x - 0) \text{ or } \underline{f}(x + 0) \neq \underline{f}(x - 0)$$

Proof

i. Let A be the set of points at which f assumes a strict maximum and for each positive integer n let A_n be the set of all points x of A such that $f(t) < f(x)$ holds for each point $t \neq x$ such that $t \in (x - \frac{1}{n}, x + \frac{1}{n})$. Then each point x of A_n is an isolated point of A_n. For, if $x \in A_n$ then there is no other point of A_n in $(x - \frac{1}{n}, x + \frac{1}{n})$. So, by Theorem 6.3 A_n is countable. Since $A = \bigcup_{n=1}^{\infty} A_n$, A is countable. Similarly, if B is the set of all points at which f assumes a strict minimum is countable.

ii. Let $E = \{x : \overline{f}(x - 0) > \overline{f}(x + 0)\}$. For each rational number r let $E_r = \{x \in E : \overline{f}(x - 0) > r > \overline{f}(x + 0)\}$. Then each point of E_r is an isolated point of E_r from the right. For, let $x \in E_r$. Since $r > \overline{f}(x + 0)$ there is a right neighborhood $(x, x + \delta)$ of x such that $f(t) < r_1$ for $t \in (x, x + \delta)$ where $r > r_1 > \overline{f}(x + 0)$. So, if $x_0 \in (x, x + \delta)$ then $\overline{f}(x_0 - 0) \leq r_1 < r$ and so $x_0 \notin E_r$. Therefore $E_r \cap (x, x + \delta) = \emptyset$ and so x is an isolated point of E_r from the right. Hence by Theorem 6.3, the set E_r is countable. Since $E = \bigcup E_r$, where the union is taken over all rational numbers, E is countable. Similarly the set $F = \{x : \overline{f}(x - 0) < \overline{f}(x + 0)\}$ is countable. So, the first part of (ii) is proved. The proof of the second part is similar. □

6.2 Derivates of a Function and Their Properties

Let f be defined in some right neighborhood $[a, a+\delta)$ of a. The right-hand upper and lower derivates of f at a are defined by

$$D^+ f(a) = \limsup_{h \to 0+} \frac{f(a+h) - f(a)}{h} \text{ and } D_+ f(a) = \liminf_{h \to 0+} \frac{f(a+h) - f(a)}{h}$$

respectively. Similarly if f is defined in some left neighborhood $(a - \delta, a]$ of a then the left-hand upper and lower derivates of f at a are defined by

$$D^- f(a) = \limsup_{h \to 0-} \frac{f(a+h) - f(a)}{h} \text{ and } D_- f(a) = \liminf_{h \to 0-} \frac{f(a+h) - f(a)}{h}$$

These four derivates are called Dini derivates of f at a. If all these four derivates are equal then f is said to have derivative at a and the common value is called the derivative of f at a and is denoted by $f'(a)$. If $f'(a)$ exists and is finite then f is said to be differentiable at a.

Theorem 6.5 *The set of points x at which*

$$D^+ f(x) < D_- f(x) \text{ or } D^- f(x) < D_+ f(x)$$

is countable.

Proof Let $A = \{x : D^+ f(x) < D_- f(x)\}$ and for each rational number r let $A_r = \{x \in A : D^+ f(x) < r < D_- f(x)\}$. Let $f_r(x) = f(x) - rx$. Then $D^+ f_r(x) < 0 < D_- f_r(x)$ for each $x \in A_r$. Since $D^+ f_r(x) < 0$ implies that there is a right neighborhood $[x, x + \delta)$ of x such that $f_r(t) < f_r(x)$ for all $t \in (x, x + \delta)$ and $D_- f_r(x) > 0$ implies that there is a left neighborhood $(x - \delta, x]$ of x such that $f_r(t) < f_r(x)$ for all $t \in (x - \delta, x)$, the function f_r assumes a strict maximum at each point of A_r. So, by Theorem 6.4 (i) the set A_r is countable. Since $A = \bigcup A_r$ where the union is taken over all rational numbers, the set A is countable. The proof of the second part is similar. □

Remark In Theorem 6.4 (ii) we get $\overline{f}(x+0) = \overline{f}(x-0)$ except on a countable set but in Theorem 6.5 we get $D^+ f(x) \geq D_- f(x)$ except on a countable set. Naturally question will arise whether $D^+ f(x) = D^- f(x)$ except on a countable set. The answer is in the negative unless some restriction is given on the function f. First A. Denjoy and then many others proved the following theorem Known as Denjoy Theorem on derivatives :

Theorem 6.6 (Denjoy) *Let f be any finite function. Then except for points x on a set E of measure zero the Dini derivates of f satisfy one of the following four relations :*

6.2 Derivates of a Function and Their Properties

1. $D^+f(x) = D^-f(x) = +\infty$, $D_+f(x) = D_-f(x) = -\infty$
2. $D^+f(x) = D_-f(x) \neq \pm\infty$, $D_+f(x) = -\infty$, $D^-f(x) = +\infty$
3. $D_+f(x) = D^-f(x) \neq \pm\infty$, $D^+f(x) = +\infty$, $D_-f(x) = -\infty$
4. $D^+f(x) = D_+f(x) = D^-f(x) = D_-f(x) \neq \pm\infty$, i.e $f'(x)$ exists finitely.

The remaining possibilities, namely

$$D^+f(x) = -\infty, \ D_+f(x) = +\infty, \ D^-f(x) = -\infty \text{ and } D_-f(x) = \infty$$

occur at points $x \in E$.

This will not be proved here. The proof can be found in [Saks p. 271; Jeffery p. 186].

The following theorem which will be used later is relevant here.

Theorem 6.7 *For every set E of measure zero on the closed interval $[a,b]$ there exists a continuous non-decreasing and non-negative function ψ on $[a,b]$ such that $\psi'(x) = +\infty$ for all $x \in E$.*

Proof Since $\mu(E) = 0$, for every positive integer n there is an open set G_n such that $E \subset G_n$ and $\mu(G_n) < \frac{1}{2^n}$. We may suppose that $G_{n+1} \subset G_n$ for all n. Define

$$\psi_n(x) = \mu(G_n \cap [a,x]) \text{ for all } x \in [a,b].$$

Then ψ_n is continuous, non-negative, and non-decreasing and satisfies the inequality $\psi_n(x) < \frac{1}{2^n}$ for all n and for all $x \in [a,b]$. Extend ψ_n by setting $\psi_n(x) = \psi_n(a)$ for $x < a$ and $\psi_n(x) = \psi_n(b)$ for $x > b$. Define

$$\psi(x) = \sum_{n=1}^{\infty} \psi_n(x).$$

Then ψ is also continuous, non-decreasing, and non-negative. Let $x_0 \in E$. We show that $\psi'(x_0) = +\infty$. Choose any positive integer N and fix it. Choose $h, k > 0$ such that $(x_0 - k, x_0 + h) \subset G_N$. Since $G_N \subset G_n$ for $n \leq N$, $(x_0 - k, x_0 + h) \subset G_n$ for all $n \leq N$ and so if $n \leq N$ then

$$\psi_n(x_0 + h) = \mu(G_n \cap [a, x_0]) + \mu(G_n \cap (x_0, x_0 + h]) = \psi_n(x_0) + h$$

$$\psi_n(x_0 - k) = \mu(G_n \cap [a, x_0]) - \mu(G_n \cap (x_0, -k, x_0]) = \psi_n(x_0) - k$$

So,

$$\frac{\psi_n(x_0 + h) - \psi_n(x_0)}{h} = 1 = \frac{\psi_n(x_0 - k) - \psi_n(x_0)}{-k}.$$

Since each ψ_n is non-decreasing,

$$\frac{\psi(x_0+h)-\psi(x_0)}{h} = \sum_{n=1}^{\infty} \frac{\psi_n(x_0+h)-\psi_n(x_0)}{h} \geq \sum_{n=1}^{N} \frac{\psi_n(x_0+h)-\psi_n(x_0)}{h}$$

$$= \sum_{n=1}^{N} 1 = N$$

Similarly
$$\frac{\psi(x_0-k)-\psi(x_0)}{-k} \geq N.$$

This being true for all intervals $(x_0-k, x_0+h) \subset G_n$, letting $h \to 0+$ and $k \to 0+$, $D_+\psi(x_0) \geq N$ and $D_-\psi(x_0) \geq N$. Since N is arbitrary, $D_+\psi(x_0) = +\infty = D_-\psi(x_0)$ and so $\psi'(x_0) = +\infty$. □

6.3 Measurablity of Dini Derivates

Theorem 6.8 *If f is continuous then its Dini derivates are measurable.*

Proof Let $\phi(x, \delta) = \sup_{0<h<\delta} \frac{f(x+h)-f(x)}{h}$. Since $\phi(x, \delta)$ decreases as δ decreases, $\lim_{\delta \to 0} \phi(x, \delta)$ exists and by definition this limit equals $D^+f(x)$. So

$$D^+f(x) = \lim_{n \to \infty} \phi\left(x, \frac{1}{n}\right) \tag{6.1}$$

Let n be fixed. Since f is continuous, $\frac{f(x+h)-f(x)}{h}$ is a continuous function of h in $(0, \frac{1}{n})$ and so

$$\sup_{0<h<\frac{1}{n}} \frac{f(x+h)-f(x)}{h} = \sup_{0<r<\frac{1}{n}} \frac{f(x+r)-f(x)}{r}, \quad r \text{ is rational} \tag{6.2}$$

Since the set of rational numbers is countable, we can arrange the set of rational numbers in $(0, \frac{1}{n})$ as a sequence $\{r_k\}$. Then by (6.2)

$$\phi\left(x, \frac{1}{n}\right) = \sup_{0<h<\frac{1}{n}} \frac{f(x+h)-f(x)}{h} = \sup_{k} \frac{f(x+r_k)-f(x)}{r_k} \tag{6.3}$$

Since f is continuous, $\frac{f(x+r_k)-f(x)}{r_k}$ is a continuous function of x for each k and hence it is measurable for each k by Theorem 3.12. So, by Theorem 3.9 the right side of (6.3) is measurable. Hence by (6.3), $\phi(x, \frac{1}{n})$ is also measurable. This being true for all n, $\lim_{n \to \infty} \phi(x, \frac{1}{n})$ is also measurable by Theorem 3.9. Hence by (6.1) D^+f is measurable. The proof for other derivates is similar.

6.3 Measurablity of Dini Derivates

We shall prove a more general theorem. To prove this theorem we need a theorem which is interesting in itself. □

Theorem 6.9 *If $f : E \to \mathbb{R}$ is measurable then there exists a sequence of measurable functions $\{f_n\}$ on E with the following property:*

(i) *For each n there is a countable collection $\{E_{n,i} : i = 0, \pm 1, \pm 2, ...\}$ of disjoint measurable sets $E_{n,i}$ such that $\bigcup_{i=-\infty}^{\infty} E_{n,i} = E$ and f_n is constant on $E_{n,i}$ for each i, and*

(ii) *$\{f_n\}$ converges to f uniformly on E.*

Proof For each n let

$$E_{n,i} = \{x \in E : \frac{i}{n} \leq f(x) < \frac{i+1}{n}\},$$

and

$$f_n(x) = \frac{i}{n} \text{ for } x \in E_{n,i}, \ i = 0, \pm 1, \pm 2, \ldots.$$

Clearly (i) is satisfied. Since

$$|f(x) - f_n(x)| \leq \frac{1}{n} \text{ for all } n \text{ and all } x \in E,$$

the proof is complete. □

Theorem 6.10 *If $f : I \to \mathbb{R}$, where I is an interval, is measurable then its Dini derivates are measurable.*

Proof We prove for D^+f, the proof for D_+f, D^-f and D_-f are similar. For each positive integer m let

$$H_m(f, x) = \sup_t \left\{ \frac{f(t) - f(x)}{t - x} : x < t < x + \frac{1}{m} \right\} \tag{6.4}$$

Then we have

$$D^+f(x) = \lim_{m \to \infty} H_m(f; x) \tag{6.5}$$

Let $\alpha \in \mathbb{R}$. Consider the set

$$A_m = \{x \in I : H_m(f; x) > \alpha\} \tag{6.6}$$

Let $E \subset I$ and E be measurable and let f be constant on E. Let $x_0 \in E \cap A_m$. Then $H_m(f, x_0) > \alpha$. So from (6.4) there exists $t_1 \in (x, x + \frac{1}{m})$ such that $\frac{f(t_1) - f(x_0)}{t_1 - x_0} > \alpha$. Since f is constant on E there is a neighborhood $N(x_0)$ of x_0 such that

$$\frac{f(t_1) - f(x)}{t_1 - x} > \alpha \text{ for } x \in E \bigcap N(x_0)$$

and hence

$$\sup \left\{ \frac{f(t) - f(x)}{t - x} : x < t < x + \frac{1}{m} \right\} > \alpha \text{ for } x \in E \bigcap N(x_0).$$

So, from (6.4) $H_m(f; x) > \alpha$ for $x \in E \bigcap N(x_0)$. Therefore, from (6.6), if $x \in E \bigcap N(x_0)$ then $x \in A_m$. Hence

$$E \bigcap N(x_0) = E \bigcap N(x_0) \bigcap A_m.$$

Since E is measurable, $E \cap A_m$ is measurable. Hence from (6.6), $H_m(f; x)$ is measurable on E. We thus proved that if f is constant on a measurable set $E \subset I$ then $H_m(f; x)$ is measurable on E.

By Theorem 6.9 there exists a sequence of measurable functions $\{f_n\}$ on I such that for each n there is a countable collection $\{E_{n,i} : i = 0, \pm 1, \pm 2, ...\}$ of disjoint measurable sets $\{E_{n,i}\}$ such that $I = \bigcup_{i=-\infty}^{\infty} E_{n,i}$ and f is constant on $\{E_{n,i}\}$ for each i and $\{f_n\}$ converges to f uniformly on I. Applying the above argument on $\{f_n\}$, $H_m(f_n; x)$ is measurable on E_{ni} for each i, $i = 0, \pm 1, \pm 2, \ldots$. and so $H_m(f_n; x)$ is measurable on $\bigcup_{i=-\infty}^{\infty} E_{ni} = I$. This being true for all n, $\lim_{n \to \infty} H_m(f_n; x)$ is measurable on I. Since $\{f_n\}$ converges to f uniformly on I,

$$\frac{f_n(t) - f_n(x)}{t - x} \to \frac{f(t) - f(x)}{t - x} \text{ as } n \to \infty \text{ uniformly for } t \in \left(x, x + \frac{1}{m} \right)$$

So, for every $\epsilon > 0$ there is N such that

$$\frac{f(t) - f(x)}{t - x} - \epsilon < \frac{f_n(t) - f_n(x)}{t - x} < \frac{f(t) - f(x)}{t - x} + \epsilon$$

for $n \geq N$ and for all $t \in \left(x, x + \frac{1}{m} \right)$. Taking 'sup' over t we have by (6.4)

$$H_m(f; x) - \epsilon \leq H_m(f_n; x) \leq H_m(f; x) + \epsilon \text{ for } n \geq N.$$

Hence taking limit as $n \to \infty$, since ϵ is arbitrary

$$\lim_{n \to \infty} H_m(f_n; x) = H_m(f; x).$$

So, $H_m(f; x)$ is measurable. This being true for all m, applying (6.5) $D^+ f$ is measurable. $\square$

6.3 Measurablity of Dini Derivates

Example 6.1 Let $\theta : [0, 1] \to [0, 1]$ be the Cantor function and let P be the Cantor ternary set (see Sect. 4.9 of Chap. 4 for the definition of Cantor function). Then show that for every open interval I_k^n which are removed from $[0, 1]$ to construct P

(i) $\theta'(x) = 0$ whenever $x \in I_k^n$ for all n and all k,
(ii) $D^-\theta(x) = 0 = D_-\theta(x)$, $D^+\theta(x) = \infty = D_+\theta(x)$, if x is the right-hand endpoint of I_k^n for all n and all k,
(iii) $D^+\theta(x) = 0 = D_+\theta(x)$, $D^-\theta(x) = \infty = D_-\theta(x)$, if x is the left-hand endpoint of I_k^n for all n and all k,
(iv) $D^-\theta(x) = \infty = D^+\theta(x)$ if $x \in P$ but x is neither a right-hand endpoint nor a left-hand endpoint of any I_k^n.

Solution Since θ is constant on every I_k^n, (i) is obvious. For (ii) let x be the right-hand endpoint of $I_{k_0}^{n_0}$, $1 \leq k_0 \leq 2^{n_0-1}$. Then $\theta(t) = \frac{2k_0-1}{2^{n_0}}$ if t belongs to the closure of $I_{k_0}^{n_0}$ and so the first part of (ii) is clear. To prove the last part of (ii) it may be noted that $I_{k_0}^{n_0}$ is removed in the n_0 th step and so its length is $\frac{1}{3^{n_0}}$. There is an open interval $I_{k_1}^{n_0+1}$ obtained in the $(n_0 + 1)$th step such that $I_{k_1}^{n_0+1}$ lies in the right of $I_{k_0}^{n_0}$ and between $I_{k_0}^{n_0}$ and $I_{k_0+1}^{n_0}$, the length of $I_{k_1}^{n_0+1}$ being $\frac{1}{3^{n_0+1}}$ and the distance between $I_{k_0}^{n_0}$ and $I_{k_1}^{n_0+1}$ is $\frac{1}{3^{n_0+1}}$. In the $(n_0 + 2)$th step we get another interval $I_{k_2}^{n_0+2}$ in the right of $I_{k_0}^{n_0}$ and lying between $I_{k_0}^{n_0}$ and $I_{k_1}^{n_0+1}$ such that the length of $I_{k_2}^{n_0+2}$ is $\frac{1}{3^{n_0+2}}$ and the distance between $I_{k_0}^{n_0}$ and $I_{k_2}^{n_0+2}$ is also $\frac{1}{3^{n_0+2}}$. In the $(n_0 + r)$ th step we get the interval $I_{k_r}^{n_0+r}$, $1 \leq k_r \leq 2^{n_0+r-1}$, such that the length of $I_{k_r}^{n_0+r}$ and the distance between $I_{k_0}^{n_0}$ and $I_{k_r}^{n_0+r}$ are $\frac{1}{3^{n_0+r}}$. Thus we get the sequence $\{I_{k_r}^{n_0+r} : r = 1, 2, \ldots\}$ of open intervals from the collection $\{I_k^n : 1 \leq k \leq 2^{n-1}; n = 1, 2, \ldots\}$ such that the length of $I_{k_r}^{n_0+r}$ is $\frac{1}{3^{n_0+r}}$ and this sequence converges to x from the right. Note that the difference of value of θ on $I_{k_0}^{n_0}$ and on $I_{k_0+1}^{n_0}$ is $\frac{1}{2^{n_0-1}}$ and so the difference of value of θ on $I_{k_0}^{n_0}$ and on $I_{k_1}^{n_0+1}$ is $\frac{1}{2^{n_0}}$ and the difference between the values of θ on $I_{k_0}^{n_0}$ and $I_{k_2}^{n_0+2}$ is $\frac{1}{2^{n_0+1}}$. The difference between the values of θ on $I_{k_0}^{n_0}$ and $I_{k_r}^{n_0+r}$ is $\frac{1}{2^{n_0+r-1}}$. Let $I_{k_r}^{n_0+r} = (a_r, b_r)$. Then $a_r - x =$ distance between $I_{k_0}^{n_0}$ and $I_{k_r}^{n_0+r} = \frac{1}{3^{n_0+r}}$. So, θ being continuous

$$\frac{\theta(a_r) - \theta(x)}{a_r - x} = \frac{1}{2^{n_0+r-1}} \cdot 3^{n_0+r} = 2\left(\frac{3}{2}\right)^{n_0+r} \to \infty \text{ as } r \to \infty \quad (6.7)$$

Also since $b_r - a_r = \frac{1}{3^{n_0+r}}$, $b_r - x = \frac{2}{3^{n_0+r}}$ and so

$$\frac{\theta(b_r) - \theta(x)}{b_r - x} = \frac{1}{2^{n_0+r-1}} \frac{3^{n_0+r}}{2} = \left(\frac{3}{2}\right)^{n_0+r} \to \infty \text{ as } r \to \infty \quad (6.8)$$

Let $\{t_i\}$ be any sequence of points which converges to x from the right. We show that

$$\frac{\theta(t_i) - \theta(x)}{t_i - x} \to \infty \text{ as } i \to \infty \quad (6.9)$$

146 6 Differentiation of Functions

We consider two cases:

Case(i) Let $\{t_i\}$ converge to x through $G = \bigcup_{n=1}^{\infty} \bigcup_{k=1}^{2^n-1} I_k^n$. Then $t_i \in G$ for all i and so for each i there is r such that $t_i \in I_{k_r}^{n_0+r} = (a_r, b_r)$. Then $a_r - x < t_i - x < b_r - x$. Since θ is continuous $\theta(t_i) = \theta(a_r) = \theta(b_r)$. Also $a_r - x = \frac{1}{3^{n_0+r}}$ and $b_r - x = \frac{2}{3^{n_0+r}}$ and so

$$\frac{\theta(t_i) - \theta(x)}{t_i - x} = \frac{\theta(a_r) - \theta(x)}{a_r - x} \cdot \frac{a_r - x}{b_r - x} \cdot \frac{b_r - x}{t_i - x} > \frac{\theta(a_r) - \theta(x)}{a_r - x} \cdot \frac{1}{2} \cdot 1 \quad (6.10)$$

Since $r \to \infty$ as $i \to \infty$, letting $i \to \infty$, we get (6.9) from (6.10) and (6.7)

Case(ii) Let $\{t_i\}$ converges to x through $P = [0, 1] \sim G$. Then $t_i \in P$ for all i and so for each i there is r such that t_i lies between two consecutive intervals $I_{k_{r+1}}^{n_0+r+1} = (a_{r+1}, b_{r+1})$ and $I_{k_r}^{n_0+r} = (a_r, b_r)$ and so $t_i \in (b_{r+1}, a_r)$. Then $b_{r+1} - x < t_i - x < a_r - x$. Since θ is non decreasing, $\theta(b_{r+1}) \le \theta(t_i)$ and since $b_{r+1} - x = \frac{2}{3^{n_0+r+1}}$ and $a_r - x = \frac{1}{3^{n_0+r}}$,

$$\frac{\theta(t_i) - \theta(x)}{t_i - x} \ge \frac{\theta(b_{r+1}) - \theta(x)}{b_{r+1} - x} \cdot \frac{b_{r+1} - x}{a_r - x} \cdot \frac{a_r - x}{t_i - x} > \frac{\theta(b_{r+1}) - \theta(x)}{b_{r+1} - x} \cdot \frac{2}{3} \cdot 1 \quad (6.11)$$

Since $r \to \infty$, as $i \to \infty$ letting $i \to \infty$ we get (6.9) form (6.11) and (6.8).

Now (6.9) being true for all sequences $\{t_i\}$ which converge to x from the right we have applying Corollary 6.2, $D^+\theta(x) = \infty = D_+\theta(x)$; completing the proof of (ii)

Case(iii) The proof of (iii) is similar.

Case(iv) Before proving (iv) note that if $x \in P$ then using ternary scale $x = .a_1 a_2 a_3 \ldots$ where $a_i = 0$ or 2 and using binary scale $\theta(x) = .b_1 b_2 b_3 \ldots$ where $b_i = 0$ or 1 for all i. Then $b_i = \frac{a_i}{2}$ for all i. In fact consider for simplicity $x = \frac{2}{9}$, the right-hand endpoint of I_1^2. Then $x = .0200\ldots$ (in ternary scale) and $\theta(x) = \frac{1}{4} = .0100\ldots$ (in binary scale).

Now we come to prove (iv). Let $x \in P$ and x is neither a right-hand endpoint nor a left-hand endpoint of any removed open interval. Then using respectively the ternary and binary scale $x = .a_1 a_2 a_3 \ldots = \sum_{k=1}^{\infty} \frac{a_k}{3^k}$ and $\theta(x) = .b_1 b_2 b_3 \ldots = \sum_{k=1}^{\infty} \frac{b_k}{2^k} = \sum_{k=1}^{\infty} \frac{a_k}{2^{k+1}}$.

Let $\beta_i = \sum_{k=1}^{i} \frac{a_k}{3^k}$. Then β_i is the right-hand endpoint of a sequence of open intervals (α_i, β_i) obtained from the collection $\{I_k^n : 1 \le k \le 2^{n-1}; n = 1, 2, \ldots\}$ which converges to x from the left. So, $\theta(\beta_i) = \sum_{k=1}^{i} \frac{a_k}{2^{k+1}}$ and hence

$$\theta(x) - \theta(\beta_i) = \sum_{k=i+1}^{\infty} \frac{a_k}{2^{k+1}} \quad \text{and} \quad x - \beta_i = \sum_{k=i+1}^{\infty} \frac{a_k}{3^k}.$$

6.3 Measurablity of Dini Derivates

Therefore

$$\frac{\theta(x) - \theta(\beta_i)}{x - \beta_i} = \frac{\sum_{k=i+1}^{\infty} \frac{a_k}{2^{k+1}}}{\sum_{k=i+1}^{\infty} \frac{a_k}{3^k}},$$

Let N be the smallest integer $k \geq i+1$ such that $a_k \neq 0$.
Then

$$\frac{\theta(x) - \theta(\beta_i)}{x - \beta_i} \geq \frac{\frac{1}{2^N}}{2 \sum_{k=N}^{\infty} \frac{1}{3^k}} = \frac{3^{N-1}}{2^N}$$

since $N \to \infty$ as $i \to \infty$, letting $i \to \infty$, $D^-\theta(x) = \infty$. To prove $D^+\theta(x) = \infty$ note that $x \in P$ if and only if $1 - x \in P$ and so from above $D^-\theta(1-x) = \infty$. Also $\theta(x) = 1 - \theta(1-x)$ and so

$$D^+\theta(x) = \limsup_{h \to 0+} \frac{\theta(x+h) - \theta(x)}{h} = \limsup_{h \to 0+} \frac{\theta(1-x-h) - \theta(1-x)}{-h} = D^-\theta(1-x) = \infty.$$

This completes the proof.

Example 6.2 Let the function ϕ be defined by

$$\phi(x) = ax \sin \frac{1}{x} + bx \cos \frac{1}{x} \text{ for } x \neq 0$$
$$= 0 \text{ for } x = 0.$$

Find $D^+\phi(0), D_+\phi(0), D^-\phi(0)$ and $D_-\phi(0)$, where $0 \leq a, b < \infty, (a, b) \neq (0, 0)$.

Solution Consider $f(x) = a \sin x + b \cos x$. Then $f'(x) = a \cos x - b \sin x$ and $f''(x) = -(a \sin x + b \cos x)$. So, $f'(x) = 0$ implies that $\tan x = \frac{a}{b}$ i.e. $x = \pm n\pi + \tan^{-1}(\frac{a}{b})$. Writing $x_0 = \pm n\pi + \tan^{-1}(\frac{a}{b})$

$$f''(x_0) = -(a \sin x_0 + b \cos x_0)$$
$$= -\left[a \sin\left(\pm n\pi + \tan^{-1}\left(\frac{a}{b}\right)\right) + b \cos\left(\pm n\pi + \tan^{-1}\left(\frac{a}{b}\right)\right)\right]$$

So

$$f''(x_0) = -\left[a \sin \tan^{-1}\left(\frac{a}{b}\right) + b \cos \tan^{-1}\left(\frac{a}{b}\right)\right] \text{ if } n \text{ is even}$$
$$= a \sin \tan^{-1}\left(\frac{a}{b}\right) + b \cos \tan^{-1}\left(\frac{a}{b}\right) \text{ if } n \text{ is odd}$$

Since $\tan^{-1}(\frac{a}{b}) \in [0, \frac{\pi}{2}]$, $f''(x_0) < 0$ if n is even and $f''(x_0) > 0$ if n is odd. Therefore f is maximum at $x_1 = \pm 2n\pi + \tan^{-1}(\frac{a}{b})$ and minimum at $x_2 = \pm(2n+1)\pi + \tan^{-1}(\frac{a}{b})$. Therefore, since $\tan^{-1}(\frac{a}{b}) \in [0, \frac{\pi}{2}]$, considering the interval $[2n\pi, (2n+2)\pi]$ f is maximum at $x_1 = 2n\pi + \tan^{-1}(\frac{a}{b})$ and minimum at $x_2 = (2n+1)\pi + \tan^{-1}(\frac{a}{b})$, the maximum and minimum value being respectively

$$f(x_1) = a \sin \tan^{-1}(\frac{a}{b}) + b \cos \tan^{-1}(\frac{a}{b}), \text{ and}$$

$$f(x_2) = -a \sin \tan^{-1}(\frac{a}{b}) - b \cos \tan^{-1}(\frac{a}{b}).$$

Now we consider the problem. Here

$$\sup_{0<x<\delta} \frac{\phi(x) - \phi(0)}{x - 0} = \sup_{0<x<\delta} \left(a \sin \frac{1}{x} + b \cos \frac{1}{x} \right) \quad (6.12)$$

Since $\sin \frac{1}{x}$ and $\cos \frac{1}{x}$ take all their values in any interval $J_n = \left(\frac{1}{(2n+2)\pi}, \frac{1}{2n\pi} \right]$, to determine (6.12) it is enough to take one interval J_n when $n > \frac{1}{2\delta\pi}$. So, supposing $n > \frac{1}{2\delta\pi}$,

$$\sup_{0<x<\delta} \left(a \sin \frac{1}{x} + b \cos \frac{1}{x} \right) = \sup_{x \in J_n} \left(a \sin \frac{1}{x} + b \cos \frac{1}{x} \right)$$

$$= \sup_{t \in [2n\pi, (2n+2)\pi]} (a \sin t + b \cos t)$$

$$= a \sin \tan^{-1}\left(\frac{a}{b}\right) + b \cos \tan^{-1}\left(\frac{a}{b}\right).$$

Hence from (6.12)

$$D^+\phi(0) = a \sin \tan^{-1}\left(\frac{a}{b}\right) + b \cos \tan^{-1}\left(\frac{a}{b}\right).$$

Similarly $D_+\phi(0) = -a \sin \tan^{-1}(\frac{a}{b}) - b \cos \tan^{-1}(\frac{a}{b})$. The proof for $D^-\phi(0)$ and $D_-\phi(0)$ are similar.

6.4 Differentiability of Monotone Functions

Theorem 6.11 *If f is monotone in $[a,b]$ then f has finite derivative almost everywhere in $[a,b]$.*

Proof Let $f : [a,b] \to \mathbb{R}$ be non-decreasing. Let

$$E = \{x \in (a,b); D_- f(x) < D^+ f(x)\}.$$

6.4 Differentiability of Monotone Functions

For each pair of rational numbers $u, v, u < v$, let

$$E_{uv} = \{x \in (a,b); D_-f(x) < u < v < D^+f(x)\}.$$

Then

$$E = \bigcup E_{uv} \qquad (6.13)$$

where the union is taken over all pair of rational numbers u, v. If possible, suppose $\mu^*(E_{uv}) = \alpha > 0$. Since f is non-decreasing, $D_-f(x) \geq 0$ for all x and hence $0 < u < v$. Choose $\epsilon, 0 < \epsilon < \frac{\alpha(v-u)}{u+2v}$. Let G be an open set such that $E_{uv} \subset G$ and $\mu(G) < \alpha + \epsilon$. Then for each $x \in E_{uv}$ there are intervals $[x-h, x]$ of arbitrary small length such that $[x-h, x] \subset G$ and

$$f(x) - f(x-h) < uh \qquad (6.14)$$

So, E_{uv} is covered by such intervals in the sense of Vitali (see Definition 2.4 of Chap. 2. Hence by Valali's Covering Theorem (Theorem 2.27) there is a finite collection of disjoint intervals $\{I_1, I_2, \ldots I_N\}$ where

$$I_n = [x_n - h_n, x_n], \; n = 1, 2, \ldots, N \qquad (6.15)$$

such that $\mu^*(E_{uv} \sim \bigcup_{n=1}^{N} I_n) < \epsilon$. Let $A = E_{uv} \cap (\bigcup_{n=1}^{N} I_n)$ then since $E_{uv} = A \bigcup (E_{uv} \sim \bigcup_{n=1}^{N} I_n)$,

$$\alpha = \mu^*(E_{uv}) \leq \mu^*(A) + \mu^*(E_{uv} \sim \bigcup_{n=1}^{N} I_n) < \mu^*(A) + \epsilon$$

and hence $\mu^*(A) > \alpha - \epsilon$. Since each interval in (6.15) satisfies (6.14) summing over the intervals in (6.15)

$$\sum_{n=1}^{N} [f(x_n) - f(x_n - h_n)] < u \sum_{n=1}^{N} h_n < u\mu(G) < u(\alpha + \epsilon) \qquad (6.16)$$

Again for each $\xi \in A$ there are intervals $[\xi, \xi + k]$ of arbitrary small length which are contained in some I_n, $n = 1, 2, \ldots, N$, such that

$$f(\xi + k) - f(\xi) > vk. \qquad (6.17)$$

So, A is covered by these intervals in the sense of Vitali and hence by Vitali's Covering Theorem (see Theorem 2.27) there is a finite collection of disjoint intervals $\{J_1, J_2, \ldots J_M\}$ where

$$J_i = [\xi_i, \xi_i + k_i], \quad i = 1, 2, \ldots, M. \tag{6.18}$$

such that $\mu^*(A \sim \bigcup_{i=1}^{M} J_i) < \epsilon$. Let $B = A \cap \bigcup_{i=1}^{M} J_i$. Then since $A = B \bigcup (A \sim \bigcup_{i=1}^{M} J_i)$,

$$\alpha - \epsilon < \mu^*(A) \le \mu^*(B) + \mu^*(A \sim \bigcup_{i=1}^{M} J_i) < \mu^*(B) + \epsilon$$

and so $\mu^*(B) > \alpha - 2\epsilon$. Since each interval in (6.19) satisfies (6.17), summing over the intervals in (6.19)

$$\sum_{i=1}^{M} [f(\xi_i + k_i) - f(\xi_i)] > v \sum_{i=1}^{M} k_i > v \mu^*(B) > v(\alpha - 2\epsilon). \tag{6.19}$$

Since f is non-decreasing, if $J_i \subset I_n$ then $f(\xi_i + k_i) - f(\xi_i) \le f(x_n) - f(x_n - h_n)$. Since each J_i is contained in some I_n, summing over those i for which $J_i \subset I_n$.

$$\sum [f(\xi_i + k_i) - f(\xi_i)] \le f(x_n) - f(x_n - h_n).$$

Hence

$$\sum_{i=1}^{M} [f(\xi_i + k_i) - f(\xi_i)] \le \sum_{n=1}^{N} [f(x_n) - f(x_n - h_n)]$$

and so from (6.16) and (6.19)

$$v(\alpha - 2\epsilon) < u(\alpha + \epsilon)$$

which is a contradiction to our choice of ϵ. So, $\mu^*(E_{uv}) = 0$ for each pair of u, v. So, by (6.13) $\mu^*(E) = 0$.

Similarly the set

$$F = \{x \in [a, b] : D_+ f(x) < D^- f(x)\}$$

is also such that $\mu^*(F) = 0$. Hence

$$D_+ f(x) = D^+ f(x) = D^- f(x) = D_- f(x) \text{ almost everywhere in } (a, b).$$

Thus f' exists a.e. in $[a, b]$. It remains to show that the set

$$G = \{x \in (a, b) : f'(x) = \infty\}$$

6.4 Differentiability of Monotone Functions

is of measure zero. If possible, suppose that $\mu^*(G) = \beta > 0$. Choose N such that $N > \frac{2[f(b)-f(a)]}{\beta}$. Then for each $x \in G$ there are intervals $[x, x+h]$ of arbitrary small length such that $[x, x+h] \subset (a, b)$ and

$$f(x+h) - f(x) > Nh \tag{6.20}$$

So, G is covered by such intervals in the sense of Vitali, and so by Vitali's Covering Theorem there exists a finite collection of disjoint intervals $\{I_1, I_2, \ldots, I_p\}$ where

$$I_n = [x_n, x_n + h_n], \quad n = 1, 2, \ldots, p$$

such that $\mu^*(G \sim \bigcup_{n=1}^{p} I_n) < \frac{1}{2}\beta$. Let $B = G \cap (\bigcup_{n=1}^{p} I_n)$. Then since $G = B \cup (G \sim \bigcup_{n=1}^{p} I_n)$.

$$\beta = \mu^*(G) \leq \mu^*(B) + \mu^*(G \sim \bigcup_{n=1}^{p} I_n) < \mu^*(B) + \frac{1}{2}\beta$$

and so $\mu^*(B) > \frac{1}{2}\beta$. Hence by (6.20)

$$\frac{1}{2}\beta N < N\mu^*(B) < N \sum_{n=1}^{p} h_n < \sum_{n=1}^{p} [f(x_n + h_n) - f(x_n)] \leq f(b) - f(a).$$

But this contradicts the choice of N. So, $\mu^*(G) = 0$. This completes the proof of the theorem when f is non-decreasing. When f is non-increasing the proof follows by considering $-f$. □

Theorem 6.12 *If f is monotone in $[a, b]$ then its derivative f' is Lebesgue integrable in $[a, b]$ and*

$$0 \leq \int_a^b f' \leq f(b) - f(a) \text{ or } f(b) - f(a) \leq \int_a^b f' \leq 0$$

according as f is non-decreasing or non-increasing.

Proof Let f be non-decreasing in $[a, b]$. Define $f(x) = f(b)$ for $x > b$ and set

$$g_n(x) = \frac{f(x + \frac{1}{n}) - f(x)}{\frac{1}{n}}, \quad n = 1, 2, \ldots, x \in [a, b].$$

Then $\{g_n\}$ is a sequence of non-negative measurable functions and by Theorem 6.11 $\lim_{n \to \infty} g_n(x) = f'(x)$ a.e. in $[a, b]$ and so by Theorem 5.27

$$\int_a^b f' \le \liminf_{n\to\infty} \int_a^b g_n = \liminf_{n\to\infty} n \int_a^b \left[f(x+\tfrac{1}{n}) - f(x) \right]$$

$$= \liminf_{n\to\infty} \left[n \int_b^{b+\frac{1}{n}} f(t) - n \int_a^{a+\frac{1}{n}} f(t) \right].$$

$$\le \liminf_{n\to\infty} \left[n \int_b^{b+\frac{1}{n}} f(b) - n \int_a^{a+\frac{1}{n}} f(a) \right] = f(b) - f(a).$$

Since f is non-decreasing $f' \ge 0$ a.e. in $[a,b]$ and so $\int_a^b f' \ge 0$. So, the theorem is proved when f is non-decreasing. If f is non-increasing the proof follows by considering $-f$. □

Remarks

(i) Theorem 6.12 is not true if the closed interval $[a,b]$ is replaced by the open inter (a,b). For, consider the function $f(x) = \log x$ for $x \in (0,1)$. Then f is non-decreasing in $(0,1)$ and $f'(x) = \tfrac{1}{x}$ for $x \in (0,1)$. But f' is not integrable in $(0,1)$. However, Theorem 6.11 remains true if $[a,b]$ is replaced by (a,b).

(ii) Theorem 6.12 is not true if Lebesgue integrability is replaced by Riemann integrability. That is there exists a monotone function on $[a,b]$ whose derivatives exist a.e. in $[a,b]$ but is not Riemann integrable in $[a,b]$. For, consider the function f introduced by Voltera which is such that f has bounded derivative f' but f' is not Riemann integrable in $[a,b]$ (see Theorem 5.23). Since f' is bounded there exists M such that $f'(x) \ge M$ for all $x \in [a,b]$. Let $\phi(x) = f(x) - Mx$. Then since $\phi' > 0$, ϕ is non-decreasing in $[a,b]$. But since f' is not Riemann integrable, $\phi' = f - M$ is not Riemann integrable.

(iii) The inequality $\int_a^b f' \le f(b) - f(a)$ in Theorem 6.12 may be strict inequality. for consider the Cantor function θ on $[0,1]$, which is non-decreasing and $\theta' = 0$ a.e on $[0,1]$ and $\theta(0) = 0$, $\theta(1) = 1$, so $\int_0^1 \theta' = 0 < 1 = \theta(1) - \theta(0)$.

Theorem 6.13 *Let $\{f_n\}$ be a sequence of non-decreasing functions defined in $[a,b]$ and let $\sum_{n=1}^{\infty} f_n$ converges to a finite function f in $[a,b]$. Then*

$$\left(\sum_{n=1}^{\infty} f_n(x) \right)' = f'(x) = \sum_{n=1}^{\infty} f_n'(x) \, a.e. \, on \, [a,b]$$

Proof Since each f_n is non-decreasing, f is non-decreasing. We may suppose that each f_n (and hence f) is non-negative. For, consider the sequence $\{f_n(x) - f_n(a)\}$ instead of $\{f_n\}$. Let

6.4 Differentiability of Monotone Functions

$$S_n(x) = \sum_{k=1}^{n} f_k(x).$$

Then each S_n is non-decreasing and non-negative. Also if $h > 0$ then

$$\frac{S_n(x+h) - S_n(x)}{h} - \frac{f(x+h) - f(x)}{h} = \frac{1}{h}\left[\sum_{k=n+1}^{\infty} f_k(x) - \sum_{k=n+1}^{\infty} f_k(x+h)\right]$$

$$= \frac{1}{h}\sum_{k=n+1}^{\infty}[f_k(x) - f_k(x+h)] \leq 0.$$

Since S_n and f are non-decreasing, S_n' and f' exist a.e. by Theorem 6.11. So, letting $h \to 0$

$$S_n'(x) \leq f'(x) \, a.e. \text{ in } [a, b]. \tag{6.21}$$

Since $S_{n+1}'(x) = S_n'(x) + f_{n+1}'(x) \geq S_n'(x)$, the sequence $\{S_n'(x)\}$ is non-decreasing and by (6.21) it is bounded for almost all $x \in [a, b]$ and hence letting $n \to \infty$ in (6.21)

$$\sum_{n=1}^{\infty} f_n'(x) \leq \left(\sum_{n=1}^{\infty} f_n(x)\right)' = f'(x) < \infty \, a.e. \text{ in } [a, b]. \tag{6.22}$$

Since $\lim_{n \to \infty} S_n(b) = f(b)$, for each j there is n_j such that $|f(b) - S_{n_j}(b)| < 2^{-j}$ and hence

$$\sum_{j=1}^{\infty}[f(b) - S_{n_j}(b)] < \infty \tag{6.23}$$

Now for $a \leq x < y \leq b$,

$$0 \leq f(x) - S_{n_j}(x) = \sum_{k=n_j+1}^{\infty} f_k(x) \leq \sum_{k=n_j+1}^{\infty} f_k(y) = f(y) - S_{nj}(y).$$

So, each member $f - S_{n_j}$ of the sequence $\{f - S_{nj}\}$, $j = 1, 2, \ldots$, is non-decreasing.

Hence by (6.23)

$$\sum_{J=1}^{\infty}[f(x) - S_{n_j}(x)] < \infty.$$

Therefore the sequence $\{f - S_{n_j}\}$, $j = 1, 2, \ldots$, satisfies the hypotheses of the theorem. So, just as we obtain (6.22) for the sequence $\{f_n\}$, we get for the sequence $\{f - S_{n_j}\}$ analogous to (6.22) the following relation :

$$\sum_{j=1}^{\infty}[f'(x)-S'_{n_j}(x)] \le \left(\sum_{J=1}^{\infty}[f(x)-S_{n_j}(x)]\right)' < \infty \text{ a.e. in } [a,b] \quad (6.24)$$

Hence
$$\lim_{j\to\infty} S'_{n_j}(x) = f'(x) \text{ a.e.}$$

So, a subsequence of the sequence $\{S'_n(x)\}$ converges to $f'(x)$ a.e. in $[a,b]$. Since the sequence $\{S'_n(x)\}$ is non-decreasing $\lim_{n\to\infty} S'_n(x) = f'(x)$ a.e. in $[a,b]$, showing that $\sum_{n=1}^{\infty} f'_n(x) = f'(x)$ a.e. in $[a,b)$. □

6.5 Functions of Bounded Variation and Their Properties

Definition 6.1 Let f be defined in $[a,b]$ and let $D: a = x_0 < x_1 < \cdots < x_n = b$ be any partition of $[a,b]$. Let

$$V_D(f;a,b) = \sum_{r=1}^{n} |f(x_r) - f(x_{r-1})|.$$

Then $V_D(f;a,b)$ is called the variation of f on $[a,b]$ corresponding to D. Let

$$V(f;a,b) = \sup_D V_D(f;a,b)$$

where the supremom is taken over all partition D of $[a,b]$. Then $V(f;a,b)$ is called the total variation of f on $[a,b]$. If $V(f;a,b)$ is finite then f is called a function of bounded variation or a function of finite variation in $[a,b]$.

Theorem 6.14 *If f is of bounded variation then f is bounded.*

Proof Let f be of bounded variation in $[a,b]$. If $x \in (a,b)$ then $D: a < x < b$ is a partition of $[a,b]$ and so

$$|f(x)| - |f(a)| \le |f(x) - f(a)| \le |f(x) - f(a)| + |f(b) - f(x)|$$
$$= V_D(f;a,b) \le V(f;a,b)$$

which gives
$$|f(x)| \le V(f;a,b) + |f(a)| + |f(b)|,$$

and if $x = a$ or b then this relation is also true. □

6.5 Functions of Bounded Variation and Their Properties 155

Theorem 6.15 *If f is monotone in $[a, b]$ then f is of bounded variation in $[a, b]$.*

Proof Let $D : a = x_0 < x_1 < \cdots < x_n = b$ be any partition of $[a, b]$. Then $V_D(f; a, b) = \sum_{r=1}^{n} |f(x_r) - f(x_{r-1})|$ and so $V_D(f; a, b) = \sum_{r=1}^{n} [f(x_r) - f(x_{r-1})] = f(b) - f(a)$ when f is non-decreasing and $V_D(f; a, b) = \sum_{r=1}^{n} [f(x_{r-1}) - f(x_r)] = f(a) - f(b)$ when f is non-increasing. So $V(f; a, b)$ is $f(b) - f(a)$ or $f(a) - f(b)$ according as f is non-decreasing or non-increasing. □

Theorem 6.16 *If f and g are of bounded variation in $[a, b]$ and α is any real number then $|f|$, $f + g$, αf and fg are of bounded variation in $[a, b]$. If there is $k > 0$ such that $|f(x)| \geq k$ for all $x \in [a, b]$ then $\frac{1}{f}$ is of bounded variation, if f is so.*

Proof The proof for $|f|$ is easy. For any partition $D : a = x_0 < x_1 < \cdots < x_n = b$. of $[a, b]$

$$V_D(f + g; a, b) = \sum_{r=1}^{n} |f(x_r) + g(x_r) - f(x_{r-1}) - g(x_{r-1})|$$

$$\leq \sum_{r=1}^{n} |f(x_r) - f(x_{r-1})| + \sum_{r=1}^{n} |g(x_r) - g(x_{r-1})|$$

$$= V_D(f; a, b) + V_D(g; a, b) \leq V(f; a, b) + V(g; a, b) \quad (6.25)$$

and

$$V_D(\alpha f; a, b) = |\alpha| \sum_{r=1}^{n} |f(x_r) - f(x_{r-1})| = |\alpha| V_D(f; a, b) \leq |\alpha| V(f; a, b) \quad (6.26)$$

Also by Theorem 6.14 f and g are bounded and so there are M_1 and M_2 such that $|f(x)| \leq M_1$ and $|g(x)| \leq M_2$ for all $x \in [a, b]$ and so

$$V_D(fg; a, b) = \sum_{r=1}^{n} |f(x_r)g(x_r) - f(x_{r-1})g(x_{r-1})|$$

$$\leq \sum_{r=1}^{n} |g(x_r)| |f(x_r) - f(x_{r-1})| + \sum_{r=1}^{n} |f(x_{r-1})| |g(x_r) - g(x_{r-1})|$$

$$\leq M_2 V_D(f; a, b) + M_1 V_D(g; a, b)$$

$$\leq M_2 V(f; a, b) + M_1 V(g; a, b) \quad (6.27)$$

Finally, since $\frac{1}{|f(x)|} \leq \frac{1}{k}$ for all $x \in [a, b]$,

$$V_D\left(\frac{1}{f};a,b\right) = \sum_{r=1}^{n}\left|\frac{1}{f(x_r)} - \frac{1}{f(x_{r-1})}\right| = \sum_{r=1}^{n}\frac{|f(x_r) - f(x_{r-1})|}{|f(x_{r-1})f(x_r)|}$$

$$\leq \frac{1}{k^2}\sum_{r=1}^{n}|f(x_r) - f(x_{r-1})|$$

$$= \frac{1}{k^2}V_D(f;a,b)$$

So

$$V_D\left(\frac{1}{f};a,b\right) \leq \frac{1}{k^2}V(f;a,b) \tag{6.28}$$

The proof of the first part of the theorem follows from (6.25), (6.26) and (6.27) while the last part follows from (6.28) □

Remark The existence of $k > 0$ in the last part of the above theorem is necessary. For let $f(x) = x$ for $0 < x \leq 1$ and $f(0) = 1$. Then $\frac{1}{f(x)} = \frac{1}{x}$ for $0 < x \leq 1$ and $\frac{1}{f(0)} = 1$. Clearly f is of bounded variation in $[0, 1]$ but $\frac{1}{f}$ is not.

Theorem 6.17 *Let $a < c < b$. Then f is of bounded variation in $[a, b]$ if and only if f is of bounded variation in both $[a, c]$ and $[c, b]$. In this case.*

$$V(f;a,b) = V(f;a,c) + V(f;c,b)$$

Proof Let f be of bounded variation in $[a, b]$ and let $D_1 : a = x_0 < x_1 < \cdots < x_n = c$ be any partition of $[a, c]$. Then since $D : a = x_0 < x_1 < \cdots < x_n = c < b$ is a partition of $[a, b]$ and $V_D(f;a,b) = V_{D_1}(f;a,c) + |f(b) - f(c)|$, $V(f;a,b) \geq V_{D_1}(f;a,c)$ showing that $V(f;a,b) \geq V(f;ac)$. Similarly $V(f;a,b) \geq V(f;c,b)$. Conversely, let f be of bounded variation in $[a, c]$ and in $[c, b]$ and let $D : a = x_0 < \cdots < x_n = b$ be any partition of $[a, b]$. Let $x_{i-1} \leq c < x_i$. Then $D_1 : a = x_0 < x_1 < \cdots < x_{i-1} \leq c$ is a partition of $[a, c]$ and $D_2 : c < x_i < \cdots < x_n = b$ is a partition of $[c, b]$. Hence $V_D(f;a,b) = V_{D_1}(f;a,c) + V_{D_2}(f;c,b) \leq V(f;a,c) + V(f;c,b)$. Since D is any partition

$$V(f;a,b) \leq V(f;a,c) + V(f;c,b) \tag{6.29}$$

So, f is of bounded variation in $[a, b]$, proving the first part.

For the last part let $\epsilon > 0$ be arbitrary. Then there exist partitions $D_1 : a = x_0 < x_1 < \cdots < x_m = c$ of $[a, c]$ and $D_2 : c = y_0 < y_1 < \cdots < y_n = b$ of $[c, b]$ such that

$$V_{D_1}(f;a,c) > V(f;a,c) - \frac{1}{2}\epsilon \text{ and } V_{D_2}(f;c,b) > V(f;c,b) - \frac{1}{2}\epsilon$$

Since D_1 and D_2 together gives a partition D of $[a, b]$

$$V(f;a,b) \geq V_D(f;a,b) = V_{D_1}(f;a,c) + V_{D_2}(f;c,b) > V(f;a,c) + V(f;c,b) - \epsilon.$$

6.5 Functions of Bounded Variation and Their Properties

Since ϵ is arbitrary,

$$V(f; a, b) \geq V(f; a, c) + V(f; c\, b) \tag{6.30}$$

The relation (6.29) and (6.30) completes the proof. □

Definition 6.2 Let f be of bounded variation in $[a, b]$. Then the function

$v_f : [a, b] \to \mathbb{R}$ defined by
$v_f(x) = 0$ if $x = a$
$\quad\quad = V(f; a, x)$ if $a < x \leq b$

is called the variation function of f.

Clearly, by Theorem 6.17, the function v_f is well defined and non-decreasing in $[a, b]$.

Theorem 6.18 *A function f is of bounded variation in $[a, b]$ if and only if f can be expressed as the difference of two non-decreasing (or two non-increasing) functions.*

Proof Let f be of bounded variation in $[a, b]$ and let $g = v_f - f$, where v_f is the variation function of f. Then for $a \leq x_1 < x_2 \leq b$,

$$\begin{aligned} g(x_2) - g(x_1) &= v_f(x_2) - f(x_2) - v_f(x_1) + f(x_1) \\ &= V(f; a, x_2) - V(f; a, x_1) - [f(x_2) - f(x_1)] \\ &\geq V(f; x_1, x_2) - |f(x_2) - f(x_1)| \geq 0. \end{aligned}$$

So, g is non-decreasing. Since v_f is non-decreasing in $[a, b]$, f is the difference of two non-decreasing functions. Conversely, if f is the difference of two non-decreasing functions then by Theorems 6.15 and 6.16, f is bounded variation in $[a, b]$. The non-increasing case is clear. □

Theorem 6.19 *A function f is of bounded variation in $[a, b]$ if and only if there is a non-decreasing function ϕ on $[a, b]$ such that*

$$|f(x_2) - f(x_1)| \leq \phi(x_2) - \phi(x_1) \text{ for all } x_1, x_2, a \leq x_1 < x_2 \leq b.$$

Proof Let f be of bounded variation in $[a, b]$. Then taking $\phi = v_f$, where v_f is the variation function of f, we have for $a \leq x_1 < x_2 \leq b$,

$$\begin{aligned} |f(x_2) - f(x_1)| &\leq V(f; x_1, x_2) = V(f; a, x_2) - V(f; a, x_1) \\ &= v_f(x_2) - v_f(x_1) = \phi(x_2) - \phi(x_1). \end{aligned}$$

Conversely, suppose that there is a non-decreasing function ϕ satisfying the given condition. Let $D : a = x_0 < x_1 < \cdots < x_n = b$, be any partition of $[a, b]$. Then

$$V_D(f;a,b) = \sum_{i=1}^{n} |f(x_i) - f(x_{i-1})| \leq \sum_{i=1}^{n} [\phi(x_i) - \phi(x_{i-1})]$$
$$= \phi(b) - \phi(a)$$

and so $V(f;a,b) \leq \phi(b) - \phi(a)$, completing the proof. □

The decomposition of a function of bounded variation as the difference of two non-decreasing functions obtained in Theorem 6.18 is not unique. The following theorem gives a unique decomposition.

Theorem 6.20 *Let f be of bounded variation in $[a,b]$ and let the functions P_f and Q_f be defined by*

$$P_f = \frac{1}{2}[v_f + f - f(a)] \text{ and } Q_f = \frac{1}{2}[v_f - f + f(a)]$$

where v_f is the variation function of f on $[a,b]$. Then P_f and Q_f are non-decreasing in $[a,b]$ and

$$P_f(a) = 0 = Q_f(a), \ P_f - Q_f = f - f(a).$$

Moreover, if g and h are any two non-decreasing functions on $[a,b]$ such that $g(a) = 0 = h(a)$, $g - h = f - f(a)$ then $P_f \leq g$ and $Q_f \leq h$.

Proof Let $a \leq x_1 < x_2 \leq b$. Then

$$P_f(x_2) - P_f(x_1) = \frac{1}{2}[V(f;x_1,x_2) + f(x_2) - f(x_1)] \text{ and}$$

$$Q_f(x_2) - Q_f(x_1) = \frac{1}{2}[V(f;x_1,x_2) - f(x_2) + f(x_1)].$$

Since $V(f;x_1,x_2) \geq |f(x_2) - f(x_1)|$, P_f and Q_f are non-decreasing. The rest is clear.

Let g and h have the stated property. Let $x \in (a,b]$ and let $D : a = x_0 < x_1 < \cdots < x_n = x$ be any partition of $[a,x]$. Then

$$V_D(f;a,x) = \sum_{i=1}^{n} |g(x_i) - g(x_{i-1}) - h(x_i) + h(x_{i-1})| \leq g(x) + h(x)$$

Since this is true for every partition D,

$$P_f(x) - Q_f(x) = v_f(x) \leq g(x) + h(x) \tag{6.31}$$

Also
$$P_f(x) - Q_f(x) = f(x) - f(a) = g(x) - h(x) \tag{6.32}$$

From (6.31) and (6.32) $P_f(x) \leq g(x)$, $Q_f(x) \leq h(x)$ □

6.5 Functions of Bounded Variation and Their Properties

Definition 6.3 If f is of bounded variation in $[a, b]$ and if v_f is the variation function of f then the functions defined by $P_f = \frac{1}{2}[v_f + f - f(a)]$ and $Q_f = \frac{1}{2}[v_f - f + f(a)]$ are called positive and negative variation of f on $[a, b]$ respectively.

Theorem 6.21 Let f be of bounded variation in $[a, b]$. Then f is continuous at $c \in [a, b]$ if and only if the variation functions v_f is continuous at c.

Proof Suppose that f is continuous at c. Let $\epsilon > 0$ be arbitrary. Then there is $\delta > 0$ such that
$$|f(x) - f(c)| < \frac{1}{2}\epsilon \text{ for } |x - c| < \delta. \tag{6.33}$$

Since f is of bounded variation, there is a division $D : c = x_0 < x_1 < \cdots < x_n = b$ of $[c, b]$ such that
$$\sum_{i=1}^{n} |f(x_i) - f(x_{i-1})| > V(f; c, b) - \frac{1}{2}\epsilon. \tag{6.34}$$

Choose x such that $c < x < \min[x_1, c + \delta]$. Since
$$|f(x_1) - f(c)| \leq |f(x) - f(c)| + |f(x_1) - f(x)|,$$

from (6.33) and (6.34)
$$V(f; c, b) - \frac{\epsilon}{2} < |f(x) - f(c)| + |f(x_1) - f(x)| + \sum_{i=2}^{n} |f(x_i) - f(x_{i-1})| < \frac{\epsilon}{2} + V(f; x, b)$$

and so, $0 \leq V(f; c, b) - V(f; x, b) < \epsilon$ which gives
$$0 \leq v_f(b) - v_f(c) - v_f(b) + v_f(x) = v_f(x) - v_f(c) < \epsilon$$

which shows that $\lim_{x \to c+} v_f(x) = v_f(c)$. Similarly $\lim_{x \to c-} v_f(x) = v_f(c)$.

Now suppose that v_f is continuous at $c \in [a, b]$. Then for every $\epsilon > 0$ there is $\delta > 0$ such that
$$|v_f(x) - v_f(c)| < \epsilon \text{ when } |x - c| < \delta.$$

Since $|f(x) - f(c)| \leq |v_f(x) - v_f(c)|$, the proof is complete. □

Theorem 6.22 A function f is a continuous function of bounded variation if and only if f can be expressed as the difference of two continuous non-decreasing functions. The proof follows from Theorems 6.18 and 6.21 and 6.16.

Theorem 6.23 Let $\{f_n\}$ be a sequence of functions of bounded variation in $[a, b]$ which converges to g in $[a, b]$. If there is K such that
$$V(f_n; a, b) \leq K \text{ for all } n$$

then g is of bounded variation in $[a, b]$ and $V(g; a, b) \leq K$.

Proof Let $D : a = x_0 < x_1 < \cdots < x_k = b$ be a partition of $[a, b]$. Then for each n.

$$\sum_{i=1}^{k} |f_n(x_i) - f_n(x_{i-1})| \leq V(f_n; a, b) \leq K.$$

Since $f_n(x_i) \to g(x_i)$ as $n \to \infty$ for $i = 1, 2, \ldots k$, the result follows. □

Remark The condition $V(f_n; a, b) \leq K$ for all n is the Theorem 6.23 is necessary. For, consider.

Example 6.3 Let $r_1, r_2, \ldots r_n, \ldots$ be an enumeration of the set of rational numbers in $[0, 1]$. Define for each n,

$$f_n(x) = 1 \text{ if } x = r_1, r_2, \ldots, r_n$$
$$= 0 \text{ otherwise}.$$

Show that each f_n is of bounded variation in $[0, 1]$ and that the sequence $\{f_n\}$ converges to a function ψ on $[0, 1]$ which is not of bounded variation.

Solution Let $D : x_0 < x_1 < \cdots < x_k = 1$ be any partition of $[0, 1]$. Then for each i, $|f_n(x_i) - f_n(x_{i-1})|$ is either 1 or 0 and so

$$V_D(f_n; 0, 1) = \sum_{i=1}^{k} |f_n(x_i) - f_n(x_{i-1})| \leq n$$

and hence $V(f_n; 0, 1) \leq n$ showing that f_n is of bounded variation in $[0, 1]$.

For the second part let x be any rational number in $[0, 1]$. Then there is m such that $x = r_m$ and so $f_n(x) = f_n(r_m) = 1$ for $n \geq m$ and so $\lim_{n \to \infty} f_n(x) = 1$. Let x be any irrational number in $[0, 1]$. Then $f_n(x) = 0$ for all n and hence $\lim_{n \to \infty} f_n(x) = 0$. So $\lim_{n \to \infty} f_n(x) = \psi(x)$ where $\psi(x) = 1$ if x is rational and $\psi(x) = 0$ if x is irrational. We show that ψ is not of bounded variation in $[a, b]$. Let G be any positive number. Choose a positive integer $k > G$. Consider a partition $D : 0 = x_0 < x_1 < \cdots < x_{2k} = 1$ such that $x_1, x_3, \ldots, x_{2k-1}$ are irrational and $x_2, x_4, \ldots, x_{2k-2}$ are rational. Then

$$V_D(\psi; 0, 1) = \sum_{i=1}^{2k} |\psi(x_i) - \psi(x_{i-1})| = 2k > G$$

Since G is arbitrary, ψ is not of bounded variation in $[a, b]$.

In view of Theorem 6.18 it is natural to think that a function of bounded variation should be monotone in some sub-interval. But this is not the case. Consider.

6.6 Absolutely Continuous Functions

Example 6.4 Let $r_1, r_2, \ldots r_k \ldots$ be an enumeration of the rational numbers in $[0, 1]$ and let $a, 0 < a < 1$, be a fixed real number. A function f is defined by

$$f(x) = a^k \text{ if } x = r_k$$
$$= 0 \text{ if } x \text{ is irrational}$$

Show that f is of bounded variation in $[0, 1]$ and there is no sub-interval of $[0, 1]$ where f is monotone.

Solution Let $D : 0 = x_0 < x_1 < \cdots < x_n = 1$ be any partition of $[0, 1]$. Then for any i, $i = 0, 1, \ldots, n$

$$|f(x_i) - f(x_{i-1})| = 0 \text{ if both } x_{i-1} \text{ and } x_i \text{ are irrational.}$$

If one of x_{i-1} and x_i is rational say x_i is rational and x_{i-1} is irrational then there is k such that $x_i = r_k$ and so in this case

$$|f(x_i) - f(x_{i-1})| = |a^k - 0| = a^k$$

Similar is the case when x_i is irrational and x_{i-1} is rational. If both x_{i-1} and x_i are rational then $x_{i-1} = r_k$ for some k and $x_i = r_l$ for some l and so

$$|f(x_i) - f(x_{i-1})| = |a^l - a^k| \le a^l + a^k < 2a^m \text{ where } m = \min[l, k]$$

Thus in any case $|f(x_i) - f(x_{i-1})| < 2a^m$ for some m and if $i \ne j$ then $|f(x_j) - f(x_{j-1})| < 2a^p$ for some p and $p \ne m$. Since $m \le p$ implies $a^m \ge a^p$,

$$V_D(f, 0, 1) = \sum_{i=1}^{n} |f(x_i) - f(x_{i-1})| \le 2 \sum_{k=1}^{n} a^k = 2a \frac{1 - a^n}{1 - a} < \frac{2a}{1 - a}.$$

Since D is any partition, f is of bounded variation in $[0, 1]$. Clearly f is nowhere monotone in $[0, 1]$, since in every sub-interval of $[0, 1]$ there are irrational points x_1, x_2 and rational point x_3 such that $x_1 < x_3 < x_2$ and so $0 = f(x_1) < f(x_3) > f(x_2) = 0$.

6.6 Absolutely Continuous Functions

Definition 6.4 Let $f : [a, b] \to \mathbb{R}$. Then f is said to be absolutely continuous if for every $\epsilon > 0$ there is $\delta > 0$ such that for every finite collection $\{(x_i, x_i')\}_{i=1}^{n}$ of non-overlapping sub-intervals of $[a, b]$

$$\sum_{i=1}^{n} |f(x_i') - f(x_i)| < \epsilon \text{ whenever } \sum_{i=1}^{n} |x_i' - x_i| < \delta.$$

Definition 6.5 A function $f : [a, b] \to \mathbb{R}$ is said to satisfy Luzin condition (N) if for every set $E \subset [a, b]$ of measure zero the image set $f(E)$ is also of measure zero.

Theorem 6.24 *Let $f : [a, b] \to R$ be absolutely continuous. Then*

(i) f is continuous in $[a, b]$
(ii) f is of bounded variation in $[a, b]$
(iii) f satisfies the Luzin condition (N) on $[a, b]$

Proof (i) This is obvious since for every $\epsilon > 0$ there is $\delta > 0$ such that

$$|f(x) - f(c)| < \epsilon \text{ whenever } x, c \in [a, b] \text{ and } |x - c| < \delta.$$

(ii) By absolute continuity for every $\epsilon > 0$ there is $\delta > 0$ such that for every collection $\{(x_i, x_i')\}_{i=1}^{n}$ of non-overlapping sub-intervals (x_i, x_i') of $[a, b]$

$$\sum_{i=1}^{n} |f(x_i') - f(x_i)| < \epsilon \text{ whenever } \sum_{i=1}^{n} |x_i' - x_i| < \delta \qquad (6.35)$$

Take $\epsilon = 1$. Divide the interval $[a, b]$ by point c_j, $a = c_0 < c_1 < \cdots < c_k = b$ such that $c_j - c_{j-1} < \delta$ for $j = 1, 2, \ldots k$. Let $D : c_{j-1} = x_0 < x_1 < \cdots < x_n = c_j$ be any partition of $[c_{j-1}, c_j]$. Then since $\sum_{i=1}^{n} |x_i - x_{i-1}| = c_j - c_{j-1} < \delta$ by (6.35)

$$V_D(f; c_{j-1}, c_j) = \sum_{i=1}^{n} |f(x_i) - f(x_{i-1})| < 1.$$

Since D is any partition, f is of bounded variation in $[c_{j-1}, c_j]$ for each $j = 1, 2, \ldots, k$, and so by Theorem 6.17, f is of bounded variation in $[a, b]$.
(iii) Let $E \subset [a, b]$ be any set of measure zero. We show that $f(E)$ is also of measure zero. We may suppose that $E \subset (a, b)$. Let $\epsilon > 0$ be arbitrary. Then there is $\delta > 0$ such that for every collection $\{(x_i, x_i')\}_{i=1}^{n}$ of non overlapping sub-intervals of $[a, b]$, the relation (6.35) holds. Since $\mu(E) = 0$, there is an open set G such that $E \subset G \subset (a, b)$ and $\mu(G) < \delta$. Let the component intervals of G be $\{(a_i, b_i)\}$. Then since $\mu(G) < \delta$, $\sum_i (b_i - a_i) < \delta$. Since f absolutely continuous, by (i) f is continuous and hence there are points x_i and x_i' in $[a_i, b_i]$ such that

$$f(x_i) = \inf\{f(x) : x \subset [a_i, b_i]\} \text{ and } f(x_i') = \sup\{f(x) : x \in [a_i, b_i]\}.$$

6.6 Absolutely Continuous Functions

Then $\sum_i |x'_i - x_i| \leq \sum_i (b_i - a_i) < \delta$ and so by (6.35) $\sum |f(x'_i) - f(x_i)| < \epsilon$. Therefore, since $E \subset G \subset \bigcup_i (a_i, b_i)$, we have

$$f(E) \subset f(G) = f(\bigcup_i (a_i b_i)) = \bigcup_i f((a_i, b_i)) \subset \bigcup_i f([a_i, b_i]).$$

So,

$$\mu^*(f(E)) \leq \sum_i \mu(f[a_i b_i]) = \sum_i |f(x'_i) - f(x_i)| < \epsilon.$$

Since ϵ is arbitrary, $\mu^*(f(E)) = 0$, completing the proof. □

The converse of Theorem 6.24 is also true and we study it after the following theorems.

Theorem 6.25 *Let $f : [a, b] \to \mathbb{R}$ and let $E \subset [a, b]$. If all the four Dini deviates $D^+ f, D_+ f, D^- f$ and $D_- f$ are bounded on E then there is a constant $M > 0$ such that*

$$\mu^*(f(E)) \leq M \mu^*(E).$$

Proof Since the Dini derivates are bounded on E there is $M > 0$ such that the upper and lower bounds of each of these derivates lie within $(-M, M)$. Let $\epsilon > 0$ be arbitrary. For each n let

$$E_n = \left\{ x \in E : |f(t) - f(x)| < (M + \epsilon)|t - x| \text{ whenever } |t - x| \leq \frac{1}{n} \right\}.$$

For each E_n we associate a sequence of intervals $\{I_k^n\}_{k=1,2,\ldots}$ such that

$$E_n \subset \bigcup_k I_k^n, \sum_k |I_k^n| < \mu^*(E_n) + \epsilon \text{ and } |I_k^n| \leq \frac{1}{n} \text{ for } k = 1, 2, \ldots \quad (6.36)$$

Then for every pair of points x_1 and x_2 of $E_n \cap I_k^n$ we have

$$|f(x_2) - f(x_1)| < (M + \epsilon)|x_2 - x_1| \leq (M + \epsilon)|I_k^n|$$

and hence

$$\mu^*(f(E_n \cap I_k^n)) \leq (M + \epsilon)|I_k^n| \quad (6.37)$$

From (6.36) and (6.37)

$$\mu^*(f(E_n)) \leq \sum_k \mu^*(f(E_n \cap I_k^n)) \leq (M + \epsilon) \sum_k |I_k^n| < (M + \epsilon)(\mu^*(E_n) + \epsilon)$$

Letting $n \to \infty$ since ϵ is arbitrary

$$\lim_n \mu^*(f(E_n)) \le M \lim_n \mu^*(E_n) \tag{6.38}$$

Since $E_n \subset E_{n+1}$ for all n and $E = \bigcup_{n=1}^{\infty} E_n$, $\mu^*(E) = \lim_n \mu^*(E_n)$. also $f(E_n) \subset f(E_{n+1})$ for all n and $f(E) = \bigcup_{n=1}^{\infty} f(E_n)$ and hence $\mu^*(f(E)) = \lim_n \mu^*(f(E_n))$. So from (6.38) $\mu^*(f(E)) \le M\mu^*(E)$, completing the proof. □

Theorem 6.26 *Let $f : [a, b] \to \mathbb{R}$ and let f' exist finitely at each point of a measurable set $D \subset [a, b]$. If f' is Lebesgue integrable on D then*

$$\mu^*(f(D)) \le \int_D |f'(t)|\, dt$$

Proof Let $\epsilon > 0$ be arbitrary. For each n let

$$D_n = \{x \in D : (n-1)\epsilon \le |f'(x)| < n\epsilon\}. \tag{6.39}$$

Then the sets D_n are measurable. pairwise disjoint and $D = \bigcup_{n=1}^{\infty} D_n$. Hence by Theorem 6.25 and (6.39)

$$\mu^*(f(D)) = \mu^*\left(\bigcup_{n=1}^{\infty} f(D_n)\right) \le \sum_{n=1}^{\infty} \mu^*(f(D_n))$$

$$\le \sum_{n=1}^{\infty} n\epsilon \mu(D_n) \le \sum_{n=1}^{\infty} \left[\int_{D_n} |f'| + \epsilon \mu(D_n)\right]$$

$$= \int_D |f'| + \epsilon \mu(D).$$

Since ϵ is arbitrary the result follows. □

Now we prove the converse of Theorem 6.24 in the following

Theorem 6.27 (Banach-Zarecki) *Let $f : [a, b] \to \mathbb{R}$ be such that*

(i) *f is continuous in $[a, b]$*
(ii) *f is of bounded variation in $[a, b]$, and*
(iii) *f satisfies the Luzin condition (N) on $[a, b]$*
    Then f is absolutely continuous in $[a, b]$

Proof Since f is of bounded variation by Theorems 6.18, 6.11 and 6.12, f' exists almost every where and is Lebesgue integrable in $[a, b]$. Let $\epsilon > 0$ be arbitrary. Then by Theorem 5.25 (vii) there is $\delta > 0$ such that for every measurable set $E \subset [a, b]$

6.6 Absolutely Continuous Functions

$$\int_E |f'(x)|\,dx < \epsilon \text{ whenever } \mu(E) < \delta \tag{6.40}$$

Let $\{[x_i, x_i']\}$ be any countable collection of non-overlapping sub-intervals of $[a, b]$ such that $\sum_i |x_i' - x_i| < \delta$. Then by (6.40)

$$\sum_i \int_{x_i}^{x_i'} |f'(x)|\,dx < \epsilon \tag{6.41}$$

Let $D = \{x \in [x_i, x_i'] : f'(x) \text{ exists and is finite}\}$ and $H = [x_i, x_i'] \sim D$.

Then $\mu(H) = 0$. Therefore since f satisfies Luzin condition (N), $\mu(f(H)) = 0$. Since f is continuous, by Theorem 6.26

$$|f(x_i') - f(x_i)| \leq \mu(f([x_i, x_i'])) = \mu(f(D \cup H))$$
$$= \mu(f(D)) + \mu(f(H)) = \mu(f(D))$$

$$\leq \int_D |f'(x)|\,dx = \int_{x_i}^{x_i'} |f'(x)|\,dx \tag{6.42}$$

So from (6.41) and (6.42)

$$\sum_i |f(x') - f(x_i)| \leq \sum_i \int_{x_i}^{x_i'} |f'(x)|\,dx < \epsilon$$

completing the proof. $\square$

Theorem 6.28 *Let f and g be absolutely continuous in $[a, c]$ and let α be any real number. Then $|f|$, $f + g$, αf and fg are also absolutely continuous in $[a, b]$. If there is $k > 0$ such that $|f(x)| \geq k$ for all $x \in [a, b]$ then $\frac{1}{f}$ is also absolutely continuous in $[a, b]$.*

The proof is similar to that of Theorem 6.16.

Theorem 6.29 *A function f is absolutely continuous if and only if its variation function v_f is absolutely continuous.*

Proof Let f be absolutely continuous in $[a, b]$. Let $\epsilon > 0$ be arbitrary. Then there is $\delta > 0$ such that for every collection $\{[x_i\, x_i']\}$ of non-overlapping sub-intervals of $[a, b]$

$$\sum_i |f(x_i') - f(x_i)| < \epsilon \text{ whenever } \sum_i |x_i' - x_i| < \delta \tag{6.43}$$

Let $\{[x_j, x'_j]\}_{j=1}^{N}$ be any collection of non-overlapping sub-intervals of $[a, b]$ satisfying $\sum_{j=1}^{N} |x'_j - x_j| < \delta$. For each j consider any partition $D_j : x_j = x_{j0} < x_{j1} < \cdots < x_{jn_j} = x'_j$. Then the collection $\{[x_{j,k-1}, x_{j,k}] : k = 1, 2, \ldots n_j;\ j = 1, 2, \ldots, N\}$ of non-overlapping sub-intervals of $[a, b]$ is such that

$$\sum_{j=1}^{N} \sum_{k=1}^{n_j} |x_{j,k} - x_{j,k-1}| = \sum_{j=1}^{N} |x'_j - x_j| < \delta.$$

So, by (6.43)

$$\sum_{j=1}^{N} \sum_{k=1}^{n_j} |f(x_j, k) - f(x_{j,k-1})| < \epsilon.$$

Since this is true for every partition D_j of $[x_j\ x'_j]$, $j = 1, 2, \ldots N$.

$$\sum_{j=1}^{N} V(f; x_j, x'_j) \leq \epsilon \ \text{i.e} \ \sum_{j=1}^{N} |v_f(x'_j) - v_f(x_j)| \leq \epsilon.$$

This shows that v_f is absolutely continuous in $[a, b]$. Conversely, suppose that v_f is absolutely continuous. Let $\epsilon > 0$ be arbitrary. Then there is $\delta > 0$ such that for every collection $\{[x_i, x'_i]\}$ of non-overlapping sub-intervals of $[a, b]$

$$\sum |v_f(x'_i) - v_f(x_i)| < \epsilon \ \text{whenever} \ \sum |x'_i - x_i| < \delta.$$

Let $\{[x_i, x'_i]\}$ be such a collection. Then

$$\sum |f(x'_i) - f(x_i)| \leq \sum V(f; x_i, x'_i) = \sum |v_f(x'_i) - v_f(x_i)| < \epsilon$$

which shows that f is absolutely continuous. $\square$

Theorem 6.30 *A function f is absolutely continuous if and only if f can be expressed as the difference between two non-decreasing absolutely continuous functions.*

Proof If f is absolutely continuous then by Theorem 6.29, v_f is absolutely continuous. So the function $g = v_f - f$ is absolutely continuous. Since $f = v_f - g$ the first part follows. The second part is obvious by Theorem 6.28. $\square$

Example 6.5 The Cantor function θ defined in Sect 4.9 is not absolutely continuous although it is continuous and of bounded variation.

Solution If is known that θ is continuous and non-decreasing in $[0, 1]$. We show that θ does not satisfy Lusin continuous $(\mathcal{N})$ which will prove the assertion by Theorem 6.24.

Let P be the Cantor ternary set in $[0, 1]$ and let $G = [0, 1] \sim P$. Then G is the union of a countable collection of open intervals $\{I_1^1; I_1^2, I_2^2; \ldots; I_k^n, 1 \leq k \leq 2^{n-1}, n = 1, 2, \ldots\}$. The function θ is constant in each interval I_k^n assuming the value $\frac{2k-1}{2^n}$, $1 \leq k \leq 2^{n-1}$, $n = 1, 2, \ldots$. Therefore $\theta(G)$ is countable. Since $\theta(P) \bigcup \theta(G) = \theta(P \bigcup G) = \theta([0, 1]) = [0, 1]$ and since $\theta(G)$ is of measure zero, $\theta(P)$ cannot be of measure zero. Since P is of measure zero, θ does not satisfy the property $(\mathcal{N})$.

6.7 Monotonicity Theorems and Their Consequences

The following theorem generalizes the well-known theorem that if f is differentiable and if the derivative f' is non-negative then f is monotone non-decreasing.

Theorem 6.31 (Zygmund) *Let f be defined in an interval I and let*
(i) $\overline{f}(x - 0) \leq f(x) \leq \overline{f}(x + 0)$ *for all $x \in I$ and*
(ii) *the image $f(E)$ of E where $E = \{x \in I : D^+ f(x) \leq 0\}$ does not contain any interval.*
Then f is non-decreasing in I.

Proof Suppose, if possible, that there are two points $a, b \in I, a < b$, such that $f(a) > f(b)$. Since by the condition (ii) $f(E)$ cannot contain any interval, there is y_0, $f(b) < y_0 < f(a)$ such that $y_0 \notin f(E)$. Let

$$S = \{x \in [a, b] : f(x) \geq y_0\} \text{ and } x_0 = \sup S.$$

Then $a \in S$, $b \notin S$. Since by (i) $\overline{f}(a + 0) \geq f(a) > y_0$ and $y_0 > f(b) \geq \overline{f}(b - 0)$ we have $x_0 \neq a$ and $x_0 \neq b$. In, fact, if $x_0 = a$ then $f(x) < y_0$ for all $x \in (a, b]$ which gives $\overline{f}(a + 0) \leq y_0 < f(a)$ which contradicts (i) and if $x_0 = b$ then since $b \notin S$ there is a sequence $\{x_n\} \in S$ such that $x_n < b$ for all n and $\lim_{n \to \infty} x_n = b$ and so $\overline{f}(b - 0) \geq y_0 > f(b)$ which also contradicts (i). So, $a < x_0 < b$. Also $f(x_0) = y_0$. For, if $f(x_0) > y_0$ then $y_0 < \overline{f}(x_0 + 0)$ and so there are points x in the right neighborhood of x_0 for which $f(x) > y_0$ which is a contradiction since $x_0 = \sup S$; and if $f(x_0) < y_0$ then $\overline{f}(x_0 - 0) < y_0$ and so there is a left neighborhood of x_0 in which there is no point x for which $f(x) \geq y_0$ which is also a contradiction since $x_0 = \sup S$. Clearly $f(x) < y_0$ for $x_0 < x < b$ and hence

$$\frac{f(x_0 + h) - f(x_0)}{h} < 0 \text{ for all } h, 0 < h < b - x_0$$

and so $D^+ f(x_0) \leq 0$. Hence $x_0 \in E$ i.e $y_0 = f(x_0) \in f(E)$. But this is a contradiction since $y_0 \notin f(E)$. So f is non-decreasing on I. □

Theorem 6.32 *Let f be defined in an interval I and let*
(i) $\overline{f}(x-0) \leq f(x) \leq \overline{f}(x+0)$ for $x \in I$, and
(ii) $D^+ f(x) \geq 0$ for all $x \in I$ except possibly on an enumerable set in I.
Then f is non-decreasing on I.

Proof Let $\epsilon > 0$ be arbitrary and let $g(x) = f(x) + \epsilon x$. Clearly $D^+ g(x) = D^+ f(x) + \epsilon$ and hence $D^+ g(x) > 0$ for all $x \in I$ except on an enumerable set in I. Let $E = \{x \in I : D^+ g(x) \leq 0\}$. Then E is enumerable and so $g(E)$ is enumerable. Hence $g(E)$ cannot contain any interval. Also g satisfies the inequalities in (i). Hence by Theorem 6.31 g is non-decreasing on I. Since ϵ is arbitrary, f is non decreasing on I. □

Remark In condition (ii) of the above two theorems $D^+ f$ may be replaced by any of the three other Dini derivates, by suitable modification of the proof.

Theorem 6.33 *If any one of the four Dini derivates of a continuous function f on I is continuous at a point x_0 in I then the three other derivates are also continuous at x_0 and the derivative $f'(x_0)$ exists.*

Proof Let us consider $D^+ f$, and let $D^+ f$ be continuous at x_0. Let $\epsilon > 0$ be arbitrary. Then there is $\delta > 0$ such that

$$D^+ f(x_0) - \epsilon < D^+ f(x) < D^+ f(x_0) + \epsilon \text{ for } x_0 - \delta < x < x_0 + \delta. \quad (6.44)$$

Putting $k_1 = D^+ f(x_0) - \epsilon$ and $k_2 = D^+ f(x_0) + \epsilon$ we see by Theorem 6.32 that the function $f(x) - k_1 x$ is non-decreasing and $f(x) - k_2 x$ is non-increasing in $(x_0 - \delta, x_0 + \delta)$. Hence

$$D^+ f(x_0) - \epsilon \leq D_+ f(x) \leq D^+ f(x_0) + \epsilon \text{ for } x_0 - \delta < x < x_0 + \delta. \quad (6.45)$$

From (6.44) and (6.45)

$$|D^+ f(x_0) - D_+ f(x_0)| < 2\epsilon$$

Since ϵ is arbitrary, $D^+ f(x_0) = D_+ f(x_0)$. So, from (6.45) $D_+ f$ is continuous at x_0. Similarly $D_- f$ are also continuous at x_0 and $D^+ f(x_0) = D^- f(x_0) = D_- f(x_0)$, completing the proof. □

Theorem 6.34 *If f is continuous is an interval I, the upper and lower bounds of each of the four Dini derivates of f are respectively equal to the upper and lower bounds of the set $\{\frac{f(x_2)-f(x_1)}{x_2-x_1} : x_1, x_2 \in I; x_1 \neq x_2\}$.*

Proof We consider $D^+ f$. Let m be the lower bound of $D^+ f$ in I and let $m > -\infty$. Let $g(x) = f(x) - mx$. Then g is continuous and $D^+ g(x) = D^+ f(x) - m$ on I. So, by Theorem 6.32 the function g is non-decreasing on I and so $f(x_2) - f(x_1) \geq m(x_2 - x_1)$ i.e. $\frac{f(x_2)-f(x_1)}{x_2-x_1} \geq m$ for $x_1, x_2 \in I, x_2 > x_1$. If $m = -\infty$, the proof is trivial. The proof for the other three derivates and for the upper bound are similar. □

6.7 Monotonicity Theorems and Their Consequences

The last two theorems are due to Dini. Theorem 6.32 can further be generalized as follows :

Theorem 6.35 *Let f be defined in an interval I and let*
(i) $\overline{f}(x-0) \leq f(x) \leq \overline{f}(x+0)$ for all $x \in I$
(ii) $D^+ f(x) \geq 0$ for almost all $x \in I$
(iii) $D^+ f(x) > -\infty$ except possibly on an enumerable set in I.
Then f is non-decreasing in I.

Proof Let

$$E = \{x \in I : -\infty < D^+ f(x) < 0, \} \text{ and } E_0 = \{x \in I : D^+ f(x) = -\infty\}.$$

Then E is of measure zero and E_0 is enumerable. By Theorem 6.7 there is a continuous non-decreasing non-negative function ψ on I such that $\psi'(x)$ exists and $\psi'(x) = \infty$ for all $x \in E$. Let n be any positive integer and let $F_n(x) = f(x) + \frac{1}{n}\psi(x)$ for $x \in I$. Then since ψ is continuous F_n satisfies (i). Also if $x \in I \sim (E \bigcup E_0)$ then $D^+ f(x) \geq 0$ and since ψ is non-decreasing, $D^+ \psi(x) \geq 0$ for all $x \in I$. So, $D^+ F_n(x) \geq D^+ f(x) + \frac{1}{n} D_+ \psi(x) \geq 0$ for all $x \in I \sim (E \bigcup E_0)$. If $x \in E$ then since $\psi'(x) = \infty$, $D^+ F_n(x) = \infty$. Therefore, $D^+ F_n(x) \geq 0$ for all $x \in I \sim E_0$. Hence by Theorem 6.32 F_n is non-decreasing in I. Since n is arbitrary F_n is non-decreasing in I for all n and hence $\lim_{n\to\infty} F_n$ is non-decreasing on I. Since $\lim_{n\to\infty} F_n = f$ the proof is complete.

The following theorem which is due to Goldowski and Tonelli is interesting since it determines the behavior of the derivative. □

Theorem 6.36 *If a continuous function f has a derivative, finite or infinite, at each point of an interval I except possibly on an enumerable set in I and if this derivative is non-negative almost everywhere in I then f is non-decreasing in I.*

Proof Let E_0 be the set of points in I where f' does not exist. Then E_0 is countable. Let $E_0 = \{x_1, x_2, x_3, \ldots\}$. Writing $E_n = \{x_1, x_2, \ldots x_n\}$ we have $E_0 = \bigcup_{n=1}^{\infty} E_n$.

Let E be the set of points x in I such that f is not non-decreasing in any neighborhood of x. Then E is closed. Also E cannot have any isolated point. We prove that $E = \emptyset$.

Suppose, if possible that $E \neq \emptyset$. For each positive integer n let

$$P_n = \{x \in I : f(t) - f(x) \leq -(t-x) \text{ for } 0 < t - x < \frac{1}{n}\}$$

and $Q_n = \{x \in I : f(t) - f(x) \geq -2(t-x) \text{ for } 0 < t - x < \frac{1}{n}\}.$

Then P_n and Q_n are closed in I and they cover the interval I except possibly the set E_0 where the derivative of f does not exist. So, $I = \bigcup_{n=1}^{\infty} (P_n \bigcup Q_n \bigcup E_n)$ and hence

$$E = \bigcup_{n=1}^{\infty} \left((P_n \cap E) \bigcup (Q_n \cap E) \bigcup E_n\right).$$

Since E is closed, by Baire's Category theorem (Theorem 4.13 of Chapter 4) at least one of the sets $P_n \cap E$, $Q_n \cap E$ and E_n is everywhere dense in a portion of E. In other words, E contains a portion which is contained in one of the sets P_n, Q_n and E_n. Since E has no isolated points E cannot contain any portion which is contained in E_n for $n = 1, 2, 3, \ldots$. So E contains a portion which is either (i) contained in one of the sets P_n or (ii) contained in one of the sets Q_n. Suppose (i). Then there is a positive integer n_0 and an open interval (a, b) such that $\emptyset \neq E \cap (a, b) \subset P_{n_0}$. We may suppose that $b - a < \frac{1}{n_0}$. Since the derivative of f is non-negative almost everywhere and P_{n_0} is closed, P_{n_0} is nowhere dense. So $E \cap (a, b)$ is nowhere dense. Let (c, d) be any interval contiguous to $E \cap (a, b)$. Then f is non-decreasing in (c, d). Since f is continuous f is non-decreasing in $[c, d]$. But $c, d \in P_{n_0}$ and $d - c < \frac{1}{n_0}$ and so $f(d) - f(c) \leq -(d - c) < 0$ which is a contradiction.

Now suppose (ii). Then as in case (i) $\emptyset \neq E \cap (a, b) \subset Q_{n_0}$ for some positive integer n_0 and some open interval (a, b), where $b - a < \frac{1}{n_0}$. So, if $x \in E \cap (a, b)$ then $x \in Q_{n_0}$ and hence $f(t) - f(x) \geq -2(t - x)$ for $0 < t - x < \frac{1}{n_0}$ and so $D^+ f(x) \geq -2$. Also if $x \in (a, b) \sim E$ then there is a neighborhood if x in which f is non-decreasing and hence $D^+ f(x) \geq 0$. Therefore $D^+ f(x) \geq -2$ for all $x \in (a, b)$. Since $f' \geq 0$ almost everywhere by Theorem 6.35 f is non decreasing in (a, b). But this is a contradiction since (a, b) contains points of E in its interior. Thun $E = \emptyset$ and this completes the proof. $\square$

From Theorem 6.36 we can deduce the very interesting property of a derivative in the following.

Theorem 6.37 *If a continuous function f has a derivative f' finite or infinite in an interval I then for every real numbers M the sets $\{x \in I : f'(x) > M\}$ and $\{x \in I : f'(x) < M\}$ are either void or is of positive measure.*

Proof Let $E = \{x \in I : f'(x) > M\}$. Suppose $E \neq \emptyset$. We show that $\mu(E) > 0$. If possible suppose $\mu(E) = 0$. Then $f' \leq M$ a.e. in I. Let $F(x) = Mx - f(x)$. Then $F'(x) = M - f'(x) \geq 0$ for almost all $x \in I$. So, by Theorem 6.36 the function F is non-decreasing in I. Hence $F' \geq 0$ everywhere in I i.e $f' \leq M$ everywhere in I and so $E = \emptyset$ which is a contradiction. The other part has similar proof. $\square$

The next theorem is also interesting.

Theorem 6.38 *If f is absolutely continuous in an interval I and if $D^+ f \geq 0$ almost everywhere in I then f is non-decreasing in I.*

Proof Let $\epsilon > 0$ be arbitrary. Let $F(x) = f(x) + \epsilon x$. Then F is absolutely continuous in I and $D^+ F = D^+ f + \epsilon > 0$ a.e. in I. Let

$$H = \{x : x \in I; D^+F(x) \leq 0\}.$$

Then $\mu(H) = 0$. Since F is absolutely continuous, by Theorem 6.24, F is continuous and has the property $(\mathcal{N})$ and so $\mu(F(H)) = 0$. Hence $F(H)$ cannot contain any interval. So, by Theorem 6.31 the function F is non-decreasing in I. Since ϵ is arbitrary, f is non-decreasing in I. □

6.8 The Indefinite Lebesgue Integral

Let f be Lebesgue integrable on a bounded measurable set E and let $E \subset [a, b]$. Then for each $x \in [a, b]$, f is Lebesgue integrable on $E \cap [a, x]$. If $F(x) = \int_{E \cap [a,x]} f$ then F is a function of x defined on $[a, b]$ which is called indefinite Lebesgue integral of f on $[a, b]$. If $E \neq [a, b]$, then defining $f(x) = 0$ for $x \in [a, b] \sim E$ we write $F(x) = \int_a^x f$. So, we shall henceforth consider Lebesgue integrable functions on $[a, b]$.

Theorem 6.39 *The indefinite Lebesgue integral $F(x) = \int_a^x f\,dx$ is absolutely continuous on $[a, b]$.*

Proof Let $\epsilon > 0$ be arbitrary. Since f is integrable, $|f|$ is integrable by Theorem 5.25 (vi) and so by Theorem 5.25 (vii) there is $\delta > 0$ such that

$$\int_E |f|\,dx < \epsilon \text{ for every measurable set } E \subset [a, b] \text{ with } \mu(E) < \delta. \qquad (6.46)$$

Let $\{[a_n, b_n]\}$ be any countable collection of pairwise disjoint sub-intervals of $[a, b]$ such that $\sum_n (b_n - a_n) < \delta$. So, by (6.46)

$$\sum_n |F(b_n) - F(a_n)| = \sum_n \left| \int_{a_n}^{b_n} f\,dx \right| \leq \sum_n \int_{a_n}^{b_n} |f|\,dx < \epsilon$$

which shows that F is absolutely continuous in $[a, b]$. □

Theorem 6.40 *Let f be integrable in $[a, b]$ and let*

$$F(x) = \int_a^x f \, dx \quad x \in [a, b].$$

Then $F'(x)$ exists and $F'(x) = f(x)$ for almost all $x \in [a, b]$.

Proof By Theorems 6.39, 6.24 (ii) and 6.18, F is the difference of two non-decreasing functions and so by Theorem 6.11 F has finite derivative a.e. Therefore we are to show that $F' = f$ a.e.

We first suppose that f is bounded. Let $|f(x)| \leq K$ for all $x \in [a, b]$. Define $f(x) = f(b)$ for $x > b$. Let for each n

$$f_n(x) = n\left[F\left(x + \frac{1}{n}\right) - F(x)\right] = n \int_x^{x+\frac{1}{n}} f \, dx, \quad x \in [a, b].$$

Hence $|f_n(x)| \leq k$ for all $x \in [a, b]$ and all n. Since F' exists a.e. $\lim_{n \to \infty} f_n = F'$ a.e. So, by Dominated convergence theorem (Theorem 5.33), for any point $c \in [a, b]$

$$\int_a^c F'(t) dt = \lim_{n \to \infty} \int_a^c f_n(t) dt = \lim_{n \to \infty} n \int_a^c \left[F(t + \frac{1}{n}) - F(t)\right] dt$$

$$= \lim_{n \to \infty} \left[n \int_c^{c+\frac{1}{n}} F(t) dt - n \int_a^{a+\frac{1}{n}} F(t) dt \right]$$

$$= F(c) - F(a) = \int_a^c f(t) dt.$$

Since $c \in [a, b]$ is arbitrary, this shows that

$$\int_a^x [F'(t) - f(t)] dt = 0 \text{ for all } x \in [a, b].$$

Hence by Theorem 5.25 (x), $F' = f$ a.e. in $[a, b]$. So, the theorem is true when f is bounded.

Next we suppose that $f \geq 0$. Let for each n and each $x \in [a, b]$

$$f_n(x) = f(x) \text{ if } f(x) \leq n$$
$$= n \text{ if } f(x) > n.$$

6.8 The Indefinite Lebesgue Integral

Then f_n is bounded and measurable for all n. Hence by the above

$$\frac{d}{dx} \int_a^x f_n(t)dt = f_n \quad a.e. \text{ in } [a, b]. \tag{6.47}$$

Since $f - f_n \geq 0$ the function G_n where

$$G_n(x) = \int_a^x [f - f_n]dx$$

is non-decreasing and hence G_n has finite derivative *a.e.* in $[a, b]$. Since

$$F(x) = \int_a^x f dx = \int_a^x [f - f_n]dx + \int_a^x f_n dx = G_n(x) + \int_a^x f_n dx$$

we have by (6.47) $F' = G'_n + f_n$ *a.e.* in $[a, b]$ and so $F' \geq f_n$ *a.e.* in $[a, b]$. This being true for all n, $F' \geq f$ *a.e.* in $[a, b]$. Hence

$$\int_a^b F' dx \geq \int_a^b f dx = F(b) - F(a).$$

Since $f \geq 0$, F is non-decreasing and hence by Theorem 6.12

$$\int_a^b F' dx \leq F(b) - F(a).$$

The last two inequalities give

$$\int_a^b F' dx = F(b) - F(a) = \int_a^b f dx.$$

Hence

$$\int_a^b [F' - f]dx = 0.$$

Since $F' \geq f$ *a.e.* in $[a, b]$, by Theorem 5.25 (xi), $F' = f$ *a.e.* in $[a, b]$. So, the theorem is proved in this case.

For the general case we have $f = f_+ - f_-$ and so

$$F(x) = \int_a^x f_+ - \int_a^x f_- \quad \text{for } x \in [a, b].$$

Since f_+ and f_- are non-negative integrable functions on $[a, b]$ applying the above on f_+ and f_- we have $F' = f_+ - f_- = f$ a.e. in $[a, b]$ completing the proof. □

6.9 Characterization of Indefinite Lebesgue Integral and Indefinite Riemann Integral

Let F be a function defined on a finite interval $[a, b]$. If there is a Lebesgue integrable function f on $[a, b]$ such that $F(x) = (L) \int_a^x f \, dx$ for all $x \in [a, b]$ then for arbitrary constant c, $c + F$ is called an indefinite Lebesgue integral of f. Similarly, if there is a Riemann integrable function f on $[a, b]$ such that $F(x) = (R) \int_a^x f \, dx$ for all $x \in [a, b]$ then $c + F$ is called an indefinite Riemann integral of f. In this section we give necessary and sufficient condition on F under which F is an indefinite Lebesgue integral and indefinite Riemann integral.

Theorem 6.41 *A function $F : [a, b] \to \mathbb{R}$ is an indefinite Lebesgue integral if and only if F is absolutely continuous. In any case*

$$F(x) = c + (L) \int_a^x F' dx, \quad x \in [a, b]$$

where c is a constant.

Proof Let F be an indefinite Lebesgue integral. Then there is a Lebesgue integrable function f such that

$$F(x) = c + (L) \int_a^x f \, dx, \quad x \in [a, b].$$

By Theorem 6.39, F is absolutely continuous in $[a, b]$ and by Theorem 6.40, $F' = f$ a.e. on $[a, b]$, proving the first part.

Conversely suppose that F is absolutely continuous in $[a, b]$. Then by Theorem 6.24 F is of bounded variation and so by Theorem 6.18 F is the difference of two non-decreasing functions. Applying Theorems 6.11 and 6.12 F has finite derivative F' a.e. in $[a, b]$ which is Lebesgue integrable. Let

6.9 Characterization of Indefinite Lebesgue Integral and Indefinite Riemann Integral

$$G(x) = (L) \int_a^x F' dx, \ x \in [a, b].$$

Then G is absolutely continuous. So, the function $\phi = F - G$ is absolutely continuous in $[a, b]$. Also by Theorem 6.40 $G' = F'$ a.e. and hence $\phi' = F' - G' = 0$ a.e. in $[a, b]$. So by Theorem 6.38 ϕ is a constant. Let $\phi = c$. So

$$F(x) = \phi(x) + G(x) = c + (L) \int_a^x F', \ x \in [a, b]$$

completing the proof. □

Corollary 6.42 *If $F : [a, b] \to \mathbb{R}$ is absolutely continuous and $F' = 0$ a.e. then F is a constant.*

By the above $F(x) = F(a) + \int_a^x F' dx = F(a)$ for all $x \in [a, b]$, completing the proof.

Theorem 6.43 *Let $F : [a, b] \to \mathbb{R}$ be strictly increasing absolutely continuous function. Then for any measurable set $E \subset [a, b]$*

$$\int_E F' dx = \mu(F(E))$$

where $F(E) = \{y : y = F(x) \text{ for some } x \in E\}$.

Proof If E is an open subinterval of $[a, b]$, say $E = (x_1, x_2)$ then

$$\int_E F' dx = \int_{x_1}^{x_2} F' dx = F(x_2) - F(x_1) = \mu(F(E))$$

and so in this case the theorem is proved. Suppose $E = G \subset [a, b]$ where G is any open subset of $[a, b]$. Then $G = \bigcup_{i=1}^{\infty} I_i$ where $\{I_i\}$ is a countable collection of open sub-intervals of $[a, b]$. Then by the first case, since F is strictly increasing,

$$\int_G F' dx = \sum_{i=1}^{\infty} \int_{I_i} F' dx = \sum_{i=1}^{\infty} \mu(F(I_i)) = \mu(\bigcup_{i=1}^{\infty} F(I_i)) = \mu(F(\bigcup_{i=1}^{\infty} I_i)) = \mu(F(G)).$$

(6.48)

Finally, let E be any measurable subset of $[a, b]$. Then by Theorem 2.15 there is a sequence $\{G_n\}$ of open sets such that $E \subset G_{n+1} \subset G_n$ for all n and $\mu(E) = \mu(\bigcap_{n=1}^{\infty} G_n)$ and so by Theorem 2.9

$$\mu(E) = \mu(\cap G_n) = \lim \mu(G_n). \tag{6.49}$$

Let $\epsilon > 0$ be arbitrary. Then by Theorem 5.25 (vii) there is $\delta > 0$ such that

$$\int_e F' dx < \epsilon \text{ whenever } \mu(e) < \delta. \tag{6.50}$$

By (6.49) there is N such that

$$\mu(G_n \sim E) = \mu(G_n) - \mu(E) < \delta \text{ for } n \geq N. \tag{6.51}$$

From (6.50) and (6.51)

$$0 \leq \int_{G_n} F' dx - \int_E F' dx = \int_{G_n \sim E} F' < \epsilon \text{ for } n \geq N.$$

Hence

$$\lim_{n \to \infty} \int_{G_n} F' dx = \int_E F' dx \tag{6.52}$$

Since the result is proved when E is an open set and since G_n is open using (6.48) we have

$$\int_{G_n} F' dx = \mu(F(G_n)).$$

Hence from (6.52)

$$\int_E F' dx = \lim_{n \to \infty} \mu(F(G_n)). \tag{6.53}$$

Since F is strictly increasing and $G_{n+1} \subset G_n$, $F(G_{n+1}) \subset F(G_n)$ for all n, and so

$$\lim_{n \to \infty} \mu(F(G_n)) = \mu(\bigcap_{n=1}^{\infty} F(G_n)) = \mu(F(\bigcap_{n=1}^{\infty} G_n)) = \mu(F(E)). \tag{6.54}$$

The proof is complete by (6.53) and (6.54). □

Theorem 6.44 *A function $F : [a, b] \to \mathbb{R}$ is an indefinite Riemann integral if and only if all the Dini derivates D^+F, D_+F, D^-F and D_-F are bounded and a.e. continuous in $[a, b]$. In any case $F(x) = c + (R) \int_a^x DF dx$ where DF stands for any one of D^+F, D_+F, D^-F, and D_-F.*

6.9 Characterization of Indefinite Lebesgue Integral and Indefinite Riemann Integral

Proof Let F be an indefinite Riemann integral. Then there is a bounded Riemann integrable function f on $[a, b]$ such that $F(x) = c + (R) \int_a^x f \, dx$ for $x \in [a, b]$. Let $|f(t)| \leq M$ for all $t \in [a, b]$. So,

$$\left| \frac{F(x+h) - F(x)}{h} \right| = \left| \frac{1}{h} \int_x^{x+h} f \, dx \right| \leq M \text{ for } x, x + h \in [a, b], h \neq 0. \quad (6.55)$$

This shows that all the Dini derivates of F are bounded in $[a, b]$.

Since f is Riemann integrable it is *a.e.* continuous on $[a, b]$. So there is a set $E \subset [a, b]$ such that $\mu(E) = b - a$ and f is continuous on E. Let $\xi \in E$ and $\epsilon > 0$. Since f is continuous at ξ, there is $\delta > 0$ such that

$$|f(x) - f(\xi)| < \epsilon \text{ for } |x - \xi| < \delta. \quad (6.56)$$

Let $x \in [a, b]$ be such that $|x - \xi| < \frac{\delta}{3}$. Then if $t \in [a, b]$ and $|t - x| < \frac{\delta}{3}$ we have $|t - \xi| \leq |t - x| + |x - \xi| < \delta$ so by (6.56)

$$|f(t) - f(x)| \leq |f(t) - f(\xi)| + |f(x) - f(\xi)| < 2\epsilon. \text{ So,}$$

$$\left| \frac{1}{h} \int_x^{x+h} [f(t) - f(x)] dt \right| \leq 2\epsilon \text{ for } 0 < |h| < \frac{\delta}{3}.$$

Therefore

$$\sup_{0 < h < \frac{\delta}{3}} \left| \frac{1}{h} \int_x^{x+h} [f(t) - f(x)] dt \right| \leq 2\epsilon$$

and hence

$$| \sup_{0 < h < \frac{\delta}{3}} \frac{1}{h} \int_x^{x+h} [f(t) - f(x)] dt | \leq 2\epsilon.$$

Letting $\delta \to 0$

$$|D^+ F(x) - f(x)| \leq 2\epsilon \quad (6.57)$$

Since f is continuous at ξ, $F'(\xi)$ exists and $F'(\xi) = f(\xi)$. So, by (6.56) and (6.57)

$$|D^+ F(x) - D^+ F(\xi)| = |D^+ F(x) - f(\xi)| \leq |D^+ F(x) - f(x)| + |f(x) - f(\xi)| \leq 3\epsilon. \quad (6.58)$$

Since x is any point in $[a, b]$ such that $|x - \xi| < \frac{\delta}{3}$, we have from (6.58)

$$|D^+ F(x) - D^+ F(\xi)| \leq 3\epsilon \text{ for } |x - \xi| < \frac{\delta}{3}$$

which shows that D^+F is continuous at ξ. Since ξ is any point in E and $\mu(E) = b - a$, D^+F is continuous $a.e.$ in $[a, b]$. From (6.55) it follows that all the derivates are bounded and that F is continuous in $[a, b]$. So, by Theorem 6.33 other derivates are also continuous $a.e.$ in $[a, b]$, proving the first part.

Conversely, suppose that all the derivates are bounded and $a.e.$ continuous in $[a, b]$. Let $M > 0$ be such that all the derivates lie within the interval $(-M, M)$. Let $x \in [a, b]$. Then there is $\delta > 0$ such that

$$-M - 1 < \frac{F(x+h) - F(x)}{h} < M + 1 \text{ for } 0 < |h| < \delta.$$

So, F is continuous at x and since x is arbitrary F is continuous in $[a, b]$. Let $\phi(x) = Mx - F(x)$. Then ϕ is continuous in $[a, b]$ and $D^+\phi(x) = M - D^+F(x) \geq 0$ for all $x \in [a, b]$. So, by Theorem 6.32 ϕ is non-decreasing in $[a, b]$ and hence F is of bounded variation in $[a, b]$. Also by Theorem 6.25 F satisfies Luzin condition $(\mathcal{N})$. Hence by Theorem 6.27 F is absolutely continuous in $[a, b]$. On the other hand D^+F being bounded and $a.e.$ continuous it is Riemann integrable in $[a, b]$. Let

$$G(x) = (R) \int_a^x D^+F dx, \ x \in [a, b].$$

Since every Riemann integral is a Lebesgue integral, G is absolutely continuous. Also D^+F being continuous $a.e.$ $G' = D^+F$ $a.e.$ and by Theorem 6.33 $G' = F'$ $a.e.$ So, if $H = F - G$ then H is absolutely continuous and $H' = 0$ $a.e.$ Hence H is constant by Theorem 6.38. Let $H = c$. Then

$$F(x) = c + G(x) = c + (R) \int_a^x D^+F dx.$$

So, F is an indefinite Riemann integral. Finally note that if f and g are Riemann integrable and $f = g$ $a.e.$ then $(R) \int_a^x f dx = (R) \int_a^x g dx$ for all x and so applying Theorem 6.33 the proof is complete. $\square$

If a function $f : [a, b] \to \mathbb{R}$ is differentiable and if the derivative f' is Lebesgue integrable in $[a, b]$ then whether the relation

$$\int_a^b f' dx = f(b) - f(a)$$

holds (which is the analog of the fundamental theorem for Riemann integral) is an interesting question. The answer is yes but the proof is not simple since it needs a more general notion of absolute continuity and we shall not prove it. For proof see [Saks, Gordon].

6.9 Characterization of Indefinite Lebesgue Integral and Indefinite Riemann Integral

Example 6.6 For every set $E \subset [a, b]$ of measure zero there exists a non-decreasing absolutely continuous function $F : [a, b] \to \mathbb{R}$ such that $\underline{D}F(x) = \infty$ for all $x \in E$, where $\underline{D} = \min[D_+, D_-]$.

Solution For each positive integer k let G_k be an open set such that $E \subset G_k$ and $\mu(G_k) < \frac{1}{2^k}$. Since $\{\chi_{G_k}\}$ is a sequence of non-negative measurable functions on $[a, b]$ by Corollary 5.31

$$\int_a^b (\sum_{k=1}^\infty \chi_{G_k}) dx = \sum_{k=1}^\infty \int_a^b \chi_{G_k} dx = \sum_{k=1}^\infty \mu(G_k) < \sum_{k=1}^\infty \frac{1}{2^k}$$

and so $\sum_{k=1}^\infty \chi_{G_k}$ is a Lebesgue integrable function on $[a, b]$. Hence the function

$$F(x) = \int_a^x (\sum_{k=1}^\infty \chi_{G_k}) dx$$

is non-decreasing and absolutely continuous on $[a, b]$. For each n let

$$F_n(x) = \int_a^x (\sum_{k=1}^n \chi_{G_k}) dx. \qquad (6.59)$$

For any fixed n, the function $F - F_n$ is non-decreasing in $[a, b]$. Also

$$\underline{D}F(x) = \underline{D}[F(x) - F_n(x) + F_n(x)] \geq \underline{D}(F(x) - F_n(x)) + \underline{D}F_n(x), \quad (6.60)$$

for all $x \in [a, b]$. If $x \in E$ then $x \in \bigcap_{k=1}^n G_k$ and since each G_k is an open set, each function χ_{G_k}, $1 \leq k \leq n$, is continuous at x, So, by (6.59)

$$F_n'(x) = \sum_{k=1}^n \chi_{G_k}(x) = \sum_{k=1}^n 1 = n$$

and so by (6.60) $\underline{D}F(x) \geq n$. Since n is arbitrary, $\underline{D}F(x) = \infty$.

Theorem 6.45 Let $f : [a, b] \to \mathbb{R}$ be Lebesgue integrable and let $F(x) = \int_a^x f$ for $x \in [a, b]$. Then the total variation of F is

$$V(F)\Big|_a^b = \int_a^b |f| dx.$$

Proof Let $D : a = x_0 < x_1 < \cdots < x_n = b$ be any partition of $[a, b]$. Then

$$\sum_{k=1}^n |F(x_k) - F(x_{k-1})| = \sum \left| \int_{x_{k-1}}^{x_k} f \right| \le \sum \int_{x_{k-1}}^{x_k} |f| = \int_a^b |f|.$$

Since D is any partition,

$$\overset{b}{\underset{a}{V}}(F) \le \int_a^b |f|. \tag{6.61}$$

To prove the reverse inequality consider the function g where $g(x) = 1$ if $f(x) > 0$, $g(x) = 0$ if $f(x) = 0$ and $g(x) = -1$ if $f(x) < 0$. Then g is Lebesgue integrable in $[a, b]$. Since the class of step functions is everywhere dense in the class of Lebesgue integrable functions (see Theorem 9.9), for each positive integer m there is a step function σ_m on $[a, b]$ such that

$$\int_a^b |g - \sigma_m| < \frac{1}{m}. \tag{6.62}$$

Let

$$\sigma_m(x) = \sum_{k=1}^{k_m} \alpha_k^m \chi_{[x_{k-1}^m, x_k^m]}(x), \; a = x_0^m < x_1^m < \ldots x_{k_m}^m = b \tag{6.63}$$

where α_k^m are constants and χ_E denotes the characteristic function of E. Since $|g(x)| \le 1$ for all $x \in [a, b]$ from (6.62) and (6.63) we may suppose that $|\alpha_k^m| \le 1$ for all k and all m. From (6.62) the sequence $\{\sigma_m\}$ converges to g in the mean and so it converges to g in measure (see Theorem 9.17). Hence there is a subsequence $\{\sigma_{m_i}\}$ of $\{\sigma_m\}$ which converges to g a.e. So by dominated convergence theorem

$$\lim_{i \to \infty} \int_a^b f \sigma_{m_i} = \int_a^b fg = \int_a^b |f|. \tag{6.64}$$

Since σ_{m_i} has the form (6.63)

$$\left| \int_a^b f \sigma_{m_i} \right| = \left| \int_a^b f \sum \alpha_k^{m_i} \chi_{[x_{k-1}^{m_i}, x_k^{m_i})} \right| = \left| \sum \alpha_k^{m_i} \int_{x_{k-1}^{m_i}}^{x_k^{m_i}} f \right|$$

$$= \left| \sum \alpha_k^{m_i} (F(x_k^{m_i}) - F(x_{k-1}^{m_i})) \right| \le \sum |\alpha_k^{m_i}| |F(x_k^{m_i}) - F(x_{k-1}^{m_i})|$$

$$\le \sum |F(x_k^{m_i}) - F(x_{k-1}^{m_i})| \le \overset{b}{\underset{a}{V}}(F).$$

So using (6.64)

$$\int_a^b |f| \le \overset{b}{\underset{a}{V}}(F). \tag{6.65}$$

By (6.61) and (6.65) the proof is complete. □

Corollary 6.46 *Let F be absolutely continuous in $[a, b]$. Then the total variation of F is*
$$V(F, a, b) = \int_a^b |F'|.$$

Since $F(x) = \int_a^x F'$ the result follows from the above theorem.

Theorem 6.47 *Let $F : [a, b] \to \mathbb{R}$. Then the following statements are equivalent.*
(a) F is absolutely continuous in $[a, b]$.
(b) For every $\epsilon > 0$ there is $\delta > 0$ such that $|\sum_{i=1}^{n}[F(x_i') - F(x_i)]| < \epsilon$ for every finite collection $\{[x_i, x_i']\}_{i=1}^n$ of non-overlapping sub intervals of $[a, b]$ satisfying $\sum_{i=1}^{n}(x_i' - x_i) < \delta$.
(c) For every $\epsilon > 0$ there is $\delta > 0$ such that $\sum_{i=1}^{\infty}|F(x_i') - F(x_i)| < \epsilon$ whenever $\{[x_i, x_i']\}$ is a countable collection of non-overlapping sub-intervals of $[a, b]$ satisfying $\sum_{i=1}^{\infty}(x_i' - x_i) < \delta$.

Proof $(a) \Leftrightarrow (b)$. Clearly $(a) \Rightarrow (b)$. Suppose (b) is true. Let $\epsilon > 0$ be arbitrary. Then there is $\delta > 0$ such that

$$\left|\sum_{i=1}^{n}[F(x_i') - F(x_i)]\right| < \frac{\epsilon}{2} \text{ whenever } \sum_{i=1}^{n}(x_i' - x_i) < \delta. \quad (6.66)$$

Writing $\sum_{i=1}^{n} = \sum_{+} + \sum_{-}$ where $\sum_{+}$ and $\sum_{-}$ are summations over those i for which $F(x_i') - F(x_i) \geq 0$ and $F(x_i') - F(x_i) < 0$ respectively, we have by (6.66)

$$\sum_{i=1}^{n}|F(x_i') - F(x_i)| = \sum_{+}|F(x_i') - F(x_i)| + \sum_{-}|F(x_i') - F(x_i)|$$

$$= \left|\sum_{+}[F(x_i') - F(x_i)]\right| + \left|\sum_{-}[F(x_i') - F(x_i)]\right| < \frac{\epsilon}{2} + \frac{\epsilon}{2} = \epsilon$$

whenever $\sum_{i=1}^{n}(x_i' - x_i) < \delta$. So $(b) \Rightarrow (a)$.

$(a) \Leftrightarrow (c)$. Let $\epsilon > 0$ be arbitrary. Then by (a) there is $\delta > 0$ such that

$$\sum_{i=1}^{n}|F(x_i') - F(x_i)| < \frac{\epsilon}{2} \text{ whenever } \sum_{i=1}^{n}(x_i' - x_i) < \delta. \quad (6.67)$$

Let $\{[x_i, x_i']\}$ be any countable collection of non-overlapping sub-intervals of $[a, b]$ such that $\sum_{i=1}^{\infty}(x_i' - x_i) < \delta$. Then $\sum_{i=1}^{n}|F(x_i') - F(x_i)| < \frac{\epsilon}{2}$ for every n and so letting $n \to \infty$.

$$\sum_{i=1}^{\infty}|F(x_i') - F(x_i)| \leq \frac{\epsilon}{2} < \epsilon.$$

So, $(a) \Rightarrow (c)$. The case $(c) \Rightarrow (a)$ is obvious. $\square$

6.10 Integration by Parts for Lebesgue Integral

Integration by parts is related to the integrability of the product of two functions. It may be noted that if two functions are Lebesgue integrable then it is not necessary that their product is Lebesgue integrable. For, consider the function $f(x) = \frac{1}{\sqrt{x}}$ for $x \neq 0$, $f(0) = 0$. Then f is integrable in $[0, 1]$ but f^2 is not. However we have

Theorem 6.48 *If f is Lebesgue integrable in $[a, b]$ and g is bounded and measurable in $[a, b]$ then fg is Lebesgue integrable in $[a, b]$.*

Proof Let $M > 0$ such that $|g(x)| \leq M$ for $x \in [a, b]$. Then fg is measurable in $[a, b]$ and $|f(x)g(x)| \leq M|f(x)|$ for all $x \in [a, b]$ and so

$$\int_a^b |fg|\,dx \leq M \int_a^b |f|\,dx < \infty$$

Hence fg is Lebesgue integrable in $[a, b]$. $\square$

Theorem 6.49 *(Integration by ports)* Let $f : [a, b] \to \mathbb{R}$ be integrable and let $F(x) = \int_a^x f\,dx$. If $G : [a, b] \to \mathbb{R}$ is absolutely continuous in $[a, b]$ then fG and FG' are integrable in $[a, b]$ and

$$\int_a^b fG\,dx = F(b)G(b) - \int_a^b FG'\,dx.$$

Proof Since F is absolutely continuous in $[a, b]$ the product FG is absolutely continuous and so by Theorem 6.41, the function $(FG)'$ and G' are Lebesgue integrable in $[a, b]$. By Theorem 6.48 FG' is Lebesgue integrable. So, $(FG)' - FG'$ is integrable in $[a, b]$. Since $(FG)' = fG + FG'$ almost every where in $[a, b]$, fG is integrable in $[a, b]$ and

$$\int_a^b fG = \int_a^b (FG)' - \int_a^b FG' = F(b)G(b) - \int_a^b FG'.$$

6.11 Change of Variable for Lebesgue Integral

Theorem 6.50 (Second Mean Value Theorem) *Let $f : [a, b] \to \mathbb{R}$ be a non-decreasing absolutely continuous function and let $g : [a, b] \to \mathbb{R}$ be Lebesgue integrable. Then there exists a point ξ in $[a, b]$ such that*

$$\int_a^b fg = f(a) \int_a^\xi g + f(b) \int_\xi^b g$$

Proof Let $G(x) = \int_a^x g$. Then by Theorem 6.49,

$$\int_a^b fg = G(b)f(b) - \int_a^b Gf' \tag{6.68}$$

Since G is absolutely continuous, it is bounded. Let m and M be the lower and upper bounds of G. Then $m \leq G(x) \leq M$ for all x and so

$$m[f(b) - f(a)] = m \int_a^b f' \leq \int_a^b Gf' \leq M \int_a^b f' = M[f(b) - f(a)]$$

and so there is $\mu \in [m, M]$ such that

$$\int_a^b Gf' = \mu[f(b) - f(a)] \tag{6.69}$$

from (6.68) and (6.69), noting that there is $\xi \in [a, b]$ such that $\mu = G(\xi)$,

$$\int_a^b fg = G(b)f(b) - \mu[f(b) - f(a)] = \mu f(a) + [G(b) - \mu]f(b)$$

$$= G(\xi)f(a) + [G(b) - G(\xi)]f(b) = f(a) \int_a^\xi g + f(b) \int_\xi^b g.$$

6.11 Change of Variable for Lebesgue Integral

Theorem 6.51 *Let ϕ be strictly increasing absolutely continuous function on $[\alpha, \beta]$ and let f be integrable on $[\phi(\alpha), \phi(\beta)]$. Then $(f \circ \phi)\phi'$ is integrable on $[\alpha, \beta]$ and*

$$\int_{\phi(\alpha)}^{\phi(\beta)} f \, dx = \int_\alpha^\beta (f \circ \phi)\phi' \, dx.$$

Proof Let $\phi(\alpha) = a$, $\phi(\beta) = b$. We first suppose that f is bounded. Let $\tau = \{E_i\}_{i=1}^n$ be any measurable partition of $[a, b]$. Let $m_i = \inf_{x \in E_i} f(x)$, $M_i = \sup_{x \in E_i} f(x)$. Let $F_i =$

$\{x \in [\alpha, \beta] : \phi(x) \in E_i\}$. Then $\tau' = \{F_i\}_{i=1}^n$ is a measurable partition of $[\alpha, \beta]$. For, since E_i are disjoint and measurable and ϕ is strictly increasing, $\phi^{-1}(E_i)$ are disjoint and measurable. Also $\bigcup_{i=1}^n F_i = [\alpha, \beta]$. To see this let $t \in [\alpha, \beta]$. Then $\phi(t) \in [a, b]$ and so there exists E_i such that $\phi(t) \in E_i$ and hence $t \in \phi^{-1}(E_i) = F_i$, proving that τ' is a measurable partition of $[\alpha, \beta]$. Also since $\{f(x) : x \in E_i\} = \{f(\phi(t)) : t \in F_i\}$, $m_i = \inf_{t \in F_i}(f \circ \phi)(t)$ and $M_i = \sup_{t \in F_i}(f \circ \phi)(t)$. Since $m_i \leq f \circ \phi \leq M_i$ on F_i, $m_i \phi' \leq (f \circ \phi)\phi' \leq M_i \phi'$ on F_i. So

$$m_i \int_{F_i} \phi' \leq \underline{\int}_{F_i} (f \circ \phi)\phi' \leq \overline{\int}_{F_i} (f \circ \phi)\phi' \leq M_i \int_{F_i} \phi'.$$

Since $\phi(F_i) = E_i$ for $i = 1, 2, ..., n$ using this and Theorem 5.5 and Theorem 6.43 we have

$$s(\tau, f) = \sum_{i=1}^n m_i \mu(E_i) = \sum_{i=1}^n m_i \mu(\phi(F_i)) = \sum_{i=1}^n m_i \int_{F_i} \phi'$$
$$\leq \sum_{i=1}^n \underline{\int}_{F_i} (f \circ \phi)\phi' = \underline{\int}_\alpha^\beta (f \circ \phi)\phi' \leq \overline{\int}_\alpha^\beta (f \circ \phi)\phi' \quad (6.70)$$
$$= \sum_{i=1}^n \overline{\int}_{F_i} (f \circ \phi)\phi' \leq \sum_{i=1}^n M_i \int_{F_i} \phi' = \sum_{i=1}^n M_i \mu(\phi(F_i)) = \sum_{i=1}^n M_i \mu(E_i)$$
$$= S(\tau, f)$$

Let $\epsilon > 0$ be arbitrary. Then since f is integrable on $[a, b]$ there is partition τ of $[a, b]$ such that

$$\int_a^b f - \epsilon < s(\tau, f) \leq S(\tau, f) < \int_a^b f + \epsilon. \quad (6.71)$$

Considering this partition we have from (6.70) and (6.71)

$$\int_a^b f - \epsilon < \underline{\int}_\alpha^\beta (f \circ \phi)\phi' \leq \overline{\int}_\alpha^\beta (f \circ \phi)\phi' < \int_a^b f + \epsilon.$$

Letting $\epsilon \to 0$, the proof is complete in this case.

Suppose now that f is non-negative. Let $[f]_n(x) = f(x)$ if $f(x) \leq n$ and $[f]_n(x) = n$ if $f(x) > n$. Then $[f]_n$ is bounded and so by the above

$$\int_a^b [f]_n = \int_\alpha^\beta ([f]_n \circ \phi)\phi'. \quad (6.72)$$

Since $\lim_{n\to\infty}([f]_n \circ \phi) = f \circ \phi$ in $[\alpha, \beta]$, $\lim_{n\to\infty}([f_n] \circ \phi)\phi' = (f \circ \phi)\phi'$ a.e. in $[\alpha, \beta]$ and so by monotone convergence theorem $i.e$ Theorem 5.28

6.11 Change of Variable for Lebesgue Integral

$$\lim_{n\to\infty} \int_\alpha^\beta ([f]_n \circ \phi)\phi' = \int_\alpha^\beta (f \circ \phi)\phi'. \qquad (6.73)$$

By (6.72) and (6.73)

$$\int_a^b f = \lim_{n\to\infty} \int_a^b [f]_n = \int_\alpha^\beta (f \circ \phi)\phi'$$

completing the proof in this case.

For the general case writing $f = f^+ - f^-$, since

$$(f^+ \circ \phi)(t) - (f^- \circ \phi)(t) = f^+(\phi(t)) - f^-(\phi(t)) = f(\phi(t)) = (f \circ \phi)(t),$$

we have applying the second case

$$\int_a^b f = \int_a^b f^+ - \int_a^b f^- = \int_\alpha^\beta (f^+ \circ \phi)\phi' - \int_\alpha^\beta (f^- \circ \phi)\phi' = \int_\alpha^\beta [(f^+ \circ \phi) - (f^- \circ \phi)]\phi'.$$

$$= \int_\alpha^\beta [(f^+ - f^-) \circ \phi]\phi' = \int_\alpha^\beta (f \circ \phi)\phi'.$$

completing the proof. □

Corollary 6.52 *Let f be integrable in $[a+h, b+h]$ where h is a fixed real number. Then the function $f(x+h)$ is integrable in $[a, b]$ and*

$$\int_{a+h}^{b+h} f(x)dx = \int_a^b f(x+h)dx$$

Proof Putting $\phi(x) = x + h$ for $x \in [a, b]$ the result follows from the above theorem. □

Definition 6.6 Let $f : [a, b] \to \mathbb{R}$. Then f is said to satisfy Lipschitz condition if there exists $k > 0$ such that $|f(x_1) - f(x_2)| \leq k|x_1 - x_2|$ for all $x_1, x_2 \in [a, b]$.

Theorem 6.53 *Let $f : [a, b] \to \mathbb{R}$. If f satisfies one of the following conditions then f is absolutely continuous.*

(i) f satisfies Lipschitz condition in $[a, b]$.

(ii) f has bounded derivative in $[a, b]$; more generally if f has bounded derivates in $[a, b]$.

Proof The proof of (i) is easy. To prove (ii) note that for any two points $x_1, x_2 \in [a, b]$ with $x_1 < x_2$ there is ξ, $x_1 < \xi < x_2$, such that $f(x_2) - f(x_1) = (x_2 - x_1)f'(\xi)$ and so $|f(x_2) - f(x_1)| \leq |x_2 - x_1|M$ where $M = \sup_{x\in[a,b]} f'(x)$. Therefore f satisfies a Lipschitz condition and so the proof follows from (i).

Finally, if f has bounded derivates then there is $M > 0$ such that $|D^+ f(x)| \leq M$, $|D_+ f(x)| \leq M$, $|D^- f(x)| \leq M$ and $|D_- f(x)| \leq M$ for all $x \in [a, b]$. So it is easy prove that f is continuous in $[a, b]$. So, by Theorem 6.34 $|f(x_2) - f(x_1)| \leq M|x_2 - x_1|$ for $x_1, x_2 \in [a, b]$. Hence the proof follows by (i) □

Theorem 6.54 Let $f : [a, b] \to \mathbb{R}$ and $g : [c, d] \to [a, b]$ be absolutely continuous. If one of the following conditions are satisfied then the composite function $f \circ g$ is absolutely continuous
(i) $f \circ g$ is of bounded variation in $[c, d]$.
(ii) f satisfies Lipschitz condition.
(iii) f has bounded derivates.
(iv) g is monotone.

Proof (i) Suppose that $f \circ g$ is of bounded variation in $[c, d]$. Let $E \subset [c, d]$ be any set of measure zero. Since g satisfies Luzin condition (N), $g(E)$ is of measure zero and hence $f(g(E))$ is of measure zero. So $f \circ g$ satisfies the Luzin condition (N). Also $f \circ g$ is continuous in $[c, d]$. Therefore by Theorem 6.27, $f \circ g$ is absolutely continuous in $[c, d]$
(ii) Suppose that f satisfies Lipschitz condition. Then there is K such that $|f(x') - f(x)| \leq K|x' - x|$ for all $x, x' \in [a, b]$. Let $\{[c_i, d_i]\}_{i=1}^n$ be any finite collection of non-overlapping intervals in $[c, d]$.Then

$$\sum_{i=1}^n |f(g(d_i)) - f(g(c_i))| \leq \sum_{i=1}^n K|g(d_i) - g(c_i)| \leq K \vee (g; c, d)$$

where $\vee(g; c, d)$ is the total variation of g in $[c, d]$. So, $f \circ g$ is of bounded variation in $[c, d]$ and hence applying the argument of (i) $f \circ g$ is absolutely continuous.
(iii) Suppose that all the derivates $D^+ f, D_+ f, D^- f$ and $D_- f$ are bounded in $[a, b]$. Then by Theorem 6.34 f satisfies Lipschitz condition and so the proof follows by applying (ii).
(iv) Suppose that g is non-decreasing. Let $\{[c_i, d_i]\}_{i=1}^n$ be any finite collection of non-overlapping intervals in $[c, d]$. Then $\{[g(c_i), g(d_i)]\}_{i=1}^n$ is a collection of non-overlapping intervals in $[a, b]$. (If $g(c_i) = g(d_i)$ for some i we exclude it) So,

$$\sum_{i=1}^n |f(g(d_i)) - f(g(c_i))| \leq \vee(f; a, b).$$

Since f is absolutely continuous, $V(f; a, b) < \infty$. Hence $f \circ g$ is of bounded variation in $[c, d]$. Applying (i) the proof is complete.
The condition (i) of the above theorem is necessary as the following example shows. □

Example 6.7 Show that the functions $F(x) = \sqrt{x}$ and $G(x) = \left|x^2 \sin \frac{\pi}{x}\right|$ for $x \neq 0$, $G(0) = 0$ are absolutely continuous in $[0, 1]$ but the composite function $F \circ G$ is not absolutely continuous.

Solution Let $\epsilon > 0$ be arbitrary. Choose α, $0 < \alpha < 1$, such that $\sqrt{\alpha} < \frac{\epsilon}{2}$. Since F has bounded derivative in $[\alpha, 1]$, by Theorem 6.53 (ii) F is absolutely continuous in $[\alpha, 1]$. So there is $\delta > 0$ such that $\sum_{i=1}^n |F(x_i') - F(x_i)| < \frac{\epsilon}{2}$ for every collection

6.12 Lebesgue Set

$\{[x_i, x'_i]\}_{i=1}^n$ of sub-intervals of $[\alpha, 1]$ satisfying $\sum_{i=1}^n |x'_i - x_i| < \delta$. Let $\{[c_i, d_i]\}_{i=1}^m$ be any collection of sub-intervals of $[0, 1]$. We may suppose that for each i either $[c_i, d_i] \subset [0, \alpha]$ or $[c_i, d_i] \subset [\alpha, 1]$. Then since F is increasing and $F(\alpha) = \sqrt{\alpha} < \frac{\epsilon}{2}$

$$\sum_{i=1}^m |F(d_i) - F(c_i)| = \sum_{d_i \leq \alpha} |F(d_i) - F(c_i)| + \sum_{c_i \geq \alpha} |F(d_i) - F(c_i)| < F(\alpha) + \frac{\epsilon}{2} < \epsilon.$$

So F is absolutely continuous in $[0, 1]$.

The function $f(x) = x^2 \sin \frac{\pi}{x}$ for $x \neq 0$ and $f(0) = 0$ has bounded derivative and so it is absolutely continuous and hence by Theorem 6.28 $|f| = G$ is absolutely continuous in $[0, 1]$.

Finally $(F \circ G)(x) = F(G(x)) = \sqrt{|x^2 \sin \frac{\pi}{x}|} = x\sqrt{|\sin \frac{\pi}{x}|}$. If possible suppose that $F \circ G$ is absolutely continuous in $[0, 1]$. Let $\epsilon > 0$ be arbitrary. Then by Theorem 6.47(c) there is $\delta > 0$ such that whenever $\{[x_i, x'_i]\}_{i=1}^\infty$ is a collection of non-overlapping sub-intervals of $[0, 1]$ satisfying $\sum_{i=1}^\infty (x'_i - x_i) < \delta$ the following relation holds:

$$\sum_{i=1}^\infty |(F \circ G)(x'_i) - (F \circ G)(x_i)| < \epsilon. \tag{6.74}$$

Consider the collection $\{[\frac{2}{4n+1}, \frac{1}{2n}]\}_{n=1}^\infty$ of non-overlapping sub-intervals of $[0, 1]$. Then $\sum_{n=1}^\infty (\frac{1}{2n} - \frac{2}{4n+1}) = \sum_{n=1}^\infty \frac{1}{2n(4n+1)}$ which is a convergent series and hence there is N such that $\sum_{n=N}^\infty \frac{1}{2n(4n+1)} < \delta$. So by (6.74)

$$\sum_{i=N}^\infty |(F \circ G)(\frac{1}{2n}) - (F \circ G)(\frac{2}{4n+1})| < \epsilon. \tag{6.75}$$

But $(F \circ G)(\frac{1}{2n}) = 0$ and $(F \circ G)(\frac{2}{4n+1}) = \frac{2}{4n+1}$ and so by (6.75) $\sum_{i=N}^\infty \frac{2}{4n+1} < \epsilon$ which is a contradiction since the series $\sum_{n=1}^\infty \frac{2}{4n+1}$ is divergent.

Note that $F \circ G$ does not satisfy the condition (i) of Theorem 6.54.

6.12 Lebesgue Set

Definition 6.7 Let $f : [a, b] \to \mathbb{R}$ to integrable. If for a point $x \in [a, b]$,

$$\lim_{h \to 0} \frac{1}{h} \int_x^{x+h} |f(t) - f(x)| dt = 0$$

then x is called a Lebesgue point of f. The set of all points x in $[a, b]$ that are Lebesgue points of f is called Lebesgue set of f.

Theorem 6.55 *Let f be integrable in $[a, b]$. Then almost all points of $[a, b]$ are Lebesgue points of f.*

Proof Let $\{r_1, r_2, \ldots r_n, \ldots\}$ be an enumeration of the set of rational numbers. Since f is integrable the function g_n where $g_n(t) = |f(t) - r_n|$ is integrable for each n. So, by Theorem 6.40 there is a set $E_n \subset [a, b]$ such that $\mu(E_n) = 0$ and

$$\lim_{h \to 0} \frac{1}{h} \int_x^{x+h} g_n(t) dt = g_n(x) \text{ for all } x \in [a, b] \sim E_n.$$

i.e

$$\lim_{h \to 0} \frac{1}{h} \int_x^{x+h} |f(t) - r_n| dt = |f(x) - r_n| \text{ for all } x \in [a, b] \sim E_n. \quad (6.76)$$

Let $E = \bigcup_{n=1}^{\infty} E_n \cup \{x \in [a, b] : |f(x)| = \infty\}$. Then $\mu(E) = 0$. Let $x \in [a, b] \sim E$ and $\epsilon > 0$ be arbitrary. Choose n such that

$$|f(x) - r_n| < \frac{\epsilon}{3}. \quad (6.77)$$

Since $x \in [a, b] \sim E \subset [a, b] \sim E_n$, by (6.76) there is $\delta > 0$ such that

$$\left| \frac{1}{h} \int_x^{x+h} |f(t) - r_n| dt - |f(x) - r_n| \right| < \frac{\epsilon}{3} \text{ for } 0 < |h| < \delta.$$

which gives by (6.77)

$$\frac{1}{h} \int_x^{x+h} |f(t) - r_n| dt < \frac{2\epsilon}{3} \text{ for } 0 < |h| < \delta. \quad (6.78)$$

from (6.77) and (6.78)

$$\frac{1}{h} \int_x^{x+h} |f(t) - f(x)| \leq \frac{1}{h} \int_x^{x+h} |f(t) - r_n| + |f(x) - r_n| < \frac{2\epsilon}{3} + \frac{\epsilon}{3} = \epsilon \text{ for } 0 < |h| < \delta$$

which shows that

$$\lim_{h \to 0} \int_x^{x+h} |f(t) - f(x)| dt = 0.$$

So x is a Lebesgue point of f. Since x is any point in $[a, b] \sim E$ and $\mu(E) = 0$, the proof is complete. □

6.12 Lebesgue Set

Theorem 6.56 *Let f be integrable in $[a, b]$. Then there exists a set $E \subset [a, b]$ such that $\mu(E) = 0$ and*

$$\lim_{h \to 0} \frac{1}{h} \int_x^{x+h} |f(t) - \xi| dt = |f(x) - \xi| \text{ for all } \xi \in \mathbb{R} \text{ and all } x \in [a, b] \sim E.$$

Proof By Theorem 6.55 there exists a set $E \subset [a, b]$ such that $\mu(E) = 0$ and

$$\lim_{h \to 0} \frac{1}{h} \int_x^{x+h} |f(t) - f(x)| dt = 0 \text{ for } x \in [a, b] \sim E. \tag{6.79}$$

Let $\xi \in \mathbb{R}$ and $x \in [a, b] \sim E$. Then for $t \in [a, b]$

$$|f(t) - \xi| \leq |f(t) - f(x)| + |f(x) - \xi|$$

and so by (6.79)

$$\lim_{h \to 0} \frac{1}{h} \int_x^{x+h} |f(t) - \xi| dt \leq |f(x) - \xi|. \tag{6.80}$$

Also

$$|f(x) - \xi| \leq |f(t) - f(x)| + |f(t) - \xi|$$

and so by (6.79)

$$|f(x) - \xi| \leq \lim_{h \to 0} \frac{1}{h} \int_x^{x+h} |f(t) - \xi| dt. \tag{6.81}$$

The proof is complete by (6.80) and (6.81). $\square$

Remark The Lebesgue set of f contains all the points of continuity of f. For, suppose that f is continuous at x. Let $\epsilon > 0$ be arbitrary. Then there is $\delta > 0$ such that $|f(t) - f(x)| < \epsilon$ for $|t - x| < \delta$. Let $|h| < \delta$ then

$$\frac{1}{h} \int_x^{x+h} |f(t) - f(x)| dt < \epsilon \text{ and so } \lim_{h \to 0} \frac{1}{h} \int_x^{x+h} |f(t) - f(x)| dt = 0.$$

Corollary 6.57 *Let f be integrable in $[a, b]$. Then for almost all $x \in [a, b]$*

$$\lim_{h \to 0} \frac{1}{h} \int_0^h |f(x + t) + f(x - t) - 2f(x)| dt = 0.$$

Proof For fixed $x \in [a, b]$ we have

$$\frac{1}{h}\int_0^h |f(x+t) + f(x-t) - 2f(x)|dt \le \frac{1}{h}\int_0^h [|f(x+t) - f(x)| + |f(x-t) - f(x)|]dt$$

$$= \frac{1}{h}\int_0^h |f(x+t) - f(x)|dt + \frac{1}{h}\int_0^h |f(x-t) - f(x)|dt$$

$$= \frac{1}{h}\int_x^{x+h} |f(t) - f(x)|dt + \frac{1}{h}\int_{x-h}^x |f(t) - f(x)|dt$$

by a change of variable formula (Theorem 6.51). Applying Theorem 6.55 the result follows. □

This result is useful in the theory of trigonometric series.

Theorem 6.58 *Let f be integrable in $[a, b]$. and $x_0 \in [a, b]$. Then the integral $F(x) = \int_a^x f(t)dt$ is differentiable at x_0 if x_0 is a Lebesgue point of f.*

Proof Let x_0 be a Lebesgue point of f. Then

$$\left|\frac{F(x_0 + h) - F(x_0)}{h} - f(x_0)\right| = \left|\frac{1}{h}\int_{x_0}^{x_0+h} f(t)dt - f(x_0)\right| \le \frac{1}{h}\int_{x_0}^{x_0+h} |f(t) - f(x_0)|dt$$

which proves the theorem. The converse is not true in general. □

6.13 Singular Function

Definition 6.8 A function $f : [a, b] \to \mathbb{R}$ is called singular if $f' = 0$ almost everywhere in $[a, b]$.

So every constant function is a singular function. The Cantor function is a non-constant singular function.

Theorem 6.59 *If a function $f : [a, b] \to \mathbb{R}$ is both absolutely continuous and singular then f is a constant.*

Proof Suppose f is absolutely continuous. Since f is singular $f' = 0$ a.e. Therefore by Corollary 6.42, f is constant. □

Theorem 6.60 *If $f : [a, b] \to \mathbb{R}$ is of bounded variation then $f = g + h$ where $g : [a, b] \to \mathbb{R}$ is absolutely continuous and $h : [a, b] \to \mathbb{R}$ is singular. The function g and h are unique up to additive constants.*

Proof Let $g(x) = \int_a^x f'$ and $h = f - g$. Then $h' = f' - g' = 0$ a.e. in $[a, b]$. So, h is singular. Since g is absolutely continuous and $f = g + h$ the result follows. For the uniqueness, suppose $f = g_1 + h_1$ is another decomposition. Then $g - g_1 = h_1 - h$. Since $g - g_1$ is absolutely continuous and $h_1 - h$ is singular, it follows from Theorem 6.59 that $g - g_1 = h_1 - h = $ constant which completes the proof. □

6.14 Points of Density and Approximate Continuity

Definition 6.9 Let $E \subset \mathbb{R}$ and $x \in \mathbb{R}$. For any $h > 0$ define

$$\phi(h) = \frac{\mu^*(E \cap [x, x+h])}{h}, \quad \mu^* \text{ denoting outer Lebesgue measure.}$$

Then the upper and lower limits of $\phi(h)$ as $h \to 0+$ are called the upper and lower outer right density of E at x and are denoted by $\overline{d}_+^*(E, x)$ and $\underline{d}_+^*(E, x)$ respectively. Since $0 \leq \phi(h) \leq 1$, $0 \leq \underline{d}_+^*(E, x) \leq \overline{d}_+^*(E, x) \leq 1$. If $\overline{d}_+^*(E, x_0) = \underline{d}_+^*(E, x)$ i.e if $\lim_{h \to 0+} \phi(h)$ exists then this limit is called the outer right density of E at x and is denoted by $d_+^*(E, x)$. If, in particular, $d_+^*(E, x) = 1$ then x is called a point of outer right density of E.

Left densities are similarly defined. If $d_+^*(E, x_0) = d_-^*(E, x)$ then the common value, denoted by $d^*(E, x)$ is called the outer density of E at x. The point x is called a point of outer density of E or a point of dispersion of E according as $d^*(E, x) = 1$ or 0. So, the point x is a point of outer density or a point of dispersion of E according as

$$d^*(E, x) = \lim_{\mu(I_n) \to 0} \frac{\mu^*(E \cap I_n)}{\mu(I_n)} = 1 \text{ or } 0,$$

for every sequence of intervals $\{I_n\}$ converging to x. If E is measurable, we write $d(E, x)$ instead of $d^*(E, x)$, and say density instead of outer density. So, if E is measurable then x is a point of density of E if and only if x is a point of dispersion of the complement $\widetilde{E}$ of E. For if $\{I_n\}$ is any sequence of intervals converging to x then since $(E \cap I_n) \cup (\widetilde{E} \cap I_n) = I_n$,

$$\frac{\mu(E \cap I_n)}{\mu(I_n)} + \frac{\mu(\widetilde{E} \cap I_n)}{\mu(I_n)} = 1$$

and hence

$$\lim_n \frac{\mu(E \cap I_n)}{\mu(I_n)} = 1 \text{ or } 0 \text{ according as } \lim_n \frac{\mu(\widetilde{E} \cap I_n)}{\mu(I_n)} = 0 \text{ or } 1.$$

Theorem 6.61 *If E is any measurable subset of $\mathbb{R}$ then almost all points of E are points of density of E and almost all points of the complement $\widetilde{E}$ of E are points of dispersion of E.*

Proof Let χ_E be the characteristic function of E i.e. $\chi_E(x) = 1$ if $x \in E$ and $\chi_E(x) = 0$ if $x \notin E$. Then χ_E is integrable over every interval $[a, b]$. Let

$$F(x) = \int_a^x \chi_E \, dt, \quad x \in [a, b].$$

Then by Theorem 6.40 $F' = \chi_E$ a.e. in $[a, b]$. Hence $F'(x) = 1$ a.e. in $E \bigcap [a, b]$ and $F'(x) = 0$ a.e. in $\widetilde{E} \bigcap [a, b]$. Let $x \in E \bigcap [a, b]$ be such that $F'(x) = 1$. Let $\{I_n\}$ be any sequence of intervals converging to x. Then

$$\frac{\mu(E \bigcap I_n)}{\mu(I_n)} = \frac{1}{\mu(I_n)} \int_{I_n} \chi_E dt \to F'(x) = 1 \text{ as } n \to \infty.$$

Hence x is a point of density of E. So, almost all points of E are points of density of E. Similarly if $x \in \widetilde{E} \bigcap [a, b]$ is such that $F'(x) = 0$ then taking $\{I_n\}$ as above

$$\frac{\mu(E \bigcap I_n)}{\mu(I_n)} = \frac{1}{\mu(I_n)} \int_{I_n} \chi_E dt \to F'(x) = 0 \text{ as } n \to \infty.$$

and so x is a point of dispersion of E and so almost all points of $\widetilde{E} \bigcap [a, b]$ are points of dispersion of E. Since $[a, b]$ is any interval, the result follows. □

Theorem 6.62 *If A is any subset of $\mathbb{R}$ then almost all points of A are points of outer density of A. The set A is measurable if and only if almost all points of $\widetilde{A}$ are points of dispersion of A.*

Proof Let A be any set. Since the Lebesgue outer measure μ^* is regular by Theorem 2.12 there is a measurable set E such that $A \subset E$ and $\mu^*(A) = \mu(E)$. Let J be any interval. Then

$$\mu(E \cap J) = \mu(E) - \mu(E \sim J) = \mu^*(A) - \mu(E \sim J)$$
$$\leq \mu^*(A) - \mu^*(A \sim J) \leq \mu^*(A \cap J) \leq \mu(E \cap J)$$

Hence $\mu(E \cap J) = \mu^*(A \cap J)$. This shows that the outer density of A at any point x is equal to the density of E at x. Since by Theorem 6.61 almost all points of E are points of density of E, almost all points A are points of outer density of A.

To prove the second part, let A be measurable. Then by Theorem 6.61 almost all points of $\widetilde{A}$ are points of dispersion of A. Conversely suppose that almost all points of $\widetilde{A}$ are points of dispersion of A. If possible suppose that A is nonmeasurable. Let E be a measurable set such that $A \subset E$ and $\mu^*(A) = \mu(E)$. Then by the first part of this theorem almost all points of $E \cap \widetilde{A}$ are points of outer density of $E \cap \widetilde{A}$ and hence almost all points of $E \cap \widetilde{A}$ are points of density of E and hence almost all points of $E \cap \widetilde{A}$ are points of outer density of A. So almost all points of $E \cap \widetilde{A}$ are not points of dispersion of A. But $\mu^*(E \cap \widetilde{A}) > 0$. For, if $\mu^*(E \cap \widetilde{A}) = 0$ then $E \cap \widetilde{A}$ is measurable and since $A = E \sim (E \cap \widetilde{A})$, A is measurable which contradicts our supposition. Thus $\mu^*(E \cap \widetilde{A}) > 0$. Hence the set of points of $\widetilde{A}$ which are not points of dispersion of A is of positive outer measure. This contradicts the hypotheses. So, A is measurable. □

Theorems 6.61 and 6.62 are called Lebesgue's Density Theorems.

Definition 6.10 Let $f : [a, b] \to \mathbb{R}$. Then a number $l \in \mathbb{R}$ is said to be the approximate limit of f at $x_0 \in [a, b]$ if for every $\epsilon > 0$. x_0 is a point of dispersion of the set $\{x \in [a, b] : |f(x) - l| \geq \epsilon\}$.
The function f is said to be approximately continuous at x_0 if for any $\epsilon > 0$, x_0 is a point of dispersion of the set $\{x \in [a, b] : |f(x) - f(x_0)| \geq \epsilon\}$.

Theorem 6.63 *Let $f : [a, b] \to \mathbb{R}$ be measurable. Then f is appropriately continuous at a point $x_0 \in [a, b]$ if and only if for every $\epsilon > 0$, x_0 is a point of density of the set $\{x \in [a, b] : |f(x) - f(x_0)| < \epsilon\}$.*

Proof To prove the theorem note that if $E \subset [a, b]$ is measurable and if $\widetilde{E} = [a, b] \sim E$ then for any interval $I, I \subset [a, b], (E \cap I) \cup (\widetilde{E} \cap I) = I$ and so $\frac{\mu(E \cap I)}{\mu(I)} + \frac{\mu(\widetilde{E} \cap I)}{\mu(I)} = 1$. Hence for any sequence of intervals $\{I_n\}$, $I_n \subset [a, b]$ which converges to x_0

$$\lim_{n \to \infty} \frac{\mu(E \cap I_n)}{\mu(I_n)} + \lim_{n \to \infty} \frac{\mu(\widetilde{E} \cap I_n)}{\mu(I_n)} = 1$$

So, x_0 is a point of dispersion of E if and only if x_0 is a point of density of $\widetilde{E}$. This argument will be used in the following.

Let $\epsilon > 0$ be arbitrary. Suppose that f is approximately continuous at x_0. Then x_0 is a point of dispersion of $E = \{x; x \in [a, b]; |f(x) - (x_0)| \geq \epsilon\}$. Since f is measurable, E is measurable and so x_0 is a point of density of $\widetilde{E} = \{x : x \in [a, b]; |f(x) - f(x_0)| < \epsilon\}$. Conversely, suppose that x_0 is a point of density of $\widetilde{E}$. Then x_0 is a point of dispersion of E and hence f is approximately continuous at x_0. □

Note that analogous result holds for approximate limit.

6.15 Properties of Approximately Continuous Function

Theorem 6.64 *If $f : [a, b] \to \mathbb{R}$ is continuous at $x_0 \in [a, b]$ then f is approximately continuous at x_0.*

Proof Let $\epsilon > 0$ be arbitrary. Since f is continuous at x_0, there is $\delta > 0$ such that $|f(x) - f(x_0)| < \epsilon$ whenever $|x - x_0| < \delta$. Hence writing $E = \{x \in [a, b] : |f(x) - f(x_0)| \geq \epsilon\}$, $E \cap (x_0 - \delta, x_0 + \delta) = \emptyset$. Hence if I is any interval such that $x_0 \in I \subset (x_0 - \delta, x_0 + \delta)$ then $E \cap I = \emptyset$ and so $\frac{\mu^*(E \cap I)}{\mu(E)} = 0$, showing that x_0 is a point of dispersion of E.

The converse of the above theorem is not true. For, consider the function $\psi(x) = 0$ if x is irrational and $\psi(x) = 1$ if x is rational. Then ψ is nowhere continuous but approximately continuous at every irrational point. For let x_0 be an irrational point and let $\epsilon > 0$ be arbitrary. Then $E = \{x : |f(x) - f(x_0)| \geq \epsilon\}$ is the set of all rational numbers or the null set $\emptyset$ according as $\epsilon \leq 1$ or $\epsilon > 1$. In any case $\mu(E) = 0$. So, x_0 is a point of dispersion of E. □

Theorem 6.65 *Let $f : [a, b] \to \mathbb{R}$. Then f is measurable if and only if f is approximately continuous almost everywhere in $[a, b]$.*

Proof Suppose that f is measurable. Let $\gamma > 0$ be arbitrary. By Luzin's theorem i.e Theorem 3.24 there is a closed set $F \subset [a, b]$ such that $\mu([a, b] \sim F) < \gamma$ and the restriction of f to F is continuous. Let $x_0 \in F$ be a point of density of F. Let $\epsilon > 0$ be arbitrary. Then there is $\delta > 0$ such that

$$|f(x) - f(x_0)| < \epsilon \text{ for all } x \in (x_0 - \delta, x_0 + \delta) \cap F.$$

Since x_0 is a point of density of F, x_0 is a point of dispersion of $\widetilde{F}$ and so x_0 is a point of dispersion of $\widetilde{F} \cap (x_0 - \delta, x_0 + \delta)$. Since $\{x \in [a, b] : |f(x) - f(x_0)| \geq \epsilon\} \cap (x_0 - \delta, x_0 + \delta)) \subset \widetilde{F} \cap (x_0 - \delta, x_0 + \delta)$, x_0 is a point of dispersion of $\{x : x \in [a, b]; |f(x) - f(x_0)| \geq \epsilon\} \cap (x_0 - \delta, x_0 + \delta)$ and so x_0 is a point of dispersion of $\{x \in [a, b] : |f(x) - f(x_0)| \geq \epsilon\}$. Hence f is approximately continuous at x_0. Since almost all points of F are points of density of F by Theorem 6.61, f is approximately continuous almost everywhere in F. So the set of points in $[a, b]$ where f is not approximately continuous is of measure less than γ. Since γ is arbitrary the set of points where f is not approximately continuous is of measure zero.

Conversely, suppose that f is approximately continuous almost everywhere in $[a, b]$. Let c be any real number and let $E = \{x \in [a, b] : f(x) > c\}$. We show that E is measurable. If $\mu^*(E) = 0$ then E is measurable and so we suppose that $\mu^*(E) > 0$. Let $x_0 \in E$ be such that f is approximately continuous at x_0. Let $\epsilon = f(x_0) - c$. Then $\epsilon > 0$. Since f is approximately continuous at x_0, x_0 is a point of dispersion of the set F where $F = \{x \in [a, b] : |f(x) - f(x_0)| \geq \epsilon\}$. Let $\widetilde{E}$ be the complement of E in $[a, b]$ i.e $\widetilde{E} = [a, b] \sim E$. Let $\xi \in \widetilde{E}$. Then $f(\xi) \leq c$ and hence $f(x_0) - f(\xi) \geq f(x_0) - c = \epsilon$. Hence $|f(\xi) - f(x_0)| \geq \epsilon$ and so $\xi \in F$. Thus $\widetilde{E} \subset F$. Since x_0 is a point of dispersion of F, x_0 is also a point of dispersion of $\widetilde{E}$. Since this is true for almost all $x_0 \in E$ almost all point of E are points of dispersion of $\widetilde{E}$. So, putting $A = \widetilde{E}$ and applying the last part of Theorem 6.62, the set $\widetilde{E}$ is measurable and consequently the set E is measurable. Therefore f is measurable. □

Note that a nonmeasurable function may be continuous at a point (see discussion below Theorem 3.12.

Recall that a function f is called essentially bounded if there is $M > 0$ such that $|f| \leq M$ almost everywhere. It may be noted that if an approximately continuous function is essentially bounded then it is bounded. For, if f is essentially bounded then there is M such that $|f| \leq M$ almost everywhere and so the set $\{x : |f(x)| > M\}$ is of measure zero and hence f cannot be approximately continuous at any point of this set and so $\{x : |f(x)| > M\} = \emptyset$.

Corollary 6.66 *Let $f : [a, b] \to \mathbb{R}$ be essentially bounded, then f is integrable if and only if f is mesearable.*

Proof If f is bounded then the result is true by Theorem 5.15. Otherwise, let

6.15 Properties of Approximately Continuous Function

$$g(x) = f(x) \text{ if } |f(x)| \leq M$$
$$= 0 \text{ otherwise.}$$

Then g is bounded and $g = f$ almost everywhere, and so the result is true for g. Hence it is true for f. □

Corollary 6.67 *Let $f : [a, b] \to \mathbb{R}$ be essentially bounded, then f is integrable if and only if f is approximately continuous almost everywhere.*

Proof Apply Theorem 6.65 and Corollary 6.66.
This is the analogue of Lebesgue-Vitali theorem for Lebesgue integral. (See Theorem 5.19)

Approximately continuous functions share many properties of continuous function. For example, they have intermediate value property and monotonicity property. But unlike continuous functions, they may not possess boundedness property *i.e* if f is approximately continuous in $[a, b]$ then it may not be bounded there. To see this consider. □

Example 6.8 There exists a function f which is approximately continuous in $[0, 1]$ but not bounded there.

Solution Let $E \subset [0, 1]$ be the set defined by $E = \bigcup_{n=2}^{\infty}(a_n, b_n)$, where $a_n = \frac{1}{n}, b_n = \frac{1}{n} + \frac{1}{n^2}$. For each $x > 0$ let $g(x) = \frac{\mu(E \cap (0,x))}{x}$. So if $x \in [b_{n+1}, a_n]$ then $g(x) = \frac{1}{x} \sum_{k=n+1}^{\infty}(b_k - a_k) = \frac{1}{x}\sum_{k=n+1}^{\infty}\frac{1}{k^2}$ and so $g(a_n) = \frac{1}{a_n}\sum_{k=n+1}^{\infty}\frac{1}{k^2}$. Since for $x \in [b_{n+1}, a_n], x \leq a_n, g(x) \geq g(a_n)$. But $g(a_n) = n\sum_{k=n+1}^{\infty}\frac{1}{k^2} \geq n\sum_{k=n+1}^{\infty}(\frac{1}{k} - \frac{1}{k+1}) = \frac{n}{n+1}$ and hence $g(x) \geq \frac{n}{n+1}$. Since as $x \to 0, n \to \infty$, $\liminf_{x \to 0} g(x) \geq \lim_{n \to \infty}\frac{n}{n+1} = 1$. Since $0 \leq g(x) \leq 1$, $\lim_{x \to 0} g(x) = 1$ *i.e.* $\lim_{x \to 0}\frac{\mu(E \cap (0,x))}{x} = 1$. So, 0 is a point of density of E, and hence 0 is a point of dispersion of the set $[0, 1] \sim E = \bigcup_{n=1}^{\infty}[b_{n+1}, a_n]$. Define $f(0) = 0$ and $f(x) = 0$ for $x \in E$. To define f on $(0, 1] \sim E$, consider any interval $[b_{n+1}, a_n]$ of $[0, 1] \sim E$. Let $c_n = \frac{b_{n+1}+a_n}{2}$ and define $f(b_{n+1}) = f(a_n) = 0$, $f(c_n) = n$ and f is linear and continuous in each of the intervals $[b_{n+1}, c_n]$ and $[c_n, a_n]$. Then f is continuous every where in $(0, 1]$ but discontinuous at $x = 0$ where f is approximately continuous. Thus f is approximately continuous in $[0, 1]$ but f is unbounded in every neighborhood of $x = 0$.

Theorem 6.68 *Let f be bounded and Lebesgue integrable in $[a, b]$ and let*

$$F(x) = \int_a^x f\,dx \text{ for } x \in [a, b].$$

If f is approximately continuous at $x_0 \in [a, b]$, then $F'(x_0) = f(x_0)$.

Proof Since f is bounded, there is $M > 0$ such that $|f(x)| \leq M$ for $x \in [a, b]$. Hence $|f(x) - f(x_0)| \leq 2M$ for all $x \in [a, b]$. Let $\epsilon > 0$ be arbitrary. Let

$$A = \{x \in [a, b] : |f(x) - f(x_0)| < \epsilon\}, B = [a, b] \sim A.$$

Suppose that f is approximately continuous at x_0. Then x_0 is a point of dispersion of B. Since f is measurable, the set A is measurable and hence x_0 is a point of density of A. Let $I = [x_0, x_0 + h]$. Then

$$\left| \frac{F(x_0 + h) - F(x_0)}{h} - f(x_0) \right| = \left| \frac{1}{h} \int_{x_0}^{x_0+h} (f(t) - f(x_0)) dt \right|$$

$$\leq \frac{1}{h} \int_{x_0}^{x_0+h} |f(t) - f(x_0)| dt$$

$$= \frac{1}{h} \int_{A \cap I} |f(t) - f(x_0)| dt + \frac{1}{h} \int_{B \cap I} |f(t) - f(x_0)| dt$$

$$\leq \frac{1}{h} \epsilon \mu(A \cap I) + \frac{1}{h} 2M \mu(B \cap I).$$

Since x_0 is a point of density of A and a point of dispersion of B,

$$\frac{1}{h} \mu(A \cap I) \to 1 \text{ and } \frac{1}{h} \mu(B \cap I) \to 0 \text{ as } h \to 0+$$

and since ϵ is arbitrary,

$$\lim_{h \to 0+} \frac{F(x_0 + h) - F(x_0)}{h} = f(x_0).$$

Similarly considering $J = [x_0 - h, x_0]$ we get

$$\lim_{h \to 0+} \frac{F(x_0 - h) - F(x_0)}{-h} = f(x_0).$$

and this completes the proof. $\square$

Theorem 6.69 *Let $f : [a, b] \to \mathbb{R}$ be measurable and let for $M > N$,*

$$g(x) = \begin{cases} M & \text{if } f(x) > M \\ f(x) & \text{if } N \leq f(x) \leq M \\ N & \text{if } f(x) < N \end{cases}$$

If f is approximately continuous at $x_0 \in [a, b]$ then g is also approximately continuous at x_0.

Proof Let f be approximately continuous at $x_0 \in [a, b]$. If $f(x_0) > M$, choose ϵ, $0 < \epsilon < f(x_0) - M$. Then x_0 is a point of density of the set $A = \{x \in [a, b] :$

$|f(x) - f(x_0)| < \epsilon$. Since $A \subset B = \{x \in [a,b] : f(x) > M\}$, x_0 is a point of density of B. Since $g(x) = M$ for $x \in B$, g is approximately continuous at x_0. If $f(x_0) < N$, choose $\epsilon, 0 < \epsilon < N - f(x_0)$. Applying similar argument, g is approximately continuous at x_0. If $f(x_0) = M$ then for $\epsilon > 0$, x_0 is a point of density of the set $\{x \in [a,b] : M - \epsilon < f(x) < M + \epsilon\}$. But

$$\{x \in [a,b] : M - \epsilon < f(x) < M + \epsilon\} \subset \{x \in [a,b] : M - \epsilon < g(x) \le M\}$$
$$= \{x \in [a,b] : M - \epsilon < g(x) < M + \epsilon\}$$

and so g is approximately continuous at x_0. If $f(x_0) = N$ then similarly g is approximately continuous at x_0. Finally if $N < f(x_0) < M$ then the proof is trivial. $\square$

Theorem 6.70 *If $f : [a,b] \to \mathbb{R}$ is approximately continuous then f has Darboux property i.e if $f(c) \ne f(d), a \le c < d \le b$ then for every η lying between $f(c)$ and $f(d)$ there is $\xi \in (c,d)$ such that $f(\xi) = \eta$.*

Proof Let $f(c) < f(d)$ and $\eta \in (f(c), f(d))$. Choose M, N such that $N < f(c) < f(d) < M$. Let g be defined by

$$g(x) = \begin{cases} M & \text{if } f(x) > M \\ f(x) & \text{if } N \le f(x) \le M \\ N & \text{if } f(x) < N \end{cases}$$

Since f is approximately continuous in $[a,b]$, by Theorem 6.69 g is approximately continuous in $[a,b]$. Moreover g is bounded and measurable and therefore g is Lebesgue integrable in $[a,b]$. Let $G(x) = \int_a^x g\, dt$ for $x \in [a,b]$. Then by Theorem 6.68 $G'(x) = g(x)$ for $x \in [a,b]$. So, g is a derivative function and hence it has Darboux property. Since $g(c) = f(c) < \eta < f(d) = g(d)$ there is $\xi \in (c,d)$ such that $g(\xi) = \eta$. So, $f(c) < g(\xi) < f(d)$ and hence $g(\xi) = f(\xi)$ which gives $f(\xi) = \eta$, completing the proof. $\square$

Theorem 6.71 *If $f : [a,b] \to \mathbb{R}$ is continuous and $g : [c,d] \to [a,b]$ is approximately continuous then the composite function $f \circ g$ is approximately continuous. The continuity of f cannot be replaced by approximate continuity even when g is continuous.*

Proof Let $x_0 \in [c,d]$. Since g is approximately continuous at x_0 there is a set $E \subset [c,d]$ having x_0 as a point of density of E and $g(x) \to g(x_0)$ as $x \to x_0, x \in E$. Since f is continuous at $g(x_0)$, $(f \circ g)(x) = f(g(x)) \to f(g(x_0)) = (f \circ g)(x_0)$ as $x \to x_0, x \in E$. So, $f \circ g$ is approximately continuous x_0.

For the last part consider the intervals $\{(a_n, b_n)\}$ defined in Example 6.8 where $a_n = \frac{1}{n}, b_n = \frac{1}{n} + \frac{1}{n^2}$. As proved there 0 is a point of density of the set $E = \bigcup_{n=2}^{\infty} (a_n, b_n)$ and 0 is a point of dispersion of $\widetilde{E} = \bigcup_{n=1}^{\infty} [b_{n+1}, a_n]$. Define $f(x) = 1$ if $x \in E \bigcup \{0\}$ and $f(x) = 0$ if $x \in \bigcup_{n=1}^{\infty} [b_{n+1}, a_n]$ and $g(x) = \frac{1}{n}$ if $x \in (a_n, b_n)$ and

$g(x) = 0$ if $x \in \{0\} \cup \bigcup_{n=1}^{\infty} [b_{n+1}, a_n]$. Then $(f \circ g)(x) = 0$ if $x \in E$ and $(f \circ g)(x) = 1$ if $x \in \{0\} \cup \bigcup_{n=1}^{\infty} [b_{n+1}, a_n]$. So $(f \circ g)(0) = 1$. Since 0 is a point of density of E, $f \circ g$ is not approximately continuous at 0, though f and g are approximately continuous at 0. Note that g is continuous at 0. □

Theorem 6.72 *Let $f : [a, b] \to \mathbb{R}$ and $g : [a, b] \to \mathbb{R}$ be approximately continuous at $x_0 \in [a, b]$. Then*

(i) *$f + g$ is approximately continuous at x_0*
(ii) *kf is approximately continuous at x_0, if k is a constant.*
(iii) *fg is approximately continuous at x_0.*
(iv) *if $f(x_0) \neq 0$ then $\frac{1}{f}$ is approximately continuous at x_0.*
(v) *$|f|$ is approximately continuous at x_0.*

Proof To prove the theorem, note that if x_0 is a point of dispersion of E and of F then x_0 is a point of dispersion of $E \cup F$. In fact for any interval I containing x_0,

$$\frac{\mu^*((E \cup F) \cap I)}{\mu(I)} \leq \frac{\mu^*(E \cap I)}{\mu(I)} + \frac{\mu^*(F \cap I)}{\mu(I)}.$$

(i) Let $\epsilon > 0$ be arbitrary. Since f is approximately continuous at x_0, x_0 is a point of dispersion of the set

$$\{x \in [a, b] : |f(x) - f(x_0)| \geq \epsilon\}$$
$$= \{x \in [a, b] : f(x) - f(x_0) \geq \epsilon\} \cup \{x \in [a, b] : f(x) - f(x_0) \leq -\epsilon\}.$$

So, x_0 is a point of dispersion of each of the sets

$$\{x \in [a, b] : f(x) - f(x_0) \geq \epsilon\} \text{ and } \{x \in [a, b] : f(x) - f(x_0) \leq -\epsilon\}.$$

Similarly x_0 is a point of dispersion of each of the sets

$$\{x \in [a, b] : g(x) - g(x_0) \geq \epsilon\} \text{ and } \{x \in [a, b] : g(x) - g(x_0) \leq -\epsilon\}.$$

Hence x_0 is a point of dispersion of each of the sets

$$P = \{x \in [a, b] : f(x) - f(x_0) \geq \epsilon\} \cup \{x \in [a, b] : g(x) - g(x_0) \geq \epsilon\} \text{ and}$$

$$Q = \{x \in [a, b] : f(x) - f(x_0) \leq -\epsilon\} \cup \{x \in [a, b] : g(x) - g(x_0) \leq -\epsilon\}$$

Since

$$U = \{x \in [a, b] : f(x) + g(x) - f(x_0) - g(x_0) \geq 2\epsilon\} \subset P$$

and

$$V = \{x \in [a, b] : f(x) + g(x) - f(x_0) - g(x_0) \leq -2\epsilon\} \subset Q,$$

x_0 is a point of dispersion of U and V. But

$$T = \{x \in [a, b] : |f(x) + g(x) - f(x_0) - g(x_0)| \geq 2\epsilon\} = U \cup V$$

and so x_0 is a point of dispersion of T which shows that $f + g$ is approximately continuous at x_0.

(ii) If $k = 0$, this is obvious. So we suppose $k \neq 0$. Let $\epsilon > 0$ be arbitrary. Since f is approximately continuous at x_0, x_0 is a point of dispersion of the set $\{x \in [a, b] : |f(x) - f(x_0)| \geq \frac{\epsilon}{|k|}\} = \{x : x \in [a, b]; |kf(x) - kf(x_0)| \geq \epsilon\}$. completing the proof of (ii).

(iii) Let $\epsilon > 0$ be arbitrary. Let

$$A = \{x \in [a, b] : |f(x)g(x) - f(x_0)g(x_0)| \geq \epsilon\}$$
$$B = \{x \in [a, b] : |f(x) - f(x_0)| \geq \frac{\sqrt{\epsilon}}{2}\}$$
$$C = \{x \in [a, b] : |g(x) - g(x_0)| \geq \frac{\sqrt{\epsilon}}{2}\}$$
$$D = \{x \in [a, b] : |f(x_0)g(x) - f(x_0)g(x_0)| \geq \frac{\epsilon}{4}\}$$
$$E = \{x \in [a, b] : |g(x_0)f(x) - g(x_0)f(x_0)| \geq \frac{\epsilon}{4}\}.$$

Then $A \subset B \cup C \cup D \cup E$. For if $x \notin B \cup C \cup D \cup E$ then

$$|f(x)g(x) - f(x_0)g(x_0)|$$
$$= |f(x)g(x) - f(x)g(x_0) + f(x)g(x_0) - f(x_0)g(x_0)|$$
$$\leq |f(x)||g(x) - g(x_0)| + |g(x_0)||f(x) - f(x_0)|$$
$$= (|f(x) - f(x_0) + f(x_0)|)|g(x) - g(x_0)| + |g(x_0)||f(x) - f(x_0)|$$
$$\leq |f(x) - f(x_0)||g(x) - g(x_0)| + |f(x_0)||g(x) - g(x_0)| + |g(x_0)||f(x) - f(x_0)|$$
$$< \frac{\sqrt{\epsilon}}{2}\frac{\sqrt{\epsilon}}{2} + \frac{\epsilon}{4} + \frac{\epsilon}{4} < \epsilon.$$

Hence $x \notin A$. So, $A \subset B \bigcup C \bigcup D \bigcup E$. Since f and g are approximately continuous at x_0, x_0 is a point of dispersion of B and C. Also by (ii) the functions $f(x_0)g$ and $g(x_0)f$ are approximately continuous at x_0 and so x_0 is a point of dispersion of D and E. Hence x_0 is a point of dispersion of $B \bigcup C \bigcup D \bigcup E$ and so x_0 is a point of dispersion of A.

(iv) Let $\epsilon > 0$ be arbitrary. Let

$$A = \{x : |\frac{1}{f(x)} - \frac{1}{f(x_0)}| \geq \epsilon\}$$

$$B = \{x : |f(x) - f(x_0)| \geq \frac{1}{2}|f(x_0)|\}$$

$$C = \{x : |f(x)| \leq \frac{1}{2}|f(x_0)|\}$$

$$D = \{x : |f(x) - f(x_0)| \geq \frac{1}{2}\epsilon|f(x_0)|^2\}.$$

Note that if $|f(x)| \leq \frac{1}{2}|f(x_0)|$ then $|f(x) - f(x_0)| \geq |f(x_0)| - |f(x)| \geq \frac{1}{2}|f(x_0)|$ and so $C \subset B$.

Since f is approximately continuous at x_0, x_0 is a point of dispersion of B and since $C \subset B$, x_0 is a point of dispersion of C. Also x_0 is a point of dispersion of D. So, x_0 is a point of dispersion of $C \cup D$. Also $A \subset C \cup D$. For, if $x \notin C \cup D$ then $x \notin C$ and $x \notin D$ and hence

$$|f(x)f(x_0)| > \frac{1}{2}|f(x_0)|^2 \text{ and } |f(x) - f(x_0)| < \frac{1}{2}\epsilon|f(x_0)|^2$$

So $\left|\frac{f(x)-f(x_0)}{f(x)f(x_0)}\right| < \epsilon$ which shows that $x \notin A$. Since x_0 is a point of dispersion of $C \cup D$, x_0 is a point of dispersion of A and hence $\frac{1}{f}$ is approximately continuous at x_0.

(v) This is easy.

$\square$

Theorem 6.73 Let $f : [a, b] \to \mathbb{R}$ be integrable. If $x_0 \in [a, b]$ is a Lebesgue point of f then f is approximately continuous at x_0.

Proof Let x_0 be a Lebesgue point of f. For any interval I containing x_0, write $I = (x_0 - h', x_0 + h)$. Since x_0 is a Lebesgue point of f

$$\lim_{h \to 0} \frac{1}{h} \int_{x_0}^{x_0+h} |f(x) - f(x_0)| = 0 \text{ and } \lim_{h \to 0} \frac{1}{h'} \int_{x_0-h'}^{x_0} |f(x) - f(x_0)| = 0.$$

Since

$$\frac{1}{\mu(I)} \int_I |f(x) - f(x_0)| \leq \frac{1}{h} \int_{x_0}^{x_0+h} |f(x) - f(x_0)| + \frac{1}{h'} \int_{x_0-h'}^{x_0} |f(x) - f(x_0)|,$$

we have

$$\lim_{\mu(I) \to 0} \frac{1}{\mu(I)} \int_I |f(x) - f(x_0)| = 0.$$

So, for any sequence $\{I_n\}$ of intervals converging to x_0

6.15 Properties of Approximately Continuous Function

$$\lim_{n\to\infty} \frac{1}{\mu(I_n)} \int_{I_n} |f(x) - f(x_0)| = 0 \tag{6.82}$$

Suppose that f is not approximately continuous at x_0. Then there exists $\epsilon_0 > 0$ such that x_0 is not a point of dispersion of the set $E = \{x \in [a, b] : |f(x) - f(x_0)| \geq \epsilon_0\}$. So there is a sequence $\{I_n\}$ of intervals converging to x_0 such that

$$\lim_{n\to\infty} \frac{\mu(E \cap I_n)}{\mu(I_n)} = l > 0.$$

Hence there is N such that

$$\frac{\mu(E \cap I_n)}{\mu(I_n)} > \frac{1}{2}l \text{ for } n \geq N.$$

So, for $n \geq N$,

$$\frac{1}{\mu(I_n)} \int_{I_n} |f(x) - f(x_0)| \geq \frac{1}{\mu(I_n)} \int_{E \cap I_n} |f(x) - f(x_0)| \geq \epsilon_0 \frac{\mu(E \cap I_n)}{\mu(I_n)} > \epsilon_0 \frac{1}{2}l.$$

Hence

$$\liminf_{n\to\infty} \frac{1}{\mu(I_n)} \int_{I_n} |f(x) - f(x_0)| \geq \frac{1}{2}\epsilon_0 l. \tag{6.83}$$

Since (6.82) and (6.83) are contradictory our supposition is wrong and so f is approximately continuous at x_0. □

Theorem 6.74 *Let $f : [a, b] \to \mathbb{R}$ be bounded and integrable. If f is approximately continuous at $x_0 \in [a, b]$ then x_0 is a Lebesgue point of f.*

Proof Let f be approximately continuous at x_0 and let $\phi(x) = |f(x) - f(x_0)|$. Then ϕ is bounded and integrable in $[a, b]$ and by Theorem 6.72 ϕ is approximately continuous at x_0. So, by Theorem 6.68

$$\lim_{h\to 0} \frac{1}{h} \int_{x_0}^{x_0+h} \phi = \phi(x_0) \text{ i.e. } \lim_{h\to 0} \frac{1}{h} \int_{x_0}^{x_0+h} |f(x) - f(x_0)| = 0$$

and so x_0 is a Lebesgue point of f. □

Comparing Theorems 6.73 and 6.74 it is natural to ask whether boundedness is necessary in Theorem 6.74. The answer is yes. Consider the following example.

Example 6.9 There exists an integrable function $f : [0, 1] \to \mathbb{R}$ which is approximately continuous at 0 but 0 is not a Lebesgue point of f.

Solution Consider the function $f : [0, 1] \to \mathbb{R}$ defined in Example 6.8. That function is approximately continuous at $x = 0$ but not bounded in $[0, 1]$. We show that f is integrable in $[0, 1]$. The integral of f on $[0, 1]$ is given by

$$\int_0^1 f = \sum_{n=1}^{\infty} \int_{b_{n+1}}^{a_n} f = \sum_{n=1}^{\infty} \frac{1}{2} n(a_n - b_{n+1}) = \sum_{n=1}^{\infty} \frac{n}{2} \left(\frac{1}{n} - \frac{1}{n+1} - \frac{1}{(n+1)^2} \right)$$

$$= \frac{1}{2} \sum_{n=1}^{\infty} \frac{1}{(n+1)^2} < \infty.$$

So, f is integrable in $[0, 1]$. To show that 0 is not a Lebesgue point of f consider any positive integer k : Then

$$k \int_0^{\frac{1}{k}} f = k \sum_{n=k}^{\infty} \int_{b_{n+1}}^{a_n} f = \frac{k}{2} \sum_{n=k}^{\infty} \frac{1}{(n+1)^2} > \frac{k}{2} \sum_{n=k}^{2k-1} \frac{1}{(n+1)^2} > \frac{k}{2} \cdot \frac{k}{(2k)^2} = \frac{1}{8}.$$

Therefore, since $f(0) = 0$ and $f \geq 0$ on $[0, 1]$, $k \int_0^{\frac{1}{k}} |f(x) - f(0)| > \frac{1}{8}$ for all k. So, putting $h_k = \frac{1}{k}$ we get the sequence $\{h_k\}$ such that $h_k \to 0$ as $k \to \infty$ and

$$\lim_{k \to \infty} \frac{1}{h_k} \int_0^{h_k} |f(x) - f(0)| \geq \frac{1}{8}.$$

So, 0 is not a Lebesgue point of f.

Example 6.10 If x_0 is a point of outer density of E and also of F and if one of E and F is measurable then x_0 is a point of outer density of $E \cap F$.

Solution Suppose F is measurable. Then

$$\mu^*(E) = \mu^*(E \cap F) + \mu^*(E \cap \tilde{F})$$
$$\mu^*(E \cup F) = \mu^*(F) + \mu^*(E \cap \tilde{F})$$

Subtracting and transposing

$$\mu^*(E) + \mu^*(F) = \mu^*(E \cup F) + \mu^*(E \cap F).$$

Let $\{I_n\}$ be a any sequence of intervals converging to x_0. Then from above

$$\frac{\mu^*(E \cap I_n)}{\mu(I_n)} + \frac{\mu^*(F \cap I_n)}{\mu(I_n)} = \frac{\mu^*((E \cup F) \cap I_n)}{\mu(I_n)} + \frac{\mu^*((E \cap F) \cap I_n)}{\mu(I_n)}$$

Since x_0 is a point of outer density of E and F, x_0 is a point of outer density of $E \bigcup F$. Hence

$$\lim_{n\to\infty} \frac{\mu^*((E\cap F)\cap I_n)}{\mu(I_n)} = \lim_{n\to\infty} \frac{\mu^*(E\cap I_n)}{\mu(I_n)} + \lim_{n\to\infty} \frac{\mu^*(F\cap I_n)}{\mu(I_n)}$$
$$-\lim_{n\to\infty} \frac{\mu^*((E\cup F)\cap I_n)}{\mu(I_n)}$$
$$= 1 + 1 - 1 = 1.$$

So x_0 is a point of outer density of $E\cap F$

The notion of approximate continuity being defined it is natural to define the approximate differentiability of a function f at a point x_0. The approximate limit of the function ϕ where $\phi(x) = \frac{f(x)-f(x_0)}{x-x_0}$, at x_0. is called the approximate derivative of f at x_0. We shall not discuss this. Interested readers may consult [Saks, Jeffery, Gordon, Bruckner].

6.16 Exercises

1. Show that for any two functions f and g $D^+(f+g) \leq D^+f + D^+g$ and give examples of f and g for which $D^+(f+g) < D^+f + D^+g$. Show that equality holds if one of f and g is differentiable. Also show that $D^+f + D_+g \leq D^+(f+g)$.
2. Let $f(x) = 1$ if x is rational and $f(x) = 0$ if x is irrational. Find the four Dini derivates of f at a rational point and at an irrational point.
3. Let $F(x) = \int_0^x \sin\frac{1}{t} dt$. Show that $F'(0) = 0$.
4. Let $F(x) = e^{\sin\frac{1}{x}}$ for $x \neq 0$ and $f(0) = 0$. Find the four Dini derivates of F at $x = 0$.

An extended real number α is said to be a derived number at x_0 of a function f if there exists a sequence $\{h_n\} \to 0$ such that $\lim_{n\to\infty} \frac{f(x_0+h_n)-f(x_0)}{h_n} = \alpha$.

5. Show that every real number in $[-1, 1]$ is a derived number at 0 of the function
$$f(x) = x\cos\frac{1}{x} \text{ for } x \neq 0, \ f(0) = 0.$$

6. Show that every extended real number is a derived number at $x = 0$ for the function
$$f(x) = \sqrt{|x|}\sin\frac{1}{x} \text{ for } x \neq 0, \ f(0) = 0.$$

An extended real number α is said to be a cluster point (or a limit point, or a limit number) at x_0 of a function f if there exists a sequence $\{h_n\} \to 0$ such that $\lim_{n\to\infty} f(x_0 + h_n) = \alpha$. The set of all cluster points at x_0 of f is denoted by $C(f; x_0)$.

7. Show that the set $C(f; x_0)$ is closed

8. Show that if f is continuous in (a, b) and if $\alpha = \liminf\limits_{x \to a+} f(x) < \limsup\limits_{x \to a+} f(x) = \beta$ then $C(f, a) = [\alpha, \beta]$.
9. If all four Dini derivates of a function f are finite at a point x then show that f is continuous at x.
10. Show that the function $f(x) = x \sin \frac{\pi}{x}$ for $x \neq 0$ and $f(0) = 0$ is continuous in $[0, 1]$ but is not of bounded variation in $[0, 1]$.
11. Show that the function $f(x) = x^2 \cos \frac{\pi}{x^2}$ for $x \neq 0$ and $f(0) = 0$ has finite derivative in $[0, 1]$ but is not of bounded variation in $[0, 1]$.
12. Let $\{f_n\}$ be a sequence of functions of bounded variation in $[a, b]$ such that

$$\sum_{n=1}^{\infty} V(f_n; a, b) < \infty \text{ and } \sum_{n=1}^{\infty} |f_n(a)| < \infty.$$

Then show that $\sum\limits_{n=1}^{\infty} f_n(x)$ is convergent and the sum function is of bounded variation in $[a, b]$.
13. If a function $f : [a, b] \to \mathbb{R}$ is such that f' exists and is bounded in $[a, b]$ then show that f is of bounded variation in $[a, b]$. If moreover f' is Riemann integrable then show that

$$V(f; a, b) = \int_a^b |f'| dx.$$

14. If f is a function of bounded variation in $[a, b]$ then show that f' exists a.e. and f' is Lebesgue integrable in $[a, b]$.
    [Hint : $f = g - h$ where g and h are non-decreasing. Apply Theorem 6.12].
15. Let $f : [0, 1] \to \mathbb{R}$ be defined by $f(x) = x^p \sin \frac{\pi}{x^q}$ for $x \neq 0$ and $f(0) = 0$ where $p, q > 0$. Show that f is absolutely continuous in $[0, 1]$ if $p > q$ and f is not even of bounded variation in $[0, 1]$ if $p \leq q$.
16. Let $f : [0, 1] \to \mathbb{R}$ be defined by $f(x) = x^2 |\sin \frac{1}{x}|$ for $x \neq 0$ and $f(0) = 0$ and $g : [0, 1] \to \mathbb{R}$ be defined by $g(x) = \sqrt{x}$. Show that f and g are absolutely continuous, the composite function $f \circ g$ is absolutely continuous but $g \circ f$ is not absolutely continuous.
17. Let $f : [0, 1] \to \mathbb{R}$ be defined by $f(x) = x \sin \frac{1}{x} + x \cos \frac{1}{x}$ for $x \neq 0$ and $f(0) = 0$. Show that f is absolutely continuous.
18. Show that if a sequence of absolutely continuous functions converges uniformly to a function f then f need not be absolutely continuous.
19. Let $f : [a, b] \to \mathbb{R}$ be measurable and let f be differentiable almost everywhere in $[a, b]$. If $E = \{x \in [a, b] : f'(E) = 0\}$ prove that $\mu^*(f(E)) = 0$ [Hint. Apply Theorem 6.25].
20. Let $f : [0, 1] \to \mathbb{R}$. If for every measurable set $E \subset [a, b]$, $f(E)$ is measurable then prove that f satisfies the Lusin condition $(\mathcal{N})$.

6.16 Exercises

21. Let $f : [a, b] \to \mathbb{R}$ be of bounded variation and let $v_f(x) = \vee(f; a, x)$ be the variation function of f in $[a, b]$. If v_f is absolutely continuous in $[a, b]$ then prove that f is absolutely continuous.
22. Let θ be the Cantor functions defined in $[0, 1]$ (See Sect. 4.9). Let $f(x) = \theta(x)$ if $x \in [0, 1]$ and let $f(x) = x$ for $x \in [1, 2]$. Show that f is of bounded variation in $[0, 2]$ but is neither absolutely continuous nor singular in $[0, 2]$.
23. If f integrable in $[a, b]$ and if f is continuous at a point $x \in [a, b]$ then show that x is a Lebesgue point of f.
24. If a point x_0 is a point of dispersion of E then prove that x_0 is a point of outer density of the complement $\widetilde{E}$ of E.
25. If E is measurable then prove that x_0 is a point of density of E if and only if x_0 is a point of dispersion of $\widetilde{E}$.
26. If x_0 is a point of outer density of E and $E \subset F$ then prove that x_0 is a point of outer density of F.
27. If x_0 is a point of dispersion of E and of F then prove that x_0 is a point of dispersion of $E \bigcup F$.
28. If f and g are approximately continuous in $[a, b]$ and if $f = g$ a.e. in $[a, b]$ then show that $f = g$ everywhere in $[a, b]$.
29. A sequence $\{f_n\}$ of approximately continuous function converges uniformly to a function f on $[a, b]$. Prove that f is approximately continuous on $[a, b]$.
30. Let $f : [a, b] \to \mathbb{R}$ be approximately continuous in $[a, b]$. If f is of bounded variation in $[a, b]$ then prove that f is continuous in $[a, b]$.
31. If $f : [a, b] \to \mathbb{R}$ is integrable then show that f is approximately continuous at each of its Lebesgue points.
32. If $f : [a, b] \to \mathbb{R}$ is bounded and Lebesgue integrable. Then prove that $x_0 \in [a, b]$ is a Lebesgue point of f if and only if f is approximately continuous at x_0.
33. Show that there exists a function f that is approximately continuous on $[0, 1]$ but not Lebesgue integrable on $[0, 1]$. [Hint: Consider the function f in Example 6.8 with $f(c_n) = (n + 1)^2$].

Chapter 7
Lebesgue Measure and Integration in $\mathbb{R}^N$

Lebesgue measure and integration in $\mathbb{R}$ being introduced in previous chapters, we now consider Lebesgue measure and integration in $\mathbb{R}^N$ for $N \geq 2$. Though $\mathbb{R} = \mathbb{R}^N$ for $N = 1$, $\mathbb{R}^N$ for $N \geq 2$ lacks some of the properties possessed by $\mathbb{R}$ and therefore $\mathbb{R}^N$ needs separate treatment in some cases particularly, in defining measures in $\mathbb{R}^N$. Many authors introduced measure in $\mathbb{R}^N$ by using product measure in $\mathbb{R}$. But the product measure is not complete and since the Lebesgue measure in $\mathbb{R}^N$ is complete, only product measure will not give Lebesgue measure in $\mathbb{R}^N$ unless sets of measure zero are included. Here Lebesgue measure in $\mathbb{R}^N$ is introduced directly. We shall discuss Lebesgue measure and integration in $\mathbb{R}^2$. There is no difficulty in extending these results in $\mathbb{R}^N$ for $N \geq 3$. We need the following two theorems.

7.1 Structure of Open Sets in $\mathbb{R}^2$

Theorem 7.1 *If G is an open set in $\mathbb{R}^2$ then G is a countable union of open rectangles.*

Proof Let $\mathfrak{J}$ be the family of all open rectangles $I = \{(x, y) : a < x < b; c < y < d\}$ such that a, b, c, d are rational numbers. Since the set of rational numbers is countable, $\mathfrak{J}$ is a countable family. Since G is open, every point of G is an interior point of G and so there are members of $\mathfrak{J}$ contained in G. Let $\mathfrak{J}_1$ be the collection of all members I of $\mathfrak{J}$ such that $I \subset G$. Since $\mathfrak{J}$ is countable, $\mathfrak{J}_1$ is also countable. Let $\mathfrak{J}_1 = \{I_1, I_2, I_3, ...\}$. Then since $I_n \subset G$ for all n, $\bigcup_{n=1}^{\infty} I_n \subset G$. Also if a point $p = (x, y) \in G$ then since p is an interior point of G, there is a member $I_n \in \mathfrak{J}_1$ such that $p \in I_n \subset G$. Hence $p \in \bigcup_{n=1}^{\infty} I_n$ and so $G \subset \bigcup_{n=1}^{\infty} I_n$. Therefore $G = \bigcup_{n=1}^{\infty} I_n$, completing the proof. □

Theorem 7.2 *Every open set in $\mathbb{R}^2$ is the union of a countable collection of closed squares which are pairwise without common interior points.*

Proof Consider in the first step two system of parallel straight lines $x = 0, \pm 1, \pm 2, \ldots$ and $y = 0, \pm 1, \pm 2, \ldots$. These two system of straight lines divide the entire plane into a countable number of squares. Including its boundaries with each of these squares, we get a countable collection C_1 of closed squares which are pairwise without common interior points. In the second step consider the two systems $x = 0, \pm\frac{1}{2}, \pm 1, \pm\frac{3}{2}\ldots$ and $y = 0, \pm\frac{1}{2}, \pm 1, \pm\frac{3}{2}, \ldots$ and these will give as above a countable collection C_2 of closed squares which are pairwise without common interior points. Continuing this process we get at the nth step the two systems of lines $x = \frac{k}{2^{n-1}}, y = \frac{k}{2^{n-1}}, k = 0, \pm 1, \pm 2, \pm 3\ldots$ which give a countable collection C_n of closed squares which are pairwise without common interior points. Clearly every member of C_n is the union of four members of C_{n+1}.

Let G be a non-void open set in $\mathbb{R}^2$. Since the length of the sides of the squares of $C_1, C_2, \ldots, C_n$ are becoming smaller and smaller and ultimately tends to 0 as $n \to \infty$, there is a least positive integer k such that some of the squares of C_k will lie completely within G and hence each of $C_n, n \geq k$ has some members which lie within G. Let $\mathfrak{I}_n$ be the collection of all members of $C_n, n \geq k$, which lie within G. Let $\mathcal{L}_k = \mathfrak{I}_k$ and $\mathcal{L}_{k+r} = \{Q \in \mathfrak{I}_{k+r} : Q$ is not contained in any member of $\bigcup_{i=0}^{r-1} \mathfrak{I}_{k+i}\}$, $r = 1, 2, 3, \ldots$ Let $\mathcal{L} = \bigcup_{i=k}^{\infty} \mathcal{L}_i$. Since each C_n is countable each $\mathfrak{I}_n$ is countable and so each $\mathcal{L}_n$ is countable and consequently $\mathcal{L}$ is countable. Let $\mathcal{L} = \{Q_1, Q_2, \ldots Q_n, \ldots\}$. Then $G = \bigcup_{i=1}^{\infty} Q_i$. For, since each $Q_i \subset G$, $\bigcup_{i=1}^{\infty} Q_i \subset G$. Let p be any point of G. Since p is an interior point of G there is a Q_i such that $p \in Q_i \subset G$ and so $p \in \bigcup_{i=1}^{\infty} Q_i$. Since p is arbitrary, $G \subset \bigcup_{i=1}^{\infty} Q_i$. This completes the proof. $\square$

7.2 Lebesgue Outer Measure and Measure in $\mathbb{R}^N$

We shall discuss Lebesgue outer measure and measure in $\mathbb{R}^2$. There is no difficulty to extend these in $\mathbb{R}^N$, for $N \geq 3$. The area of an open rectangle $I = \{(x, y) : a < x < b; c < y < d\}$ will be denoted by $|I| = (b - c)(d - c)$. The outer measure and measure of a set $E \subset \mathbb{R}^2$ will be denoted by $\mu_2^*(E)$ and $\mu_2(E)$ respectively while μ^* and μ will denoted outer measure and measure of a set in $\mathbb{R}$ as is (Chap. 2).

Definition 7.1 For any set $E \subset \mathbb{R}^2$ the outer measure of E is defined as

$$\mu_2^*(E) = \inf\left\{\sum_n |I_n| : E \subset \bigcup_n I_n; I_n \text{ is an open rectangle}\right\}$$

7.2 Lebesgue Outer Measure and Measure in $\mathbb{R}^N$

the infimum being taken over all countable collection $\{I_n\}$ of open rectangles such that $E \subset \bigcup I_n$. It is clear from the definition that

(i) $\mu_2^*(\phi) = 0$ where ϕ is the empty subset of $\mathbb{R}^2$,
(ii) if $E_1 \subset E_2$ and $E_1, E_2 \subset \mathbb{R}^2$ then $\mu_2^*(E_1) \leq \mu_2^*(E_2)$,

(iii) for any countable collection $\{E_n\}$ of subsets of $\mathbb{R}^2$ $\mu_2^*(\bigcup_n E_n) \leq \sum_n \mu_2^*(E_n)$.

The proof of (i) and (ii) follows from the definition. We prove (iii). Let $\{E_n\}$ be any countable collection of subset of $\mathbb{R}^2$ and let $E = \bigcup_n E_n$. Take $\epsilon > 0$ arbitrary. Then for each n there is, by definition, a countable collection $\{I_{nk}\}$ of open rectangles such that $E_n \subset \bigcup_k I_{nk}$ and $\mu_2^*(E_n) + \frac{\epsilon}{2^n} > \sum_k |I_{nk}|$. The collection $\{I_{n,k}\}, n = 1, 2, \ldots; k = 1, 2, \ldots$ is also countable and this collection covers $\bigcup_n E_n$. So,

$$\mu_2^*(\bigcup_n E_n) \leq \sum_n \sum_k |I_{nk}| < \sum_n [\mu_2^*(E_n) + \frac{\epsilon}{2^n}] = \sum_n \mu_2^*(E_n) + \epsilon.$$

Letting $\epsilon \to 0$, we have (iii).

Thus the set function μ_2^* defined on the class of all subset of $\mathbb{R}^2$ satisfies (i), (ii) and (iii) of Theorem 2.1.

Theorem 7.3 (a) For any open rectangle I, $\mu_2^*(I) = |I|$, the area of I.
(b) If E is a countable set then $\mu_2^*(E) = 0$.
(c) If $E_1 = \{(x, y) : a \leq x \leq b; y = k\}$ and $E_2 = \{(x, y) : x = k; c \leq y \leq d\}$ then $\mu_2^*(E_1) = 0$ and $\mu_2^*(E_2) = 0$.

Proof (a) follows from the definition. If $E = \{(x_0, y_0)\}$ is a singleton set then $E \subset \{(x, y) : x_0 - \epsilon < x < x_0 + \epsilon; y_0 - \epsilon < y < y_0 + \epsilon\}$ and so $\mu_2^*(E) \leq 4\epsilon^2$ for every $\epsilon > 0$ and hence $\mu_2^*(E) = 0$. So (b) follows from (iii) of the definition. Finally, $E_1 \subset \{(x, y) : a - \epsilon < x < b + \epsilon; k - \epsilon < y < k + \epsilon\}$ and so $\mu_2^*(E_1) \leq (b - a + 2\epsilon).2\epsilon$ for every $\epsilon > 0$ and hence $\mu^*(E_1) = 0$. Similarly $\mu^*(E_2) = 0$, proving (c). $\square$

Theorem 7.4 For any closed rectangle $\bar{I} = \{(x, y) : a \leq x \leq b; c \leq y \leq d\}$, $\mu_2^*(\bar{I}) = \mu_2^*(I)$, where I is the interior of $\bar{I}$.

The proof follows from (ii) and (iii) of Definition 7.1 and property (a) and (c) of Theorem 7.3.

Theorem 7.5 If $f : [a, b] \to \mathbb{R}$ is continuous and $E = \{(x, y) : a \leq x \leq b; y = f(x)\}$ then $\mu_2^*(E) = 0$. Analogously, if $g : [c, d] \to \mathbb{R}$ and $E = \{(x, y) : x = g(y) : c \leq y \leq d\}$ then $\mu_2^*(E) = 0$.

Proof Since f is continuous in $[a, b]$ it is uniformly continuous. Let $\epsilon > 0$ be arbitrary. Then there is $\delta > 0$ such that

$$|f(x_1) - f(x_2)| < \epsilon \text{ for } x_1, x_2 \in [a, b] \text{ and } |x_1 - x_2| < \delta. \tag{7.1}$$

Let $a = a_0 < a_1 < ... < a_n = b$ be a partition of $[a, b]$ such that $|a_{i+1} - a_i| < \delta$ for $i = 0, 1, 2, ..., n - 1$. Let $m_i = \inf\{f(x) : a_i \leq x \leq a_{i+1}\}$ and $M_i = \sup\{f(x) : a_i \leq x \leq a_{i+1}\}$. Then by (7.1) $M_i - m_i < \epsilon$. Consider the closed rectangle $\overline{I_i} = \{(x, y) : a_i \leq x \leq a_{i+1}; m_i \leq y \leq M_i\}$ for $i = 0, 1, 2, ..., n - 1$. Then by Theorem 7.4 $\mu_2^*(\overline{I_i}) = (a_{i+1} - a_i)(M_i - m_i) < \epsilon(a_{i+1} - a_i)$. Since $E \subset \bigcup_{i=0}^{n-1}(\overline{I_i})$, $\mu_2^*(E) \leq \sum_{i=0}^{n-1} \mu_2^*(\overline{I_i}) < \epsilon(b - a)$. Since ϵ is arbitrary $\mu_2^*(E) = 0$. The second part is similar. □

Definition 7.2 A set $E \subset \mathbb{R}^2$ is said to be measurable if for every

$$A \subset \mathbb{R}^2, \mu_2^*(A) = \mu_2^*(E \cap A) + \mu_2^*(\widetilde{E} \cap A).$$

If E is measurable then $\mu_2^*(E)$ is called the measure of E and is denoted by $\mu_2(E)$. If $\mathcal{M}_2$ is the class of all measurable sets in $\mathbb{R}^2$ and if μ_2 is the measure then $(\mathbb{R}^2, \mathcal{M}_2, \mu_2)$ is called the Lebesgue measure space on $\mathbb{R}^2$.
Note Clearly the Lebesgue measure space $(\mathbb{R}^2, \mathcal{M}_2, \mu_2)$ is complete.
Note that a measure space is said to be complete if whenever a set E is measurable and is of measure zero then every subset of E is also measurable. Also note that all the properties of measurable sets in $\mathbb{R}$ proved in Sect. 2.2 are also true for measurable sets in $\mathbb{R}^2$. Therefore to prove the completeness of $(\mathbb{R}^2, \mathcal{M}_2, \mu_2)$, let $E \subset \mathcal{M}_2$ and $\mu_2(E) = 0$ and let $E_1 \subset E$. Then $\mu_2^*(E_1) \leq \mu_2(E) = 0$ and so applying analog of Theorem 2.3 (ii), E_1 is measurable.

Theorem 7.6 *If $\{E_n\}$ is any countable collection of measurable sets in $\mathbb{R}^2$ then $\bigcup E_n$ and $\bigcap E_n$ are also measurable.*

The proof is similar to that of Theorem 2.4.

Theorem 7.7 *All rectangles in $\mathbb{R}^2$ are measurable.*

Proof Let I be any rectangle in $\mathbb{R}^2$. We first suppose that I is bounded and open. Let A be any set in $\mathbb{R}^2$. We show that

$$\mu_2^*(A) \geq \mu_2^*(I \cap A) + \mu_2^*(\widetilde{I} \cap A).$$

If $\mu_2^*(A) = \infty$ this is obvious. So, we suppose that $\mu_2^*(A) < \infty$. Let $\epsilon > 0$ be arbitrary. Then there exists a countable collection $\{I_n\}$ of open rectangles such that $A \subset \bigcup_{n=1}^{\infty} I_n$ and $\sum_{n=1}^{\infty} \mu_2^*(I_n) < \mu_2^*(A) + \epsilon$. We divide the members of $\{I_n\}$ into two classes C_1 and C_2 such that if $I_n \cap \widetilde{I} = \emptyset$ then we put I_n in C_1 and if $I_n \cap I = \emptyset$ then we put

7.2 Lebesgue Outer Measure and Measure in $\mathbb{R}^N$

I_n in C_2. If a rectangles I_n is such that $I_n \cap I \neq \emptyset$ and $I_n \cap \tilde{I} \neq \emptyset$ then we get from I_n at most five open rectangles $I_{n_1}, I_{n_2}, I_{n_3}, I_{n_4}$ and I_{n_5} (neglecting the boundaries of each of $I_{n_i}, i = 1, 2, 3, 4, 5$) such that $I_{n_1} \subset I$ and $(I_{n_2} \cup I_{n_3} \cup I_{n_4} \cup I_{n_5}) \subset \tilde{I}$. We put in this case I_{n_1} in C_1 and the remaining $I_{n_2}, I_{n_3}, I_{n_4}, I_{n_5}$ in C_2. Then C_1 and C_2 are the countable collections of open rectangles. Renaming and reindexing the members of C_1 and C_2 we have $I \cap A \subset \bigcup\{I_i : I_i \in C_1\}$ and $\tilde{I} \cap A \subset \bigcup\{I_j : I_j \in C_2\}$. (Here also we have neglected the portions of A which are covered by boundaries of the rectangles. This is permissible, since by Theorem 7.3 (c) they are of outer measure zero). So

$$\mu_2^*(I \cap A) \leq \sum_{I_i \in C_1} \mu_2^*(I_i) \text{ and } \mu_2^*(\tilde{I} \cap A) \leq \sum_{I_j \in C_2} \mu_2^*(I_j).$$

Hence

$$\mu_2^*(I \cap A) + \mu_2^*(\tilde{I} \cap A) \leq \sum_{I_i \in C_1} \mu_2^*(I_i) + \sum_{I_j \in C_2} \mu_2^*(I_j)$$

$$= \sum_{n=1}^{\infty} \mu_2^*(I_n) < \mu_2^*(A) + \epsilon.$$

Since ϵ is arbitrary. $\mu_2^*(A) \geq \mu_2^*(I \cap A) + \mu_2^*(\tilde{I} \cap A)$ which proves that I is measurable.

Next suppose that I is bounded. Let I^0 be the interior of I. Then I^0 is a bounded open rectangle and so is measurable. Let L_1, L_2, L_3, and L_4 be the sides of I. By Theorem 7.3 (c) $\mu^*(L_i) = 0$ for $i = 1, 2, 3, 4$ and so L_i is measurable for $i = 1, 2, 3, 4$. Since I is the union of I^0 and some or all of L_1, L_2, L_3 and L_4, by Theorem 7.6 the set I is measurable.

Finally if I is unbounded then we can select a countable collection $\{I_n\}$ of bounded rectangle such that $I = \cup I_n$. Since each I_n is measurable, by Theorem 7.6 the set I is measurable, completing the proof. □

Theorem 7.8 *Every open set in $\mathbb{R}^2$ is measurable. More generally all Borel sets in $\mathbb{R}^2$ are measurable.*

Proof By Theorems 7.6 and 7.1 every open set in $\mathbb{R}^2$ is measurable.
Note that as in Theorem 2.5 the class of measurable sets in $\mathbb{R}^2$ is also a $\sigma-$ algebra and this $\sigma-$ algebra contains all open sets in $\mathbb{R}^2$. Since Borel $\sigma-$ algebra is the smallest $\sigma-$ algebra containing all open sets (see Definition 1.4), all Borel sets in $\mathbb{R}^2$ are measurable. □

Theorem 7.9 (Regularity properly) *For any set $A \subset \mathbb{R}^2$ there is a G_δ—set (and hence a measurable set) E such that $A \subset E$ and $\mu_2^*(A) = \mu_2(E)$.*
The proof is similar to that of Theorem 2.12.

Corollary 7.10 *For every measurable set $E \subset \mathbb{R}^2$ there is a decreasing sequence $\{G_n\}$ of open sets such that $E \subset G_n$ for each n and $\mu_2(E) = \lim_{n \to \infty} \mu_2(G_n)$.*

Proof By Theorem 7.9 there is a $G_\delta-$ set S such that $E \subset S$ and $\mu_2(E) = \mu_2(S)$. Let $S = \bigcap_{n=1}^{\infty} Q_n$ where Q_n are open sets. Let $G_n = \bigcap_{r=1}^{n} Q_r$. Then $\{G_n\}$ is a decreasing sequence of open sets and $E \subset G_n$ for each n and $\mu_2(E) = \mu_2(S) = \mu_2(\bigcap_{n=1}^{\infty} Q_n) = \mu_2(\lim_{n\to\infty} \bigcap_{r=1}^{n} Q_r) = \mu_2(\lim_{n\to\infty} G_n) = \lim_{n\to\infty} \mu_2(G_n)$, where we have applied analogous theorem of Theorem 2.9. $\square$

Theorem 7.11 *If A and B are measurable subset of $\mathbb{R}$ then $A \times B = \{(x, y) : x \in A, y \in B\}$ is a measurable subset of $\mathbb{R}^2$ and $\mu_2(A \times B) = \mu(A).\mu(B)$.*

Proof We first suppose that A and B are bounded.

Case-*I* Let A and B be open subsets of $\mathbb{R}$. Then there are countable collection $\{A_i\}$ and $\{B_j\}$ of disjoint open intervals such that $A = \cup A_i$ and $B = \cup B_j$ and hence $\mu(A) = \sum_i \mu(A_i)$ and $\mu(B) = \sum_j \mu(B_j)$. Then $A_i \times B_j$ is an open rectangle and so is measurable for each pair (i, j) by Theorem 7.7 and hence $\bigcup_i \bigcup_j (A_i \times B_j)$ is measurable. Also

$$A \times B = (\bigcup_i A_i) \times (\bigcup_j B_j) = \bigcup_i \bigcup_j (A_i \times B_j) \quad (7.2)$$

and so $A \times B$ is measurable. Also $\{A_i \times B_j\}$ is a countable collection of disjoint rectangles and so by (7.2)

$$\mu_2(A \times B) = \sum_i \sum_j \mu_2(A_i \times B_j) = \sum_i \sum_j (\mu(A_i) \times \mu(B_j))$$
$$= \sum_i \mu(A_i) \sum_j \mu(B_j) = \mu(A)\mu(B).$$

Case-*II* Let A be of measure 0 and B is measurable. Then for arbitrary $\epsilon > 0$ there are open subsets G and H of $\mathbb{R}$ such that $A \subset G, B \subset H, \mu(G) < \epsilon$ and $\mu(H) < \mu(B) + \epsilon$. Since $A \times B \subset G \times H$, by Case - *I*, $\mu_2^*(A \times B) \leq \mu_2^*(G \times H) = \mu(G).\mu(H) < \epsilon(\mu(B) + \epsilon)$. Since ϵ is arbitrary $\mu_2^*(A \times B) = 0$ and so $A \times B$ is measurable and $\mu_2(A \times B) = 0$. Also $\mu(A) = 0$ and so $= \mu_2(A \times B) = \mu(A)\mu(B)$.

Case-*III* Let A and B be $G_\delta-$ sets. Then $A = \bigcap_i A_i$ and $B = \bigcap_j B_j$ where A_i and B_j are open. We may suppose that $A_{i+1} \subset A_i$ and $B_{j+1} \subset B_j$ for all i and all j. Since A_i and B_j are open, by Case-*I*, $A_i \times B_j$ is measurable and $\mu_2(A_i \times B_j) = \mu(A_i)\mu(B_j)$ for all pair (i, j). Since $A \times B = (\bigcap_i A_i) \times (\bigcap_j B_j) = \bigcap_i \bigcap_j (A_i \times B_j)$, $A \times B$ is measurable by Theorem 7.6. Also we can write

7.2 Lebesgue Outer Measure and Measure in $\mathbb{R}^N$

$$\bigcap_i \bigcap_j (A_i \times B_j) = (A_1 \times B_1) \cap [(A_2 \times B_1) \cap (A_1 \cap B_2)]$$
$$\cap (A_2 \times B_2) \cap [(A_3 \times B_2) \cap (A_2 \times B_3)] \cap \ldots$$

Since $\{A_i\}$ and $\{B_j\}$ are decreasing, the terms in the right side are decreasing and so applying analogous theorem of Theorem 2.9

$$\mu_2(A \times B) = \mu_2(\bigcap_i \bigcap_j A_i \times B_j) = \mu_2(\lim_{i,j \to \infty}(A_i \times B_j))$$
$$= \lim_{i,j \to \infty} \mu_2(A_i \times B_j) = \lim_{i,j \to \infty}[\mu(A_i)\mu(B_j)]$$
$$= \lim_{i \to \infty} \mu(A_i) \lim_{j \to \infty} \mu(B_j) = \mu(A)\mu(B)$$

Case-IV Let A and B be measurable sets. Then by Theorem 2.13 there are $G_\delta-$ sets G and H such that $A \subset G$, $B \subset H$, $\mu(G \sim A) = 0$ and $\mu(H \sim B) = 0$. Since

$$(G \times H) \sim (A \times B) \subset ((G \sim A) \times (H \sim B)) \cup ((G \sim A) \times B) \cup (A \times (H \sim B)),$$

by Case II

$$\mu_2^*((G \times H) \sim (A \times B)) \leq \mu_2((G \sim A) \times (H \sim B)) + \mu_2((G \sim A) \times B)$$
$$+ \mu_2((A \times (H \sim B))$$
$$= \mu(G \sim A)\mu(H \sim B)) + \mu(G \sim A)\mu(B) + \mu(A)\mu(H \sim B) = 0.$$

So, $G \times H \sim A \times B$ is measurable and is of measure 0. By Case III, $G \times H$ is measurable and $\mu_2(G \times B) = \mu(G)\mu(H)$. Hence $A \times B$ is measurable and

$$\mu_2(A \times B) = \mu_2(G \times H) = \mu(G)\mu(H) = \mu(A)\mu(B).$$

So, the theorem is proved when A and B are bounded.

To complete the proof we suppose that A or B or both are unbounded. Let $A_n = A \cap [-n, n]$, $B_n = B \cap [-n, n]$. Then using an analog of Theorem 2.8 $\mu(A) = \lim_{n \to \infty} \mu(A_n)$ and $\mu(B) = \lim_{n \to \infty} \mu(B_n)$. Also $A \times B = \lim_{n \to \infty}(A_n \times B_n)$. By Case IV $A_n \times B_n$ is measurable and $\mu_2(A_n \times B_n) = \mu(A_n)\mu(B_n)$ for all n. So $A \times B$ is measurable and $\mu_2(A \times B) = \mu_2(\lim_{n \to \infty}(A_n \times B_n)) = \lim_{n \to \infty} \mu_2(A_n \times B_n) = \lim_{n \to \infty} \mu(A_n)\mu(B_n) = \lim_{n \to \infty} \mu(A_n) \lim_{n \to \infty} \mu(B_n) = \mu(A)\mu(B)$. Completing the proof. □

Definition 7.3 Let $E \subset \mathbb{R}^2$. Then for fixed x the set $E_x = \{y : (x, y) \in E\}$ is called the $x-$ section of E and for fixed y the set $E^y = \{x : (x, y) \in E\}$ is called the $y-$ section of E.

It is clear that if the sets A, B, C are such that $A \subset B$, $A = \cup A_n$, $C = A \sim B$ then their $x-$ section and $y-$ section also satisfy these relations. It may be noted that if E is measurable then E_x may not be measurable for all x and E^y may not be measurable for all y.

Theorem 7.12 *If $E \subset \mathbb{R}^2$ is measurable then*

(a) for almost all $x \in \mathbb{R}$ E_x is measurable and for almost all $y \in \mathbb{R}$ the set E^y is measurable.

(b) if $\Delta_1 = \{x : E_x$ is measurable$\}$ then Δ_1 is measurable and the function $f : \Delta_1 \to \mathbb{R}$ defined by $f(x) = \mu(E_x)$ is measurable and if $\Delta_2 = \{y : E^y$ is measurable$\}$ then Δ_2 is measurable and the function $g : \Delta_2 \to \mathbb{R}$ defined by $g(y) = \mu(E^y)$ is measurable.

Also

$$\mu_2(E) = \int_{\Delta_1} f dx = \int_{\Delta_2} g dy$$

where the integrals in the right side are Lebesgue integral in $\mathbb{R}$.

Proof We shall prove for E_x and $f : \Delta_1 \to \mathbb{R}$, the proof for E^y and $g : \Delta_2 \to \mathbb{R}$ are similar.

We first suppose that E is bounded.

Case-I Let E be a closed square and let $E = Q = \{(x, y) : \alpha \le x \le \beta; \gamma \le y \le \delta\}$. Then for fixed $x \in [\alpha, \beta]$, $Q_x = \{y : (x, y) \in Q\}$ and so $Q_x = [\gamma, \delta]$. If $x \notin [\alpha, \beta]$ then $Q_x = \emptyset$. So, Q_x is measurable for all $x \in \mathbb{R}$, proving (a). For (b) note that $\Delta_1 = \mathbb{R}$ and $f : \Delta_1 \to \mathbb{R}$ is defined by $f(x) = \delta - \gamma$ if $x \in [\alpha, \beta]$ and $f(x) = 0$ for $x \notin [\alpha, \beta]$. So, f is a step function on $\mathbb{R}$ and hence is measurable and

$$\mu_2(Q) = (\beta - \alpha)(\delta - \gamma) = \int_\alpha^\beta f dx = \int_{\Delta_1} f dx$$

proving (b).

Case-II Let E be an open set in $\mathbb{R}^2$. Then by Theorem 7.2 there is a countable collection $\{Q_i\}$ of closed squares which are pairwise without common interior points such that $E = \cup Q_i$. So, for each $x \in \mathbb{R}$, $E_x = \cup(Q_i)_x$. By case I, $(Q_i)_x$ is measurable for each i and all $x \in \mathbb{R}$. So, E_x is measurable for all $x \in \mathbb{R}$, proving (a). For (b) note that here also $\Delta_1 = \mathbb{R}$. Since Q_i are pairwise without common interior points, $(Q_i)_x$ are also pairwise without common interior points for all $x \in \mathbb{R}$ and therefore, since $E_x = \cup(Q_i)_x$, $\mu(E_x) = \sum \mu((Q_i)_x)$ for all $x \in \mathbb{R}$. So, writing $f(x) = \mu(E_x)$ and $f_i(x) = \mu((Q_i)_x)$, $f(x) = \sum f_i(x)$ for all $x \in \mathbb{R}$. By case I, f_i is measurable for all i and so f is measurable. Again apply Case I and Monotone convergence Theorem 5.28 and get the relation

$$\mu_2(E) = \sum \mu_2(Q_i) = \sum \int_{\Delta_1} f_i dx = \int_{\Delta_1} (\sum f_i) dx = \int_{\Delta_1} f dx$$

proving (b).

7.2 Lebesgue Outer Measure and Measure in $\mathbb{R}^N$

Case-III Let E be a closed set in $\mathbb{R}^2$. Let $I = \{(x, y), \alpha < x < \beta; \gamma < y < \delta\}$ be an open rectangle such that $E \subset I$. Clearly E_x is also a closed set in $\mathbb{R}$ for all $x \in \mathbb{R}$ and hence E_x is measurable for all $x \in \mathbb{R}$, proving (a). To prove (b) Let $G = I \sim E$. Then G is an open set. Hence by Case II G_x is measurable for all $x \in \mathbb{R}$ and $\mu_2(G) = \int_{\Delta_1} \mu(G_x)dx = \int_\alpha^\beta \mu(G_x)dx$.
Since $E = I \sim G$, $E_x = (\gamma, \delta) \sim G_x$. Hence $\mu(E_x) = (\delta - \gamma) - \mu(G_x)$ and so

$$\int_\alpha^\beta \mu(E_x)dx = (\beta - \alpha)(\delta - \gamma) - \int_\alpha^\beta \mu(G_x)dx = \mu_2(I) - \mu_2(G) = \mu_2(E)$$

proving (b).

Case IV Let E be any measurable set in $\mathbb{R}^2$. Note that analogs of Theorems 2.13 and 2.14 of Chap. 2 are also true for sets in $\mathbb{R}^2$ (proofs are similar). So applying these theorems we get for every $\epsilon > 0$ an open set G and a closed set F in $\mathbb{R}^2$ such that $F \subset E \subset G$ and $\mu_2(E \sim F) < \epsilon$ and $\mu_2(G \sim E) < \epsilon$. Hence

$$\mu_2(E) - \mu_2(F) < \epsilon \text{ and } \mu_2(G) - \mu_2(F) < \epsilon \tag{7.3}$$

By Case II and Case III

$$\mu_2(G) = \int_a^b \mu(G_x)dx \text{ and } \mu_2(F) = \int_a^b \mu(F_x)dx \tag{7.4}$$

where $[a, b]$ is an interval taken such that $E \subset I = \{(x, y) : a \leq x \leq b; c \leq y \leq d\}$. By (7.3) and (7.4)

$$\mu_2(E) - \epsilon < \mu_2(F) = \int_a^b \mu(F_x)dx \leq \int_a^b \mu(G_x)dx = \mu_2(G) < \mu_2(E) + \epsilon \tag{7.5}$$

Hence

$$0 \leq \int_a^b [\mu(G_x) - \mu(F_x)]dx < 2\epsilon.$$

Since ϵ is arbitrary, $\mu(F_x) = \mu(G_x)$ for almost all $x \in \mathbb{R}$. Since $F_x \subset E_x \subset G_x$; E_x is measurable for almost all $x \in \mathbb{R}$ proving (a). Also $\mu(F_x) = \mu(E_x) = \mu(G_x)$ for almost all $x \in \mathbb{R}$ and so by 7.5 letting $\epsilon \to 0$, $\mu_2(E) = \int_{\Delta_1} \mu(E_x)dx$ where $\Delta_1 = \{x : E_x \text{ is measurable}\}$ proving (b) and so the theorem is proved when E is bounded.

To complete the proof let E be any measurable set. Consider the sequence $Q_n = \{(x, y) : -n \leq x \leq n; -n \leq y \leq n\}$. Then $E \cap Q_n$ is bounded and measurable. Writing $f_n(x) = \mu((E \cap Q_n)_x)$ we have by case IV

$$\mu_2(E \cap Q_n) = \int_{S_n} f_n dx \text{ where } S_n = \{x : (E \cap Q_n)_x \text{ is measurable}\}.$$

The sequence $\{f_n\}$ is non-decreasing and $f_n \to f$ where $f(x) = \mu(E_x)$ and so letting $n \to \infty$ we have

$$\mu_2(E) = \int_{\Delta_1} f\,dx \text{ where } \Delta_1 = \{x : E_x \text{ is measurable}\}$$

completing the proof. □

Theorem 7.13 *If $A \subset \mathbb{R}$ is of Lebesgue measure zero and $B \subset \mathbb{R}$ is any set then $A \times B$ is of Lebesgue measure zero.*

Proof We first suppose that B is bounded. Let $I = \{x : a \leq x \leq b\}$ be such that $B \subset I$. Since A and I are measurable sets in $\mathbb{R}$, by Theorem 7.11 $A \times I$ is measurable in $\mathbb{R}^2$ and $\mu_2(A \times I) = \mu(A)\mu(I)$. Also $A \times B \subset A \times I$ and so $\mu_2^*(A \times B) \leq \mu_2(A \times I) = \mu(A)\mu(I) = 0$ since $\mu(A) = 0$. Hence by the property of outer measure in $\mathbb{R}^2$ $A \times B$ is measurable in $\mathbb{R}^2$ and $\mu_2(A \times B) = 0$, proving the theorem in this case. If B is unbounded consider $I_n = \{x : -n \leq x \leq n\}$ and $B_n = B \cap I_n$. Then B_n is bounded for fixed n and so $\mu_2(A \times B_n) = 0$. Since this is true for all n, letting $n \to \infty$, $\mu_2(A \times B) = 0$ where we have used analog of Theorem 2.8.
The following theorem generalizes Theorem 7.5. □

Theorem 7.14 *Let $E \subset \mathbb{R}$ be a measurable set and let $f : E \to \mathbb{R}$ be measurable. Then the set $G(f, E) = \{(x, y) : x \in E; y = f(x)\}$ is measurable and is of measure zero.*

Proof Let $\epsilon > 0$ be arbitrary. For each integer k, $k = \ldots -2, -1, 1, 2, \ldots$ Let $E_k = \{x \in E : \epsilon k \leq f(x) < \epsilon(k+1)\}$. Then the sets E_k are measurable and disjoint and $E = \bigcup_{k=-\infty}^{\infty} E_k$. For each k let $G(f, E_k) = \{(x, y) : x \in E_k; y = f(x)\}$. Then

$$G(f, E) = \bigcup_{k=-\infty}^{\infty} G(f, E_k) \tag{7.6}$$

Since $G(f, E_k) \subset E_k \times \{y : \epsilon k \leq y < \epsilon(k+1)\}$ by Theorem 7.11

$$\mu_2^*(G(f, E_k)) \leq \mu(E_k).\epsilon \tag{7.7}$$

By (7.6) and (7.7)

$$\mu_2^*(G(f, E)) \leq \sum_{k=-\infty}^{\infty} \mu_2^*(G(f, E_k)) \leq \sum_{k=-\infty}^{\infty} \mu(E_k)\epsilon = \epsilon\mu(E).$$

Since ϵ is arbitrary, $\mu_2^*(G(f, E)) = 0$, proving the theorem. □

Theorem 7.15 *Let $E \subset \mathbb{R}$ be a measurable set and let f be a non-negative function on E. Then f is measurable if and only if the set $F = \{(x, y) : x \in E; 0 \leq y \leq f(x)\}$ is measurable set in $\mathbb{R}^2$.*

Proof Let f be measurable on E. Then by Theorem 3.21 there is a non-decreasing sequence of non-negative simple functions f_n such that $f_n \to f$ on E. Let n be fixed. Then there are disjoint measurable sets $E_1, E_2, \ldots, E_N$ and real numbers $a_1, a_2, \ldots a_N$ such that $f_n = \sum_{i=1}^{N} a_i \chi_i$ where χ_i is defined by $\chi_i(x) = 1$ if $x \in E_i$ and $\chi_i(x) = 0$ elsewhere. If $F_n = \{(x, y) : x \in E; 0 \le y \le f_n(x)\}$ and $S_i = \{(x, y) : x \in E_i; 0 \le y \le a_i\}$ for $i = 1, 2, \ldots, N$ then $F_n = \bigcup_{i=1}^{N} S_i$. Since E_i is measurable, by Theorem 7.11 S_i is measurable for each i, $i = 1, 2, \ldots, N$ and so F_n is measurable. Since n is arbitrary, each F_n is measurable and so $\bigcup_{n=1}^{\infty} F_n$ is measurable. Also by Theorem 7.14 $G(f, E)$ is measurable where $G(f, E) = \{(x, y) : x \in E; y = f(x)\}$. Since $F = \bigcup_{n=1}^{\infty} F_n \cup G(f, E)$, F is measurable.

Conversely, suppose that F is measurable in $\mathbb{R}^2$. We are to show that $\{x \in E : f(x) > a\}$ is measurable for all a. For y, $0 \le y < \infty$, let $E(y) = \{x \in E :: y \le f(x)\}$. Then $E(y) = \{x : (x, y) \in F\}$. Since $F \subset \mathbb{R}^2$ is measurable and since $\{x : (x, y) \in F\}$ is the $y-$ section of F, by Theorem 7.12 (a) $\{x : (x, y) \in F\}$ is measurable for almost all y and so $E(y)$ is measurable for almost all y. Choose a and a sequence $\{y_n\}$ such that $y_n > y_{n+1}$ for all n and $y_n \to a$ and $E(y_n)$ is measurable for all n. Then since $E(y_n) \subset E(y_{n+1}) \subset \ldots$ and $\{x \in E : f(x) > a\} = \bigcup_{n=1}^{\infty} E(y_n)$, the set $\{x \in E : f(x) > a\}$ is measurable and so the proof is complete. □

Theorem 7.16 *Let $E \subset \mathbb{R}$ be a measurable set of finite measure and let f be a non-negative finite function on E. Then f is integrable if and only if the set $F(E, f) = \{(x, y) : x \in E; 0 \le y \le f(x)\}$ is measurable and is of finite measure in $\mathbb{R}^2$. In either case $\int_E f dx = \mu_2(F(E, f))$.*

Proof We first suppose that f is bounded. If f is constant and $f = k$ on E then $\int_E f dx = k\mu(E) = \mu_2\{(x, y) : x \in E; 0 \le y \le k\}$ by Theorem 7.11 and so the theorem is true. Hence if f is a simple function then also the theorem is true. Let f be integrable. Then by Theorem 5.6 f is measurable and so by Theorem 3.21 there is a non-decreasing sequence of non-negative simple functions $\{f_n\}$ which converges to f uniformly on E. Let $\epsilon > 0$ be arbitrary. Then there is n_0 such that $|f(x) - f_n(x)| < \epsilon$ for all $n \ge n_0$ and all $x \in E$. So, $\left| \int_E f dx - \int_E f_n dx \right| \le \epsilon \mu(E)$ for all $n \ge n_0$. Since ϵ is arbitrary, $\lim_{n \to \infty} \int_E f_n dx = \int_E f dx$. Since the theorem is true for simple functions $\int_E f_n dx = \mu_2(F(E, f_n))$. Since

$$f_n \le f_{n+1} \le \ldots \le f, F(E, f_n) \subset F(E, f_{n+1}) \subset \ldots \subset F(E, f)$$

for all n and so $\bigcup_{n=1}^{\infty} F(E, f_n) \cup G(E, f) = F(E, f)$ where $G(E, f) = \{(x, y) : x \in E; y = f(x)\}$. Since by Theorem 7.14 $\mu_2(G(E, f)) = 0$, $\lim_{n \to \infty} \mu_2(F(E, f_n)) = \mu_2(F(E, f))$. Hence $\lim_{n \to \infty} \int_E f_n dx = \mu_2(F(E, f))$ and so $\int_E f dx = \mu_2(F(E, f))$.

Conversely, if $F(E, f)$ is measurable and is of finite measure then by Theorem 7.15 f is measurable and since f is assumed to be bounded, f is integrable by Theorem 5.6. So, by the above argument the equality follows. Thus the theorem is proved when f is bounded.

For unbounded f, suppose that f is integrable. For each positive integer N let $f_N(x) = f(x)$ if $0 \le f \le N$ and $f_N(x) = N$ if $f(x) > N$. Then $\{f_N\}$ is a non-decreasing sequence of non-negative bounded integrable functions which converges to f on E. Hence $F(E, f_N) \subset F(E, f_{N+1}) \subset \ldots \subset F(E, f)$ and so $\bigcup_{N=1}^{\infty} F(E, f_N) \cup G(E, f) = F(E, f)$ where $G(E, f) = \{(x, y) : x \in E; y = f(x)\}$. Hence as above $\lim_{N \to \infty} \mu_2(F(E, f_N)) = \mu_2(F(E, f))$. Since the theorem is true for bounded functions, $\int_E f_N dx = \mu_2(F(E, f_N))$ and have $\lim_{N \to \infty} \int_E f_N dx = \mu_2(F(E, f))$. But by definition $\lim_{N \to \infty} \int_E f_N dx = \int_E f dx$ and so $\int_E f dx = \mu_2(F(E, f))$. Conversely, if $F(E, f)$ is measurable and is of finite measure then by Theorem 7.15 f is measurable and therefore, since f is non-negative, $\int_E f dx$ exists as an extended real number. By definition $\lim_{N \to \infty} \int_E f_N dx = \int_E f dx$ where f_N is defined as above. Since f_N is bounded and since the theorem is true for a bounded function, $\int_E f_N dx = \mu_2(F(E, f_N))$. As above $\lim_{N \to \infty} \mu_2(F(E, f_N)) = \mu_2(F(E, f))$ and so $\int_E f dx = \mu_2(F(E, f))$. Since $\mu_2(F(E, f))$ is finite f is integrable. □

Example 7.1 Let $E_1 = \{(x, y) : 0 \le x \le 1; 0 \le y \le 1; x \text{ is rational and } y \text{ is irrational}\}$ $E_2 = \{(x, y) : 0 \le x \le 1; 0 \le y \le 1; x \text{ and } y \text{ are both rational}\}$ and $E_3 = \{(x, y) : 0 \le x \le 1; 0 \le y \le 1; x \text{ and y are both irrational}\}$. Then both sets E_1 and E_2 are of measure zero and $\mu_2(E_3) = 1$. The set E_2 is countable but both the sets E_1 and E_3 has the cardinal number $\underline{c}$.

To see this, let $A = \{x : 0 \le x \le 1; x \text{ is rational}\}$ and $B = \{y : 0 \le y \le 1; y \text{ is irrational}\}$. Then $\mu(A) = 0$ and $\mu(B) = 1$. Also $E_1 = A \times B$, $E_2 = A \times A$ and $E_3 = B \times B$. So, by Theorem 7.11 $\mu_2(E_1) = 0$ and $\mu_2(E_2) = 0$ and $\mu_2(E_3) = 1$. Regarding cardinal numbers of E_1 E_2 and E_3 note that A being countable, it has cardinal number $\underline{a}$ (see Art.4.2) i.e $\overline{\overline{A}} = \underline{a}$ and $\overline{\overline{B}} = \underline{c}$. Since $\underline{c} \le \underline{a}\,\underline{c} \le \underline{c}\,\underline{c} = \underline{c}$, $\underline{a}\,\underline{c} = \underline{c}$ and so $\overline{\overline{E_1}} = \overline{\overline{A \times B}} = \underline{a}\,\underline{c} = \underline{c}$. Since $\underline{a}\,\underline{a} = \underline{a}$ and $\underline{c}\,\underline{c} = \underline{c}$, $\overline{\overline{E_2}} = \underline{a}$ and $\overline{\overline{E_3}} = \underline{c}$.

7.3 Lebesgue Measurable Function in $\mathbb{R}^N$

The definition of measurable function on $\mathbb{R}^N$ is similar to that given in Chap. 3 and all the properties of measurable functions on $\mathbb{R}$ proved there are also true for the function on $\mathbb{R}^N$, $N \ge 2$. The proofs being similar we omit them. We consider $N = 2$.

Definition 7.4 Let $f : \mathbb{R} \times \mathbb{R} \to \mathbb{R}$. For $x \in \mathbb{R}$, define f_x by $f_x(y) = f(x, y)$ and for $y \in \mathbb{R}$ define f^y by $f^y(x) = f(x, y)$. The functions f_x and f^y are called the $x-$ section and $y-$ section of f respectively.

Theorem 7.17 *Let $f : \mathbb{R} \times \mathbb{R} \to \mathbb{R}$ be measurable. Then for almost all $x \in \mathbb{R}$ the function f_x is measurable on E_x and for almost all $y \in \mathbb{R}$ the function f^y is measurable on E^y, where E_x and E^y are defined in Definition 7.3.*

Proof Let α be any real number and let $E = \{(x, y) : f(x, y) > \alpha\}$. Since f is measurable, E is measurable and so by Theorem 7.12 $(a) E_x$ is measurable for almost all $x \in \mathbb{R}$ and E^y is measurable for almost all $y \in \mathbb{R}$. Since $E_x = \{y : f_x(y) > \alpha\}$ and $E^y = \{x : f^y(x) > \alpha\}$ and since α is arbitrary, the result follows. □

Theorem 7.18 *If A and B are measurable subsets of $\mathbb{R}$ and if $f : A \to \mathbb{R}$ and $g : B \to \mathbb{R}$ are measurable then the functions $F : A \times B \to \mathbb{R}$ and $G : A \times B \to \mathbb{R}$ defined by $F(x, y) = f(x)$ and $G(x, y) = g(x)$ for all $(x, y) \in A \times B$ are measurable.*

Proof Let α be any real number. Then $\{(x, y) : (x, y) \in A \times B; F(x, y) > \alpha\} = \{x \in A : f(x) > \alpha\} \times B$. Since f is measurable, by Theorem 7.11 the set $\{x \in A : f(x) > \alpha\} \times B$ is measurable. Since α is arbitrary. F is measurable on $A \times B$. The proof for G is similar. □

Corollary 7.19 *If $f : A \to \mathbb{R}$ and $g : B \to \mathbb{R}$ are measurable then the function $S(x, y) = f(x) + g(y)$ and $P(x, y) = f(x)g(x)$ are measurable on $A \times B$.*

Proof Writing $F(x, y) = f(x)$ and $G(x, y) = g(y)$ for $(x, y) \in A \times B$ and applying the above theorem. F and G are measurable on $A \times B$. Since $S = F + G$ and $P = FG$, the result follows. □

7.4 Lebesgue Integral in $\mathbb{R}^N$

The definition of Lebesgue integral in $\mathbb{R}^N$ to quite similar to that in $\mathbb{R}$ and so we shall not repeat this. All properties of the integration in $\mathbb{R}$ are also true. Here also we restrict our discussion for $N = 2$ and there is no difficulty to extend it for $N > 2$. If a set $E \subset \mathbb{R}^2$ is measurable and $f : E \to \mathbb{R}$ is integrable then the integral of f on E is denoted by $\int \int_E f(x, y) dx\, dy$, or simply by $\int \int_E f\, dx\, dy$. The properties of the integral which are true in $\mathbb{R}$ are also true in $\mathbb{R}^2$. We start with the following theorem which will be used.

Theorem 7.20 (Monotone convergence Theorem) *Let $\{f_n\}$ be a non-decreasing sequence of non-negative measurable functions defined on a measurable set $E \subset \mathbb{R}^2$ of finite measure. Then*

$$\int_E (\lim_{n \to \infty} f_n) dx\, dy = \lim_{n \to \infty} \int_E f_n dx\, dy.$$

The proof is similar to that of Theorem 5.28.

Theorem 7.21 *Let $E \subset \mathbb{R}^2$ be a measurable set and let $f : E \to \mathbb{R}$ be a non-negative measurable function. If $\Delta_1 = \{x : E_x \text{ is measurable}\}$ and $\Delta_2 = \{y : E^y \text{ is measurable}\}$ then the functions $\int_{E_x} f_x(y)dy$ and $\int_{E^y} f^y(x)dx$ are measurable on Δ_1 and Δ_2 respectively and*

$$\int\int_E f(x,y)dx\,dy = \int_{\Delta_1}\left\{\int_{E_x} f_x(y)dy\right\}dx = \int_{\Delta_2}\left\{\int_{E^y} f^y(x)dx\right\}dy. \quad (7.8)$$

Proof Extend f to the whole of $\mathbb{R}^2$ by defining $f(x, y) = 0$ if $(x, y) \notin E$. By Theorem 7.17, f_x is measurable on E_x for all $x \in \Delta_1$ and f^y is measurable on E^y for all $y \in \Delta_2$. Let $g : \Delta_1 \to \mathbb{R}$ be defined by

$$g(x) = \int_{E_x} f_x(y)dy. \quad (7.9)$$

We show that g is measurable on Δ_1 and

$$\int_{\Delta_1} g(x)dx = \int\int_E f(x,y)dx\,dy. \quad (7.10)$$

If f is the characteristic function of E then by Theorem 7.12 the function g is measurable and (7.10) is true. So, (7.10) is true for all non-negative measurable simple functions. Let f be a non-negative measurable function on E. Then by an analogous theorem of Theorem 3.21 there is a non-decreasing sequence $\{f_n\}$ of non-negative measurable simple functions such that $f_n \to f$ as $n \to \infty$ on E. For each n define

$$g_n(x) = \int_{E_x} (f_n)_x(y)dy. \quad (7.11)$$

Since (7.10) is true for non-negative measurable simple functions g_n is measurable on Δ_1 and

$$\int_{\Delta_1} g_n(x)dx = \int\int_E f_n(x,y)dxdy \text{ for all } n. \quad (7.12)$$

Since $f_n \to f$ on E as $n \to \infty$, $(f_n)_x \to f_x$ on E as $n \to \infty$ and so by monotone convergence theorem (see Theorem 5.28) we get from (7.9) and (7.11), $g_n \to g$ as $n \to \infty$. Again applying Theorems 5.28 and 7.20 respectively on the left side and right side of (7.12)

$$\int_{\Delta_1} g(x)dx = \int\int_E f(x,y)dxdy.$$

So (7.10) is proved. From (7.9) and (7.10) the first equality of (7.8) is proved. The proof of the second equality of (7.8) is similar. □

Theorem 7.22 *Let $E \subset \mathbb{R}^2$ be a measurable set and let $f : E \to \mathbb{R}$ be a non-negative measurable function. If $\Delta_1 = \{x : E_x \text{ is measurable}\}$ and $\Delta_2 = \{y : E^y \text{ is measurable}\}$ and if one of the integrals*

7.4 Lebesgue Integral in $\mathbb{R}^N$

$$\int\int_E f(x,y)dx\,dy, \quad \int_{\Delta_1}\left\{\int_{E_x} f_x(y)dy\right\}dx, \quad \int_{\Delta_2}\left\{\int_{E^y} f^y(x)dx\right\}dy$$

is finite then all are finite and equal and f is integrable on E.
The proof follows Theorem 7.21.

Theorem 7.23 (Fubini's theorem) *Let $E \subset \mathbb{R}^2$ be measurable and let $f : E \to \mathbb{R}$ be integrable. Let $\Delta_1 = \{x : E_x \text{ is measurable}\}$ and $\Delta_2 = \{y : E^y \text{ is measurable}\}$. Then f_x is integral on E_x for each $x \in \Delta_1$ and f^y is integrable on E^y for each $y \in \Delta_2$. The functions $\int_{E_x} f_x(y)dy$ and $\int_{E^y} f^y(x)dx$ are integrable on Δ_1 and Δ_2 respectively and the following relation hold :*

$$\int\int_E f(x,y)dx\,dy = \int_{\Delta_1}\left\{\int_{E_x} f_x(y)dy\right\}dx = \int_{\Delta_2}\left\{\int_{E^y} f^y(x)dx\right\}dy.$$

Proof Since f is integrable on E, f^+ and f^- are integrable on E. Since f^+ and f^- are non-negative measurable functions on E by Theorem 7.21

$$\int\int_E f^+(x,y)dx\,dy, = \int_{\Delta_1}\left\{\int_{E_x}(f^+)_x(y)dy\right\}dx$$

and

$$\int\int_E f^-(x,y)dx\,dy = \int_{\Delta_1}\left\{\int_{E_x}(f^-)_x(y)dy\right\}dx.$$

Subtracting we have

$$\int\int_E f(x,y)dx\,dy = \int_{\Delta_1}\left\{\int_{E_x}(f^+)_x(y)dy - \int_{E_x}(f^-)_x(y)dy\right\}dx$$
$$= \int_{\Delta_1}\left[\int_{E_x}\{(f^+)_x(y)-(f^-)_x(y)\}dy\right]dx$$
$$= \int_{\Delta_1}\left\{\int_{E_x} f_x(y)dy\right\}dx.$$

proving the first equality. The proof of the second equality is similar. □

Theorem 7.24 *Let $E = A \times B$ where A and B are measurable subsets of $\mathbb{R}$ and let f be integrable on E. Then f_x is integrable on B for all $x \in A$ and f^y is integrable on A for $y \in B$. The functions $\int_B f_x(y)dy$ and $\int_A f^y(x)dx$ are integrable on A and B respectively and*

$$\int\int_E f(x,y)dxdy = \int_A\left\{\int_B f_x(y)dy\right\}dx = \int_B\left\{\int_A f^y(x)dx\right\}dy.$$

Proof By Theorem 7.11 the set E is measurable. Also $E_x = B$ for all $x \in A$ and $E^y = A$ for all $y \in B$. Since A and B are measurable the result follows by Theorem 7.23. □

Example 7.2 There exist measurable functions f such that both repeated integrals exist but are not equal and f is not integrable.

Solution: Let $f(x, y) = \frac{x^2-y^2}{(x^2+y^2)^2}$, if $(x, y) \neq (0, 0)$ and $f(0, 0) = 0$. Since for fixed x, $x \neq 0$, $\frac{\partial}{\partial y}\left(\frac{y}{x^2+y^2}\right) = f(x, y)$, $\int_0^1 f(x, y)dy = \frac{1}{x^2+1}$ and so $\int_0^1\{\int_0^1 f(x, y)dy\}dx = \frac{\pi}{4}$. So, $\int_0^1\{\int_0^1 f(y, x)dx\}dy = \frac{\pi}{4}$. Since $f(y, x) = -f(x, y)$, $\int_0^1\{\int_0^1 f(x, y)dy\}dx = -\frac{\pi}{4}$. So, by Theorem 7.23, f is not integrable.

Example 7.3 There exists measurable functions f such that both repeated integrals exists and are equal and finite but f is not integrable.

Solution: Let $f(x, y) = \frac{xy}{(x^2+y^2)^2}$, if $(x, y) \neq (0, 0)$ and $f(0, 0) = 0$. If $x \neq 0$ then $f(x, y) = -f(x, -y)$ for all y and so $\int_{-1}^1 f(x, y)dy = 0$ for all $x \neq 0$. If $x = 0$ then $f(x, y) = 0$ for all y and so $\int_{-1}^1 f(x, y)dy = 0$ for $x = 0$. So $\int_{-1}^1 f(x, y)dy = 0$ for all x and hence $\int_{-1}^1\{\int_{-1}^1 f(x, y)dy\}dx = 0$. Since $f(y, x) = f(x, y)$, $\int_{-1}^1\{\int_{-1}^1 f(x, y)dx\}dy = 0$. But f is not integrable in $[-1, 1] \times [-1, 1]$. For, if so then f is integrable in $[0, 1] \times [0, 1]$ and therefore by Theorem 7.23 $\int_0^1\{\int_0^1 f(x, y)dy\}dx$ is finite. But

$$\int_0^1 f(x, y)dy = \int_0^1 \frac{xy}{(x^2+y^2)^2}dy = \frac{x}{2}\left(\frac{1}{x^2} - \frac{1}{x^2+1}\right) = \frac{1}{2x(x^2+1)}$$

and so

$$\int_0^1\left\{\int_0^1 f(x, y)dy\right\}dx = \int_0^1 \frac{1}{2x(x^2+1)}dx = \infty$$

which is a contradiction.

Example 7.4 Let $E = \{(x, y) : 0 \leq x \leq 1; 0 \leq y \leq 1\}$ and let $f : E \to \mathbb{R}$ be defined by $f(x, y) = \frac{1}{3}$ if y is rational and $f(x, y) = x^2$ if y is irrational. Then
(i) if Riemann integral is considered then the repeated Riemann integral $\int_0^1\left\{\int_0^1 f(x, y)dx\right\}dy$ exists but the other repeated integral and the double integral of f do not exist.
(ii) if the Lebesgue integral is considered then both repeated integrals and the double integral exist finitely and all are equal.

Solution (i) Considering Riemann integral $\int_0^1 f(x, y)dx = \frac{1}{3}$ for all $y \in [0, 1]$ and so $\int_0^1\{\int_0^1 f(x, y)dx\}dy = \frac{1}{3}$. But for fixed $x \in [0, 1]$, $f(x, y)$ is everywhere discontinuous in $[0, 1]$ and so for any fixed $x \in [0, 1]$ $\int_0^1 f(x, y)dy$ does not exist as a Riemann integral and so $\int_0^1\{\int_0^1 f(x, y)dy\}dx$ does not exist as Riemann integral. To see that f is not Riemann integrable in E, let $E_1 = \{(x, y) : \frac{1}{\sqrt{2}} \leq x \leq 1; 0 \leq y \leq 1\}$. Then $E_1 \subset E$. If $(x, y) \in E_1$ then $x^2 \geq \frac{1}{2}$ and so $f(x, y) \geq \frac{1}{2}$ if y is irrational and $f(x, y) = \frac{1}{3}$ if y is rational. So if $D : \frac{1}{\sqrt{2}} = x_0 < x_1 < \cdots < x_m = 1; 0 = y_0 < y_1 < \cdots < y_n = 1$ is any partition of E_1 and if M_{rs} and m_{rs} are the upper and lower bounds of $f(x, y)$ in $\{(x, y) : x_{r-1} \leq x \leq x_r; y_{s-1} \leq y \leq y_s\}$ then $M_{rs} \geq \frac{1}{2}$ and $m_{rs} = \frac{1}{3}$ and so upper and lower Riemann sums are

$$S_D(f) = \sum_{r=1}^{m} \sum_{s=1}^{n} M_{rs}(x_1 - x_{r-1})(y_s - y_{s-1}) \geq \frac{1}{2}(1 - \frac{1}{\sqrt{2}})$$

$$s_D(f) = \sum_{r=1}^{m} \sum_{s=1}^{n} m_{rs}(x_r - x_{r-1})(y_s - y_{s-1}) = \frac{1}{3}(1 - \frac{1}{\sqrt{2}})$$

which shows that the upper Riemann integral of f in E_1 is $\geq \frac{1}{2}(1 - \frac{1}{\sqrt{2}})$ while the lower Riemann integral of f in E_1 is $= \frac{1}{3}(1 - \frac{1}{\sqrt{2}})$ and so f is not Riemann integrable in E_1 and so f is not Riemann integrable in E.

(ii) Considering Lebesgue integral the function $g(x, y) = x^2$ for all $(x, y) \in [0, 1] \times [0, 1]$, is continuous and so g is Riemann integrable and hence g is Lebesgue integrable. If $S = [0, 1] \times \{y : 0 \leq y \leq 1; \ y \text{ is rational}\}$ then by Theorem 7.11, S is measurable and is of measure 0. So, $g = f$ a.e. in $[0, 1] \times [0, 1]$ and hence f is Lebesgue integrable in $[0, 1] \times [0, 1]$. So, by Theorem 7.23 the result follows. (Note that in this case for any fixed $x \in [0, 1]$ $f(x, y)$ is integrable in $[0, 1]$ and $\int_0^1 f(x, y) dy = x^2$ for all $x \in [0, 1]$ and so $\int_0^1 \{\int_0^1 f(x, y) dy\} dx = \frac{1}{3}$).

The following theorem is also known as Fubini's theorem.

Theorem 7.25 (Fubini) *Let $E \subset \mathbb{R}^2$ be measurable and let $f : E \to \mathbb{R}$ be measurable. If any one of the integrals*

$$\int\int_E |f|(x, y) dx\, dy, \int_{\Delta_1} \left\{\int_{E_x} |f_x|(y) dy\right\} dx, \int_{\Delta_2} \left\{\int_{E^y} |f^y|(x) dx\right\} dy \quad (7.13)$$

is finite then all are finite and equal, where Δ_1, Δ_2, E_x and E^y are as in Theorem 7.22.

Proof If one of the three integrals is finite then that integral is finite if f is replaced by f^+ and f^-. So, using Theorem 7.22 we have

$$\int\int_E f^+(x, y) dx\, dy = \int_{\Delta_1} \left\{\int_{E_x} (f^+)_x(y) dy\right\} dx = \int_{\Delta_2} \left\{\int_{E^y} (f^+)^y(x) dx\right\} dy \quad (7.14)$$

and

$$\int\int_E f^-(x, y) dx\, dy = \int_{\Delta_1} \left\{\int_{E_x} (f^-)_x(y) dy\right\} dx = \int_{\Delta_2} \left\{\int_{E^y} (f^-)^y(x) dx\right\} dy. \quad (7.15)$$

Since $|f| = f^+ + f^-$ and $f = f^+ - f^-$, adding (7.14) and (7.15) the proof of Theorem (7.25) is complete.

Also note that subtracting (7.15) from (7.14) we get Theorem 7.23 since if one of (7.13) is finite then f is integrable. □

Example 7.5 Let $A, B \subset \mathbb{R}$ where A is nonmeasurable and B is of measure zero and let $f : \mathbb{R}^2 \to \mathbb{R}$ be defined by $f(x, y) = 1$ if $(x, y) \in A \times B$ and $f(x, y) = 0$ if $(x, y) \notin A \times B$. Then f is measurable in $\mathbb{R}^2$ but f^y is measurable for all $y \in \mathbb{R} \sim B$ and f^y is not measurable for all $y \in B$.

Solution: By Theorem 7.13, $A \times B$ is of measure zero and so $f = 0$ a.e. and hence f is measurable in $\mathbb{R}^2$. If $y \in \mathbb{R} \sim B$ then $y \notin B$ and so $(x, y) \notin A \times B$ and hence $f(x, y) = 0$ for all x and hence $f^y(x) = 0$ for all x and so f^y is measurable. If $y \in B$, then $(x, y) \in A \times B$ for all $x \in A$ and so $f(x, y) = 1$ for all $x \in A$ and $(x, y) \notin A \times B$ for all $x \in \mathbb{R} \sim A$ and hence $f(x, y) = 0$ for all $x \in \mathbb{R} \sim A$. Therefore $f^y(x) = 1$ for all $x \in A$ and $f^y(x) = 0$ for all $x \in \mathbb{R} \sim A$. So, f^y is not measurable.

This example shows that in Theorem 7.17 'almost all' cannot be replaced by 'all'.

7.5 Exercises

1. Show that the set of points (x, y) in $[0, 1] \times [0, 1]$ where x and y are rational is measurable and is of measure, 0.
2. If A and B are bounded sets in $\mathbb{R}^2$ and $\mu_2^*(A) = 0$ then show that $\mu_2^*(A \cup B) = \mu_2^*(B)$.
3. Show that a set $E \subset \mathbb{R}^2$ is measurable if and only if for every $\epsilon > 0$ there is an open set $G \subset \mathbb{R}^2$ such that $E \subset G$ and $\mu_2^*(G \sim E) < \epsilon$.
4. Show that a set $E \subset \mathbb{R}^2$ is measurable if and only if for every $\epsilon > 0$ there is a closed set $F \subset \mathbb{R}^2$ such that $F \subset E$ and $\mu_2^*(E \sim F) < \epsilon$.
5. Show that for every Lebesgue measurable set $E \subset \mathbb{R}^2$ there are Borel sets F and G, $F, G \subset \mathbb{R}$ such that $F \subset E \subset G$ and $\mu_2(F) = \mu_2(E) = \mu_2(G)$.
6. If $A \subset \mathbb{R}$, $B \subset \mathbb{R}$, $\mu^*(A) > 0$ and B is nonmeasurable then show that $A \times B$ is nonmeasurable in $\mathbb{R}^2$.
7. If $A, B \subset \mathbb{R}$ and if B is a measurable set of positive measure then show that A is measurable if and only if $A \times B$ is measurable.
8. Let $f(x) = 0$ if x is rational and $f(x) = x^2$ if x is irrational. Let $E = \{(x, y) : 0 \leq x \leq 1; 0 \leq y \leq f(x)\}$. Show that E is measurable and find $\mu_2(E)$.
9. Let $E = \{(x, y) : 0 \leq x \leq 1; 1 \leq y < \infty\}$ and let $f(x, y) = e^{-xy} - 2e^{-2xy}$ for $(x, y) \in E$. Show that the repeated integrals are convergent but f is not integrable on E.
10. Let $A, B \subset [0, 1]$ and A and B be measurable. Let $f : [0, 1] \times [0, 1] \to \mathbb{R}$ be defined by $f(x, y) = x^2 + y^2$ if $(x, y) \in A \times B$ and $f(x, y) = 0$ if $(x, y) \notin A \times B$. Examine whether $f(x, y)$ is (i) measurable (ii) integrable.

Chapter 8
General Measure and Outer Measure

The measure and outer measure discussed in Chap. 2 are the Lebesgue measure and Lebesgue outer measure. These measures are defined using the definition of outer measure on the real line. In this chapter, we give the definition of measure and outer measure which are independent of each other in the sense that a measure can give an outer measure and an outer measure can give a measure. These definitions of measure and outer measure include Lebesgue measure and Lebesgue outer measure. The integral corresponding to this measure is the general Lebesgue integral. Note that in this chapter the notation μ and μ^* are used to mean general measure and general outer measure, not necessarily Lebesgue measure and Lebesgue outer measure respectively.

8.1 Measure

Definition 8.1 Let $\mathcal{A}$ be an algebra of sets. A function $\psi : \mathcal{A} \to [0, \infty]$ is called a measure on $\mathcal{A}$ if

(i) $\psi(\emptyset) = 0$, $\emptyset$ being the null set, and
(ii) for every countable collection of disjoint sets $\{E_i\} \subset \mathcal{A}$, $\psi(\cup E_i) = \sum \psi(E_i)$ whenever $\cup E_i \in \mathcal{A}$. (Note that $\cup E_i \in \mathcal{A}$ is true if $\mathcal{A}$ is a σ-algebra).
Thus a non-negative completely additive set function (see Definition 1.9) defined on an algebra is called a measure, which will be denoted by μ.

Definition 8.2 Let X be a non-void set and $\mathcal{A}$ be a σ-algebra of subsets of X. Then $(X, \mathcal{A})$ is called a measurable space. If μ is a measure defined on $\mathcal{A}$ then $(X, \mathcal{A}, \mu)$ is called a measure space. The members of $\mathcal{A}$ are called measurable sets. If $A \in \mathcal{A}$, then $\mu(A)$ is called the measure of A. The measure space $(X, \mathcal{A}, \mu)$ is called finite measure space if $\mu(E)$ is finite for all $E \in \mathcal{A}$ and it is called σ-finite if $E \in \mathcal{A}$

and $\mu(E) = \infty$ imply that there is a countable collection of disjoint measurable sets $E_n, n = 1, 2, \ldots$, such that $E = \bigcup_{n=1}^{\infty} E_n$ and $\mu(E_n)$ is finite for all n.

The Lebesgue measure space $(\mathbb{R}, \mathcal{M}, \mu)$ is not finite since $\mu(\mathbb{R}) = \infty$ but it is σ-finite since $\mathbb{R} = (\bigcup_{n=0}^{\infty}[n, n+1)) \cup (\bigcup_{n=0}^{\infty}[-n-1, -n))$.

Most of the results which are true for Lebesgue measure are also true for the general measure. The proofs of these results either follow from the definition or are similar to that of Lebesgue measure and so we omit them.

Theorem 8.1 *Let μ be a measure on an algebra $\mathcal{A}$ and let $\{E_n\}$ be a sequence of sets in $\mathcal{A}$. If $\cup E_n \in \mathcal{A}$ then*

$$\mu(\cup E_n) \leq \sum \mu(E_n).$$

Proof Since $\mathcal{A}$ is an algebra and $\{E_n\} \subset \mathcal{A}$ there is a collection $\{B_n\} \subset \mathcal{A}$ such that $B_m \cap B_n = \emptyset$ for $m \neq n$, $B_n \subset E_n$ for all n and $\cup B_n = \cup E_n$ (see Theorem 1.4) since $\mu(E_n) = \mu((E_n \sim B_n) \cup B_n) = \mu(E_n \sim B_n) + \mu(B_n) \geq \mu(B_n)$ we have

$$\mu(\cup E_n) = \mu(\cup B_n) = \sum \mu(B_n) \leq \sum \mu(E_n)$$

□

Theorem 8.2 *Let $(X, \mathcal{A}, \mu)$ be a measure space and let $\{E_n\} \subset \mathcal{A}$. Then*

$$\mu(\liminf_{n \to \infty} E_n) \leq \liminf_{n \to \infty} \mu(E_n).$$

Proof For each n, let $A_n = \bigcap_{k=n}^{\infty} E_k$. Then $\{A_n\}$ is an increasing sequence. So, by Theorem 1.9

$$\lim_{n \to \infty} \mu(A_n) = \mu(\lim_{n \to \infty} A_n). \tag{8.1}$$

Since $A_n \subset E_n$ for all n, $\mu(A_n) \leq \mu(E_n)$ and hence by 8.1

$$\liminf_{n \to \infty} \mu(E_n) \geq \lim_{n \to \infty} \mu(A_n) = \mu(\lim_{n \to \infty} A_n) = \mu(\bigcup_{n=1}^{\infty} A_n)$$

$$= \mu(\bigcup_{n=1}^{\infty} \bigcap_{k=n}^{\infty} E_k) = \mu(\liminf_{k \to \infty} E_k).$$

□

Theorem 8.3 *Let $(X, \mathcal{A}, \mu)$ be a measure space and let $\{E_n\} \subset \mathcal{A}$ be such that $\mu(\cup E_n) < \infty$. Then*

$$\mu(\limsup_{n \to \infty} E_n) \geq \limsup_{n \to \infty} \mu(E_n).$$

8.1 Measure

Proof Let $S = \cup E_n$ and $F_n = S \sim E_n$. Then by Theorem 8.2

$$\mu(\liminf F_n) \leq \liminf \mu(F_n) \qquad (8.2)$$

Since

$$S \sim \limsup E_n = S \sim \bigcap_{k=1}^{\infty}\bigcup_{n=k}^{\infty} E_n = \bigcup_{k=1}^{\infty}\bigcap_{n=k}^{\infty}(S \sim E_n) = \bigcup_{k=1}^{\infty}\bigcap_{n=k}^{\infty} F_n = \liminf F_n,$$

we have by (8.2)

$$\mu(S \sim \limsup E_n) \leq \liminf \mu(F_n) = \liminf \mu(S \sim E_n). \qquad (8.3)$$

Since $\limsup E_n \subset S$ and $E_n \subset S$ and since $\mu(S) < \infty$, (8.3) gives

$$\mu(S) - \mu(\limsup E_n) \leq \liminf[\mu(S) - \mu(E_n)] = \mu(S) - \limsup \mu(E_n)$$

and so $\mu(\limsup E_n) \geq \limsup \mu(E_n)$. $\square$

Theorem 8.4 *Let $(X, \mathcal{A}, \mu)$ be a measure space and let $\{E_n\}$ be a convergent sequence in $\mathcal{A}$ such that $\mu(\cup E_n) < \infty$. Then*

$$\lim_{n\to\infty} \mu(E_n) = \mu(\lim_{n\to\infty} E_n).$$

Proof Since $\{E_n\}$ is convergent by Theorems 8.2 and 8.3

$$\mu(\lim E_n) \leq \liminf \mu(E_n) \leq \limsup \mu(E_n) \leq \mu(\lim E_n)$$

and the proof follows. $\square$

Definition 8.3 A measure space $(X, \mathcal{A}, \mu)$ is said to be complete if $B \in \mathcal{A}$, $\mu(B) = 0$ and $E \subset B$ imply $E \in \mathcal{A}$.

Let $(X, \mathcal{B}, \nu)$ be a measure space which is not complete. If $(X, \mathcal{A}, \mu)$ be another measure space which is complete and such that $\mathcal{B} \subset \mathcal{A}$ and $\mu(E) = \nu(E)$ for all $E \in \mathcal{B}$ then $(X, \mathcal{A}, \mu)$ is called completion of $(X, \mathcal{B}, \nu)$.

Theorem 8.5 *There exists a measure space which is not complete.*

Proof Let $\mathcal{B}$ be the class of all Borel sets in $X = [0, 1]$. Since $\mathcal{B}$ is a σ- algebra (see Definition 1.4), $(X, \mathcal{B})$ is a measurable space. Let μ be the Lebesgue measure restricted on $\mathcal{B}$. Then $(X, \mathcal{B}, \mu)$ is a measure space. The measure space $(X, \mathcal{B}, \mu)$ is not complete. For, there are members $B \in \mathcal{B}$ such that $\mu(B) = 0$ but B contains subsets which are not Borel sets. (See consequence II of Theorem 4.19). So, there are subsets $E \subset B$ but $E \notin \mathcal{B}$. Therefore $(X, \mathcal{B}, \mu)$ is not complete. $\square$

8.2 Outer Measure

Definition 8.4 Let X be a non-void set and let S be the class of all subsets of X. An extended real-valued set function $\psi : S \to \overline{\mathbb{R}}$ is called an outer measure if it satisfies the following conditions :

(i) $\psi(\emptyset) = 0$, where $\emptyset$ is the null set
(ii) $A, B \in S$ and $A \subset B$ imply $\psi(A) \leq \psi(B)$, and
(iii) $\psi(\bigcup_n E_n) \leq \sum_n \psi(E_n)$ for every sequence $\{E_n\} \subset S$.

An outer measure is usually denoted by μ^*. It follows from (i) and (ii) that μ^* is non-negative.

Note The property stated in (ii) is called monotone non-decreasing and the property stated in (iii) is called countably sub additive.

Definition 8.5 Let μ^* be an outer measure defined on the class S of all subsets of a non-void set X. A set $E \in S$ is said to be measurable with respect to μ^*, or simply μ^*- measurable, if for every $A \in S$

$$\mu^*(A) = \mu^*(A \cap E) + \mu^*(A \sim E).$$

Theorem 8.6 *A set $E \in S$ is μ^*- measurable if and only if for every set $A \in S$*

$$\mu^*(A) \geq \mu^*(A \cap E) + \mu^*(A \sim E).$$

Proof Since $A = (A \cap E) \cup (A \sim E)$ by the conditions (iii) $\mu^*(A) \leq \mu^*(A \cap E) + \mu^*(A \sim E)$ and so if the condition of the theorem holds then $\mu^*(A) = \mu^*(A \cap E) + \mu^*(A \sim E)$ proving that E is measurable. The other part is obvious. $\square$

Theorem 8.7 *Let μ^* be an outer measure defined on the class of all subsets of a non-void set X. If $\mathcal{M}$ is the class of all μ^*- measurable sets then $\mathcal{M}$ is a $\sigma-$ algebra and the restriction of μ^* on $\mathcal{M}$ is a measure. This measure is called the measure induced by the outer measure μ^*.*

Proof Let $E \in \mathcal{M}$. Then for any set $A \subset X$,

$$\mu^*(A) = \mu^*(A \cap E) + \mu^*(A \cap \widetilde{E}) = \mu^*(A \cap \widetilde{E}) + \mu^*(A \cap E).$$

Since $E = \widetilde{\widetilde{E}}$, $\widetilde{E} \in \mathcal{M}$. Let $E_1, E_2 \in \mathcal{M}$. Then for any $A \subset X$

$$\mu^*(A) = \mu^*(A \cap E_1) + \mu^*(A \cap \widetilde{E_1}) \text{ and}$$

$$\mu^*(A \cap \widetilde{E_1}) = \mu^*(A \cap \widetilde{E_1} \cap E_2) + \mu^*(A \cap \widetilde{E_1} \cap \widetilde{E_2}).$$

8.2 Outer Measure

Hence

$$\mu^*(A) = \mu^*(A \cap E_1) + \mu^*(A \cap \widetilde{E_1} \cap E_2) + \mu^*(A \cap \widetilde{(E_1 \cup E_2)}) \quad (8.4)$$

Since $(A \cap E_1) \cup (A \cap \widetilde{E_1} \cap E_2) = A \cap (E_1 \cup E_2)$, by (8.4)

$$\mu^*(A) \geq \mu^*(A \cap (E_1 \cup E_2)) + \mu^*(A \cap \widetilde{(E_1 \cup E_2)})$$

showing that $E_1 \cup E_2 \in \mathcal{M}$. Thus $\mathcal{M}$ is an algebra.

Let $\{E_i\}$ be any countable collection of sets in $\mathcal{M}$. We show that $E = \cup E_i \in \mathcal{M}$. We first suppose that the sets E_i are disjoint. Let $S_n = \bigcup_{i=1}^{n} E_i$. Then $S_n \in \mathcal{M}$ for all n. Let $A \subset X$. Since $S_n \subset E$,

$$\mu^*(A) = \mu^*(A \cap S_n) + \mu^*(A \cap \widetilde{S_n}) \geq \mu^*(A \cap S_n) + \mu^*(A \cap \widetilde{E}) \quad (8.5)$$

Since for any k, $S_k \cap \widetilde{E_k} = S_{k-1}$, by the measurability of E_k,

$$\mu^*(A \cap S_k) = \mu^*(A \cap S_k \cap E_k) + \mu^*(A \cap S_k \cap \widetilde{E_k})$$
$$= \mu^*(A \cap E_k) + \mu^*(A \cap S_{k-1}).$$

This being true for all k, putting $k = 2, 3, \ldots n$ and adding

$$\mu^*(A \cap S_n) = \sum_{i=1}^{n} \mu^*(A \cap E_i) \quad (8.6)$$

From (8.5) to (8.6)

$$\mu^*(A) \geq \sum_{i=1}^{n} \mu^*(A \cap E_i) + \mu^*(A \cap \widetilde{E})$$

Letting $n \to \infty$,

$$\mu^*(A) \geq \sum_{i=1}^{\infty} \mu^*(A \cap E_i) + \mu^*(A \cap \widetilde{E})$$
$$\geq \mu^*(A \cap E) + \mu(A \cap \widetilde{E}).$$

So, $E \in \mathcal{M}$. If the sets $\{E_i\}$ are not disjoint then since $\mathcal{M}$ is an algebra, by Theorem 1.4 there is a countable collection $\{F_i\}$ of disjoint sets in $\mathcal{M}$ such that $\cup E_i = \cup F_i$. By the above argument $\cup F_i \in \mathcal{M}$ and so $\cup E_i \in \mathcal{M}$. So, $\mathcal{M}$ is a σ- algebra.

To show that the restriction of μ^* on $\mathcal{M}$ is a measure we need only to show that $\mu^*(\cup E_i) = \sum \mu^*(E_i)$ whenever $\{E_i\}$ is a countable collection of disjoint sets in $\mathcal{M}$.

Let $E_1, E_2, \in \mathcal{M}$ and $E_1 \cap E_2 = \emptyset$. Then since E_1 is μ^*- measurable

$$\mu^*(E_1 \cup E_2) = \mu^*((E_1 \cup E_2) \cap E_1) + \mu^*((E_1 \cup E_2) \cap \widetilde{E_1})$$
$$= \mu^*(E_1) + \mu^*(E_2).$$

Thus for any finite collection of disjoint sets $E_1, E_2, \ldots, E_n$ in $\mathcal{M}$

$$\mu^*\left(\bigcup_{i=1}^n E_i\right) = \sum_{i=1}^n \mu^*(E_i) \tag{8.7}$$

Let $\{E_i\}$ be any countable collection of disjoint sets in $\mathcal{M}$ and let $E = \bigcup_{i=1}^\infty E_i$. Then since $\bigcup_{i=1}^n E_i \subset E$ for all n, by (8.7)

$$\mu^*(E) \geq \mu^*(\bigcup_{i=1}^n E_i) = \sum_{i=1}^n \mu^*(E_i) \text{ for all } n.$$

So, letting $n \to \infty$

$$\mu^*(E) \geq \sum_{i=1}^\infty \mu^*(E_i).$$

Since μ^* is an outer measure, $\mu^*(E) \leq \sum_{i=1}^\infty \mu^*(E_i)$ and so the proof is complete. □

Theorem 8.8 *Let μ^* be an outer measure defined on the class of all subsets of a non-void set X and let $\mathcal{M}$ be the class of all μ^*- measurable sets. Then the measure space $(X, \mathcal{M}, \overline{\mu})$, where $\overline{\mu}$ is the restriction of μ^* on $\mathcal{M}$, is a complete measure space.*

Proof Let $B \in \mathcal{M}, \overline{\mu}(B) = 0$ and $E \subset B$. Then $\mu^*(E) \leq \mu^*(B) = \overline{\mu}(B) = 0$. So, for every set $A \subset X, \mu^*(A \cap E) \leq \mu^*(E) = 0$. Hence

$$\mu^*(A) \geq \mu^*(A \cap \widetilde{E}) = \mu^*(A \cap E) + \mu^*(A \cap \widetilde{E}).$$

So, by Theorem 8.6, $E \in \mathcal{M}$, proving the theorem. □

Note Since Lebesgue measure is defined by Lebesgue outer measure, Lebesgue measure is complete.

Definition 8.6 An outer measure μ^* defined on the class of all subsets of a non-void set X is said to be regular if for every set $E \subset X$ there is a μ^*- measurable set $B \subset X$ such that $E \subset B$ and $\mu^*(E) = \mu^*(B)$.

Theorem 8.9 *There exist outer measures which are not regular.*

8.2 Outer Measure

Proof Let S be the class of all subsets of $X = (0, 1)$ and let $\mu^* : S \to \overline{\mathbb{R}}$ be defined by $\mu^*(\emptyset) = 0$, $\mu^*(X) = 2$ and $\mu^*(E) = 1$ for $E \in S$, $\emptyset \neq E \neq X$. Clearly μ^* is an outer measure on S. The set $\emptyset$ and X are μ^*- measurable. For, if $A \subset X$ then $\mu^*(A) = \mu^*(A \cap \emptyset) + \mu^*(A \sim \emptyset)$ and $\mu^*(A) = \mu^*(A \cap X) + \mu^*(A \sim X)$. We observe that if $E \in S$ and $\emptyset \neq E \neq X$ then E is not μ^*- measurable. For, suppose E is μ^*- measurable. Choose a set $A \subset X$, $A \neq X$ such that $A \cap E \neq \emptyset$ and $A \sim E \neq \emptyset$. Then since E is μ^*- measurable,

$$1 = \mu^*(A) = \mu^*(A \cap E) + \mu^*(A \sim E) = 1 + 1 = 2$$

which is a contradiction. The outer measure defined above is not regular. For, if possible, suppose that μ^* is regular. Consider the interval $I = [\frac{1}{3}, \frac{2}{3}]$. Then since μ^* is regular, there is a μ^*- measurable set $E \in S$ such that $I \subset E$ and $\mu^*(I) = \mu^*(E)$. If $E \neq X$ then by our observation E is not μ^*- measurable which is a contradiction. So, $E = X$. Hence, since $\mu^*(I) = 1$, we have $1 = \mu^*(I) = \mu^*(E) = \mu^*(X) = 2$ which is again a contradiction. So, μ^* is not regular. $\square$

Theorem 8.10 *Let μ^* be a regular outer measure defined on the class of all subset of a non-void set and let $\mu^*(X) < \infty$. Then a set $E \subset X$ is μ^*- measurable if and only if*

$$\mu^*(E) + \mu^*(\widetilde{E}) = \mu^*(X).$$

Proof Let E be μ^*- measurable. Then

$$\mu^*(X) = \mu^*(X \cap E) + \mu^*(X \cap \widetilde{E}) = \mu^*(E) + \mu^*(\widetilde{E})$$

and so the condition holds.

Now suppose that the condition holds. Let $A \subset X$. Since μ^* is regular, there is a μ^*- measurable set $B \subset X$ such that $A \subset B$ and $\mu^*(A) = \mu^*(B)$. Let $E \subset X$ satisfy the given condition. Since B is μ^*- measurable

$$\mu^*(E) = \mu^*(E \cap B) + \mu^*(E \cap \widetilde{B})$$

and

$$\mu^*(\widetilde{E}) = \mu^*(\widetilde{E} \cap B) + \mu^*(\widetilde{E} \cap \widetilde{B}).$$

Therefore, since E satisfies the given condition

$$\begin{aligned}\mu^*(X) &= \mu^*(E) + \mu^*(\widetilde{E}) \\ &= \mu^*(E \cap B) + \mu^*(E \cap \widetilde{B}) + \mu^*(\widetilde{E} \cap B) + \mu^*(\widetilde{E} \cap \widetilde{B}) \\ &= [\mu^*(E \cap B) + \mu^*(\widetilde{E} \cap B)] + [\mu^*(E \cap \widetilde{B}) + \mu^*(\widetilde{E} \cap \widetilde{B})] \\ &\geq \mu^*(B) + \mu^*(\widetilde{B}) \geq \mu^*(X).\end{aligned}$$

Since $\mu^*(X)$ is finite this gives

$$[\mu^*(E \cap B) + \mu^*(\widetilde{E} \cap B)] + [\mu^*(E \cap \widetilde{B}) + \mu^*(\widetilde{E} \cap \widetilde{B})] = \mu^*(B) + \mu^*(\widetilde{B}).$$

Since $\mu^*(\widetilde{B}) \leq \mu^*(E \cap \widetilde{B}) + \mu^*(\widetilde{E} \cap \widetilde{B})$, we have

$$\mu^*(E \cap B) + \mu^*(\widetilde{E} \cap B) \leq \mu^*(B).$$

Since $A \subset B$ and $\mu^*(A) = \mu^*(B)$, this gives

$$\mu^*(E \cap A) + \mu^*(\widetilde{E} \cap A) \leq \mu^*(A).$$

Hence by Theorem 8.6 the set E is μ^* measurable. $\square$

8.3 Outer Measure Induced by a Measure

Definition 8.7 Let μ be a measure on an algebra $\mathcal{A}$ of subsets of X. For any set $E \subset X$ define

$$\mu^*(E) = \inf\{\sum \mu(A_i) : E \subset \cup A_i, A_i \in \mathcal{A}\}$$

where infimum is taken over all countable collection $\{A_i\} \subset \mathcal{A}$ satisfying $E \subset \cup A_i$. In the following theorem we shall show that μ^* is an outer measure. This outer measure is called the outer measure induced by μ.

Theorem 8.11 *The set function μ^* defined above is an outer measure. Moreover if $E \in \mathcal{A}$ then E is μ^*- measurable and $\mu^*(E) = \mu(E)$.*

Proof Since the null set $\emptyset$ is a subset of every set, $\mu^*(\emptyset) = 0$. Let $E_1 \subset E_2$. Then

$$\{\sum \mu(A_i) : E_2 \subset \cup A_i; A_i \in \mathcal{A}\} \subset \{\sum \mu(A_i) : E_1 \subset \cup A_i; A_i \in \mathcal{A}\}$$

and hence $\mu^*(E_1) \leq \mu^*(E_2)$.

Finally, let $\{E_i\}$ be any countable collection of subsets of X, and let $E = \cup E_i$. If $\mu^*(E_i) = \infty$ for some i say $i = i_0$ then

$$\mu^*(E) \leq \infty = \mu^*(E_{i_0}) \leq \sum \mu(E_i).$$

So, we suppose that $\mu^*(E_i) < \infty$ for all i. Let $\epsilon > 0$ be arbitrary. Then for each i there is $\{A_{ij}\}_{j=1,2,\ldots}$ such that $E_i \subset \bigcup_j A_{ij}, A_{ij} \in \mathcal{A}$ and

$$\sum_j \mu(A_{ij}) < \mu^*(E_i) + \frac{\epsilon}{2^i}.$$

8.3 Outer Measure Induced by a Measure

Since $E = \bigcup_i E_i \subset \bigcup_i \bigcup_j A_{ij} = \bigcup_m A_m$ say, we have

$$\mu^*(E) \leq \sum_m \mu(A_m) = \sum_i \sum_j \mu(A_{ij}) \leq \sum_i \mu^*(E_i) + \epsilon.$$

Since ϵ is arbitrary, $\mu^*(E) \leq \sum_i \mu^*(E_i)$. So μ^* is an outer measure.

Next let $E \in \mathcal{A}$ and let $A \subset X$. If $\mu^*(A) = \infty$ then

$$\mu^*(A) = \infty \geq \mu^*(A \cap E) + \mu^*(A \cap \widetilde{E}).$$

If $\mu^*(A) < \infty$, take $\epsilon > 0$ arbitrary. Then there is $\{B_i\} \subset \mathcal{A}$ such that $A \subset \cup B_i$ and

$$\sum \mu(B_i) < \mu^*(A) + \epsilon. \tag{8.8}$$

Also

$$\mu(B_i) = \mu(B_i \cap E) + \mu(B_i \cap \widetilde{E}) \tag{8.9}$$

and since $A \cap E \subset \cup(B_i \cap E)$ and $A \cap \widetilde{E} \subset \cup(B_i \cap \widetilde{E})$, from (8.8) and (8.9)

$$\mu^*(A) + \epsilon > \sum \mu(B_i \cap E) + \sum \mu(B_i \cap \widetilde{E})$$
$$\geq \mu^*(A \cap E) + \mu^*(A \cap \widetilde{E}).$$

Since ϵ is arbitrary, $\mu^*(A) \geq \mu^*(A \cap E) + \mu^*(A \cap \widetilde{E})$, showing that E is μ^*-measurable.

Finally, since $E \in \mathcal{A}$, $\mu^*(E) \leq \mu(E)$. On the other hand if $E \subset \cup B_i$, $B_i \in \mathcal{A}$, then $E = \cup(B_i \cap E)$ and so

$$\mu(E) \leq \sum \mu(B_i \cap E) \leq \sum \mu(B_i)$$

and hence taking infimum of $\sum \mu(B_i)$ over all such $\{B_i\}$ we get $\mu(E) \leq \mu^*(E)$. So, $\mu^*(E) = \mu(E)$. $\square$

Theorem 8.12 *The outer measure μ^* induced by a measure as defined in Definition 8.7 is regular.*

Proof Let $E \subset X$. If $\mu^*(E) = \infty$, then $\mu^*(X) \geq \mu^*(E) = \infty$ and hence $\mu^*(X) = \mu^*(E)$. Since $X \in \mathcal{A}$, X is μ^*-measurable and so the result follows. So, we suppose $\mu^*(E) < \infty$. For each n there is a sequence $\{A_{in}\} \subset \mathcal{A}$ such that $E \subset \bigcup_i A_{in}$ and

$$\sum_i \mu(A_{in}) < \mu^*(E) + \frac{1}{n} \tag{8.10}$$

Let $B = \bigcap_{n=1}^{\infty} \bigcup_i A_{in}$. Since $A_{in} \in \mathcal{A}$, A_{in} are μ^*- measurable and hence B is μ^*- measurable. Since $B \subset \bigcup_i A_{in}$ we get $\mu^*(B) \le \sum_i \mu(A_{in})$ for all n. So letting $n \to \infty$ we have from (8.10) $\mu^*(B) \le \mu^*(E)$. Also $E \subset B$ and so $\mu^*(E) \le \mu^*(B)$. Thus $\mu^*(E) = \mu^*(B)$, completing the proof. □

8.4 Extension of Measure. Interplay Between Measure and Outer Measure

Definition 8.8 Let X be a non-void set and let $\mathcal{S}$ and $\mathcal{J}$ be algebras of subsets of X such that $\mathcal{S} \subset \mathcal{J}$. If μ and ν are measures defined on $\mathcal{S}$ and $\mathcal{J}$ respectively and if $\mu(E) = \nu(E)$ for all $E \subset \mathcal{S}$ then ν is called an extension of μ. If moreover $\mathcal{S}$ and $\mathcal{J}$ are $\sigma-$ algebras then the measure space $(X, \mathcal{J}, \nu)$ is called an extension of the measure space $(X, \mathcal{S}, \mu)$.

Theorem 8.13 *For, every measure μ on an algebra $\mathcal{S}$ of subsets of a non-void set X there is a measure space $(X, \mathcal{J}, \nu)$ such that $\mathcal{S} \subset \mathcal{J}$ and $\mu(E) = \nu(E)$ for $E \in \mathcal{S}$. Hence ν is an extension of μ.*

Proof For $E \subset X$, let

$$\mu^*(E) = \inf\{\sum_i \mu(A_i); E \subset \bigcup_i A_i; A_i \in \mathcal{S}\}$$

where the infimum is taken over all countable collections $\{A_i\} \subset \mathcal{S}$ such that $E \subset \cup A_i$. Then by Theorem 8.11 μ^* is an outer measure and if $E \in \mathcal{S}$ then E is μ^*- measurable and $\mu^*(E) = \mu(E)$. Applying Theorem 8.7 on this μ^* the class $\mathcal{M}$ of all μ^*- measurable subsets of X is a $\sigma-$ algebra and the restriction of μ^* on $\mathcal{M}$ is a measure which we denote by $\bar{\mu}$. If $E \in \mathcal{S}$ then by Theorem 8.11 $E \in \mathcal{M}$ and $\mu(E) = \bar{\mu}(E)$. Hence $\mathcal{S} \subset \mathcal{M}$. So writing $\mathcal{J} = \mathcal{M}$ and $\nu = \bar{\mu}$, ν is an extension of μ. □

Note 1 In some cases it may be so that $\mathcal{S} = \mathcal{J}$. In these cases the extension is trivial. Let $(X, \mathcal{A}, \mu)$ be a measure space which is not complete (Such measure space exists by Theorem 8.5), and let μ^* be the outer measure obtained from μ by applying Definition 8.7. Again from μ^* applying Theorem 8.8 we get a complete measure space $(X, \mathcal{M}, \bar{\mu})$ which is such that $\mathcal{A} \subset \mathcal{M}$ and $\mu(E) = \bar{\mu}(E)$ for $E \in \mathcal{A}$ where $\mathcal{M}$ is class of all μ^*- measurable sets and $\bar{\mu}$ is the restriction of μ^* on $\mathcal{M}$. So, $(X, \mathcal{M}, \bar{\mu})$ is an extension of $(X, \mathcal{A}, \mu)$. Since $(X, \mathcal{M}, \bar{\mu})$ is complete and $(X, \mathcal{A}, \mu)$ is not, $\bar{\mu}$ is a true extension of μ.

Note 2 Let μ^* be an outer measure defined on the class of subsets of a non-void set X which is not regular (such outer measure exists by Theorem 8.9). Applying Theorem 8.7 we get a measure space $(X, \mathcal{M}, \bar{\mu})$ where $\bar{\mu}$ is the restriction of μ^* on the $\sigma-$

8.4 Extension of Measure. Interplay Between Measure and Outer Measure 235

algebra M of μ^*- measurable sets. Again using $\bar{\mu}$ and applying Theorem 8.11 we get another outer measure $\bar{\mu}^*$ and this outer measure is regular by Theorem 8.12. So if the original outer measure μ^* is not regular then $\bar{\mu}^*$ gives more measurable sets and hence applying Theorem 8.7 on $\bar{\mu}^*$ it gives a measure say ν which is an extension of $\bar{\mu}$.

Lebesgue measure can also be considered as an extension of Borel measure. For, if M is the $\sigma-$ algebra of all Lebesgue measurable sets and $\mathcal{B}$ is the $\sigma-$ algebra of all Borel measurable sets then $\mathcal{B} \subset M$. If μ is the Lebesgue measure on M and ν is the restriction of μ on $\mathcal{B}$ then μ is an extension of ν. Thus by Theorems 8.12 and 8.8 the Lebesgue measure is regular and complete which are already proved.

The following theorem gives more information regarding this.

Definition 8.9 Let μ be a measure on an algebra $\mathcal{A}$ of subsets of a non-void set X. Then μ is said to be finite if $\mu(A)$ is finite for all $A \subset \mathcal{A}$ and μ is said to be $\sigma-$ finite if there is $\{A_i\} \subset \mathcal{A}$ such that $\mu(A_i) < \infty$ for all i and $X = \cup A_i$. Clearly if μ is finite then μ is $\sigma-$ finite but the converse is not true. For, the Lebesgue measure on the real line $\mathbb{R}$ is not finite but it is a $\sigma-$ finite measure on the $\sigma-$ algebra generated by the intervals (a, b).

Theorem 8.14 (Caratheodory-Hahn Extension Theorem) *Let X be a non-void set and $\mathcal{A}$ be an algebra of subsets of X. Let μ be a measure on $\mathcal{A}$ and let μ^* be the outer measure induced by μ. Let M be the $\sigma-$ algebra of all μ^*- measurable sets. Then*
(i) the restriction $\bar{\mu}$ of μ^* in M is an extension of μ.
(ii) If μ is $\sigma-$ finite and if $\mathcal{B}$ is any $\sigma-$ algebra satisfying $\mathcal{A} \subset \mathcal{B} \subset M$ then $\bar{\mu}$ is the only measure on $\mathcal{B}$ which is an extension of μ.

Proof From Theorem 8.11 it follows that if $E \in \mathcal{A}$ then E is μ^*- measurable and $\mu(E) = \mu^*(E)$. So, $\bar{\mu}$ is an extension of μ in M and (i) is proved.

To prove (ii) let λ be a measure on $\mathcal{B}$ which is an extension of μ i.e. $\lambda(E) = \mu(E)$ for all $E \in \mathcal{A}$. Let E be any set such that $E \in \mathcal{B}$. Let $\{A_i\}$ be such that $E \subset \bigcup_i A_i$ and $A_i \in \mathcal{A}$. Such a countable collection exists since $X \in \mathcal{A}$ and $E \subset X$. Then

$$\lambda(E) \leq \lambda(\bigcup_i A_i) \leq \sum_i \lambda(A_i) = \sum_i \mu(A_i).$$

Hence by the definition of μ^*

$$\lambda(E) \leq \mu^*(E). \tag{8.11}$$

We shall show that equality holds in (8.11) for all sets $E \in \mathcal{B}$. We first suppose, as a special case, that there exists $A \in \mathcal{A}$ such that $E \subset A$ and $\mu(A) < \infty$. Since $A \sim E \in \mathcal{B}$ and since (8.11) is true for every $E \in \mathcal{B}$ we have by (8.11)

$$\lambda(A \sim E) \leq \mu^*(A \sim E). \tag{8.12}$$

Also
$$\lambda(E) + \lambda(A \sim E) = \lambda(A) = \mu(A) = \mu^*(A)$$
$$= \overline{\mu}(A) = \overline{\mu}(E) + \overline{\mu}(A \sim E)$$
$$= \mu^*(E) + \mu^*(A \sim E). \tag{8.13}$$

From (8.12) and (8.13)
$$\lambda(E) \geq \mu^*(E) \tag{8.14}$$

From (8.11) and (8.14) $\lambda(E) = \mu^*(E) = \overline{\mu}(E)$.

In the general case, since μ is $\sigma-$ finite, there exists $\{A_i\} \subset \mathcal{A}$ such that $\mu(A_i) < \infty$ for all i and $X = \cup A_i$. Since $\mathcal{A}$ is an algebra, by Theorem 1.4 there exists $\{B_i\} \subset \mathcal{A}$ such that $B_i \subset A_i$, for all i, $B_m \cap B_n = \emptyset$ for $m \neq n$ and $\cup A_i = \cup B_i$. So, $\mu(B_i) < \infty$ and $X = \cup B_i$. Since $E \cap B_i \in \mathcal{B}$, $E \cap B_i \subset B_i \in \mathcal{A}$ and $\mu(B_i) < \infty$ we have by the above special case $\lambda(E \cap B_i) = \mu^*(E \cap B_i)$ for all i. Hence

$$\lambda(E) = \sum_i \lambda(E \cap B_i) = \sum_i \mu^*(E \cap B_i) = \sum_i \overline{\mu}(E \cap B_i) = \overline{\mu}(E).$$

Since $E \in \mathcal{B}$ is arbitrary, $\lambda = \overline{\mu}$ on $\mathcal{B}$. □

Note Properties of the Lebesgue measure and Lebesgue outer measure are also true for the general measure and general outer measure and the proofs are similar so these proofs are omitted.

8.5 Construction of Outer Measure in $\mathbb{R}$

Let $\mathcal{I}$ be the family of all open intervals on the real line $\mathbb{R}$ together with the null set $\emptyset$. Let $\mathcal{T}$ be an extended non-negative real-valued function defined on $\mathcal{I}$ such that $\mathcal{T}(\emptyset) = 0$. For every set $E \subset \mathbb{R}$, define

$$\mu_\mathcal{T}^*(E) = \inf\{\sum \mathcal{T}(I_n) : E \subset \cup I_n; I_n \in \mathcal{I}\}$$

where the infimum is taken over all countable collection $\{I_n\} \subset \mathcal{I}$ such that $E \subset \cup I_n$. We show that $\mu_\mathcal{T}^*$ is an outer measure.

Theorem 8.15 *The set function $\mu_\mathcal{T}^*$ is an outer measure.*

Proof Clearly $\mu_\mathcal{T}^*(\emptyset) = 0$. Let $E_1 \subset E_2$. Then since

$$\{\sum \mathcal{T}(I_n) : E_2 \subset \cup I_n; I_n \in \mathcal{I}\} \subset \{\sum \mathcal{T}(I_n) : E_1 \subset \cup I_n; I_n \in \mathcal{I}\},$$

we have $\mu_\mathcal{T}^*(E_1) \leq \mu_\mathcal{T}^*(E_2)$. Finally, let $\{E_n\}$ be any countable collection of sets and let $E = \cup E_n$. If $\mu_\mathcal{T}^*(E_i) = \infty$ for some i then

$$\mu_\mathcal{T}^*(E) \leq \infty = \mu_\mathcal{T}^*(E_i) \leq \sum \mu_\mathcal{T}^*(E_i). \tag{8.15}$$

If $\mu_\mathcal{T}^*(E_i) < \infty$ for all i, then for arbitrary $\epsilon > 0$, there is for each i a countable collection $\{I_{in}\} \subset \mathcal{I}$ such that $E_i \subset \bigcup_{n=1}^{\infty} I_{in}$ and $\sum_{n=1}^{\alpha} \mathcal{T}(I_{in}) < \mu_\mathcal{T}^*(E_i) + \frac{\epsilon}{2^i}$. Hence

$$\sum_{i=1}^{\infty} \sum_{n=1}^{\infty} \mathcal{T}(I_{in}) \leq \sum_{i=1}^{\infty} \mu_\mathcal{T}^*(E_i) + \epsilon. \tag{8.16}$$

Since $E = \bigcup_{i=1}^{\infty} E_i \subset \bigcup_{i=1}^{\infty} \bigcup_{n=1}^{\infty} I_{in}$ and since the countable union $\bigcup_{i=1}^{\infty} \bigcup_{n=1}^{\infty} I_{in}$ can be written as $\bigcup_{k=1}^{\infty} J_k$ where $J_k = I_{in}$ for some i and some n, we have $E \subset \bigcup_{k=1}^{\infty} J_k$ where $J_k \in \mathcal{I}$. Hence by (8.16)

$$\mu_\mathcal{T}^*(E) \leq \sum_{k=1}^{\infty} \mathcal{T}(J_k) = \sum_{i=1}^{\infty} \sum_{n=1}^{\infty} \mathcal{T}(I_{in}) \leq \sum_{i=1}^{\infty} \mu_\mathcal{T}^*(E_i) + \epsilon. \tag{8.17}$$

Since ϵ is arbitrary, (8.15) and (8.17) show that $\mu_\mathcal{T}^*(E) \leq \sum_{i=1}^{\infty} \mu_\mathcal{T}^*(E_i)$. So, by Definition 8.4, $\mu_\mathcal{T}^*$ is an outer measure. □

8.6 Lebesgue Stieltje's Outer Measure and Measure

Definition 8.10 Let f be a non-decreasing function on the real line $\mathbb{R}$ which is continuous on the right throughout. Let $\mathcal{I}$ be the final of all open intervals on $\mathbb{R}$ together with the null set $\emptyset$. Define $\mathcal{T}(\emptyset) = 0$ and $\mathcal{T}((a,b)) = f(b) - f(a)$ for $(a,b) \in \mathcal{I}$. Then $\mathcal{T}$ induces an outer measure as in Sect. 8.5. This outer measure is called Lebesgue-Stieltje's outer measure and is denoted by μ_f^*. Then for any set $A \subset \mathbb{R}$

$$\mu_f^*(A) = \inf \left\{ \sum \mathcal{T}(I_n); I_n \in \mathcal{I}; A \subset \cup I_n \right\}$$
$$= \inf \left\{ \sum [f(b_n) - f(a_n)] : (a_n b_n) \in \mathcal{I}; A \subset \cup (a_n b_n) \right\}.$$

The outer measure μ_f^* generates a σ− algebra $\mathcal{M}$ and the restriction of μ_f^* on $\mathcal{M}$ is a measure. This measure is called the Lebesgue-Stieltje's measure and is denoted by μ_f.

Theorem 8.16 *For any semiclosed interval $(a, b]$*
$$\mu_f^*((a, b]) = f(b) - f(a).$$

Proof Since $(a, b + \epsilon)$ covers $(a, b]$ for every $\epsilon > 0$ we have
$$\mu_f^*((a, b]) \leq \mathcal{T}((a, b + \epsilon)) = f(b + \epsilon) - f(a).$$

Letting $\epsilon \to 0$, since f is continuous at the right
$$\mu_f^*((a, b]) \leq f(b) - f(a). \tag{8.18}$$

Let $\{(a_k, b_k)\}$ be any countable collection in $\mathcal{I}$ such that $(a, b] \subset \bigcup_k (a_k b_k)$. Choose a' such that $a < a' < b$. Then $[a', b] \subset \bigcup_k (a_k, b_k)$. So, there is a finite collection from $\{(a_k, b_k)\}$ which also covers $[a', b]$. By reindexing this finite collection, if necessary, we may suppose that $[a', b] \subset \bigcup_{n=1}^{N} (a_n, b_n)$ and $a_1 < a', b < b_N$, and $a_{n+1} < b_n$ for $n = 1, 2, \ldots, N - 1$. Hence $f(a_{n+1}) \leq f(b_n)$ for $n = 1, 2, \ldots, N - 1$, and $f(a) \leq f(a')$ and $f(b) \leq f(b_N)$. So,

$$\sum_k \mathcal{T}((a_k, b_k)) \geq \sum_{n=1}^{N} \mathcal{T}((a_n, b_n)) = \sum_{n=1}^{N} [f(b_n) - f(a_n)]$$
$$= f(b_N) - f(a_1) + \sum_{n=1}^{N-1} [f(b_n) - f(a_{n+1})]$$
$$\geq f(b_N) - f(a_1) \geq f(b) - f(a').$$

This being true for every $a', a < a' < b$, letting $a' \to a$, since f is continuous at the right,
$$\sum_k \mathcal{T}((a_k, b_k)) \geq f(b) - f(a).$$

Since this is true for all countable collections $\{(a_k, b_k)\}$ in $\mathcal{I}$ such that $(a, b] \subset \bigcup_k (a_k, b_k)$,
$$\mu_f^*((a, b]) \geq f(b) - f(a). \tag{8.19}$$

So, (8.18) and (8.19) complete the proof. □

Theorem 8.17 *All Borel sets are μ_f^*- measurable.*

Proof We first show that the interval $(-\infty, a]$ is μ_f^*- measurable for all a. Let A be any set. Put $A_1 = A \cap (-\infty, a]$, $A_2 = A \cap (a, \infty)$. We are to show that

$$\mu_f^*(A_1) + \mu_f^*(A_2) \leq \mu_f^*(A). \tag{8.20}$$

8.6 Lebesgue Stieltje's Outer Measure and Measure

This is obvious if $\mu_f^*(A) = \infty$. So, we suppose that $\mu_f^*(A) < \infty$. Let $\epsilon > 0$ be arbitrary. Then there is a countable collection of open intervals $\{I_n = (a_n, b_n)\} \subset \mathcal{I}$ such that $A \subset \cup I_n$ and

$$\sum_n \mathcal{T}(I_n) < \mu_f^*(A) + \epsilon. \tag{8.21}$$

Let $I_n' = I_n \cap (-\infty, a]$ and $I_n'' = I_n \cap (a, \infty)$.

Clearly each I_n'' is either empty or an open interval and $A_2 \subset \cup I_n''$. Hence

$$\mu_f^*(A_2) \leq \sum \mathcal{T}(I_n''). \tag{8.22}$$

Also each I_n' is either empty or is an open interval or an interval which is open in the left and closed in the right. Divide the collection $\{I_n'\}$ into two groups Q_1 and Q_2 such that Q_1 contains all I_n' which are open and Q_2 contains all I_n' which are all open in the left and closed in the right. Let $A_{11} = A_1 \cap \bigcup_{Q_1} I_n'$ and $A_{12} = A_1 \cap \widetilde{A_{11}}$.

Then $A_{11} \subset \bigcup_{Q_1} I_n'$ and so

$$\mu_f^*(A_{11}) \leq \sum_{Q_1} \mathcal{T}(I_n'). \tag{8.23}$$

If $I_n' \in Q_2$ then I_n' is of the form $(a, b]$ and since by Theorem 8.16 $\mu_f^*((a, b]) = f(b) - f(a) = \mathcal{T}((a, b))$, denoting by $I_n'^0$ the interior of I_n' we have

$$\mu_f^*(A_{12}) \leq \sum_{Q_2} \mu_f^*(I_n') = \sum_{Q_2} \mathcal{T}(I_n'^0) \tag{8.24}$$

Since $A_1 \subset A_{11} \cup A_{12}$ we have from (8.23), (8.24) and (8.22)

$$\mu_f^*(A_1) + \mu_f^*(A_2) \leq \mu_f^*(A_{11}) + \mu_f^*(A_{12}) + \mu_f^*(A_2)$$
$$\leq \sum_{Q_1} \mathcal{T}(I_n') + \sum_{Q_2} \mathcal{T}(I_n'^0) + \sum \mathcal{T}(I_n''). \tag{8.25}$$

Since for abutting interval (a, b) and (b, c)

$$\mathcal{T}((a, b)) + \mathcal{T}((b, c)) = f(b) - f(a) + f(c) - f(b) = \mathcal{T}((a, c))$$

the right-hand side of (8.25) is $\sum_n \mathcal{T}(I_n)$ and so by (8.21)

$$\mu_f^*(A_1) + \mu_f^*(A_2) \leq \mu_f^*(A) + \epsilon.$$

Since ϵ is arbitrary, (8.20) is proved. So, the set $(-\infty, a]$ is μ_f^* measurable for all a. Since $(-\infty, a) = \bigcup_{n=1}^{\infty} (-\infty, a - \frac{1}{n}]$, $(-\infty, a)$ is μ_f^* measurable for all a. Since

$(a, b) = (-\infty, b) \cap \widetilde{(-\infty, a]}$, every open interval (a, b) is μ_f^*- measurable. So, the $\sigma-$ algebra $\mathcal{M}$ of all μ_f^*- measurable sets contains all open intervals. Since the class $\mathcal{B}$ of all Borel sets is the smallest $\sigma-$ algebra containing all open intervals, $\mathcal{B} \subset \mathcal{M}$. Hence all Borel sets are μ_f^*- measurable, completing the proof. $\square$

If a set is μ_f^*- measurable, we write μ_f.

Theorem 8.18 *All intervals including the singleton sets are μ_f^*- measurable and*

$$\mu_f(\{a\}) = f(a) - f(a-), \mu_f([a, b]) = f(b) - f(a-)$$
$$\mu_f((a, b)) = f(b-) - f(a), \mu_f([a, b)) = f(b-) - f(a-).$$

Proof Let $I_n = (a - \frac{1}{n}, a]$. Since I_n is a Borel set I_n is μ_f^*- measurable for all n. Also $\{I_n\}$ is a decreasing sequence which converges to the singleton set $\{a\}$. Since $\mu_f(I_1) = f(a) - f(a - 1)$ by Theorem 8.16, $\mu_f(I_n) < \infty$, applying Theorem 8.4

$$\mu_f(\{a\}) = \mu_f(\lim I_n) = \lim \mu_f(I_n) = \lim[f(a) - f(a - \frac{1}{n})] = f(a) - f(a-)$$

Since $[a, b] = \{a\} \cup (a, b]$, applying Theorem 8.16

$$\mu_f([a, b]) = \mu_f(\{a\}) + \mu_f((a, b])$$
$$= f(a) - f(a-) + f(b) - f(a)$$
$$= f(b) - f(a-)$$

and since $(a, b) = (a, b] \sim \{b\}$,

$$\mu_f((a, b)) = \mu((a, b]) - \mu(\{b\})$$
$$= f(b) - f(a) - f(b) + f(b-)$$
$$= f(b-) - f(a)$$

and since $[a, b) = \{a\} \cup (a, b)$

$$\mu_f([a, b)) = \mu_f(\{a\} + \mu_f((a, b))$$
$$= f(a) - f(a-) + f(b-) - f(a)$$
$$= f(b-) - f(a-).$$

$\square$

Theorem 8.19 *The outer measure μ_f^* is such that given any set E there is a Borel set B such that $E \subset B$ and*

$$\mu_f^*(E) = \mu_f(B).$$

8.6 Lebesgue Stieltje's Outer Measure and Measure

Proof If $\mu_f^*(E) = \infty$ then since $E \subset \mathbb{R}$, $\mu_f^*(\mathbb{R}) = \mu_f^*(E)$. Since $\mathbb{R}$ is a Borel set, the theorem is proved. So, we suppose that $\mu_f^*(E) < \infty$. Since

$$\mu_f^*(E) = \inf\{\sum \mathcal{T}(I_n) : E \subset \cup I_n : I_n \in \mathcal{I}\},$$

for any positive integer k there is a countable collection $\{(a_n^k, b_n^k)\} \subset \mathcal{I}$ such that $E \subset \bigcup_n (a_n^k, b_n^k)$ and

$$\sum_n \mathcal{T}((a_n^k, b_n^k)) < \mu_f^*(E) + \frac{1}{k}.$$

Let $B_k = \bigcup_n (a_n^k, b_n^k)$ and $B = \bigcap_k B_k$. Then B is a Borel set. Moreover, for all k

$$\mu_f(B) \leq \mu_f(B_k) \leq \sum_n \mathcal{T}((a_n^k, b_n^k)) < \mu_f^*(E) + \frac{1}{k}$$

Letting $k \to \infty$, $\mu_f(B) \leq \mu_f^*(E)$. Since $E \subset B_k$ for all k, $E \subset B$ and so $\mu_f^*(E) \leq \mu_f(B)$. Hence $\mu_f^*(E) = \mu_f(B)$. $\square$

Corollary 8.20 *The outer measure μ_f^* is regular.*

Since every Borel set is μ_f^*- measurable, the result follows from the above theorem.

Definition 8.11 A measure μ defined on the $\sigma-$ algebra of all Borel sets such that μ is finite for bounded sets is called a Borel measure.

Definition 8.12 A non-decreasing function $f : \mathbb{R} \to \mathbb{R}$ such that f is continuous on the right throughout is called a distribution function.

Theorem 8.21 *The class of all Borel measure is in one-to-one correspondence with the class of all distribution functions, where two distribution functions are considered to be the same if they differ by a constant.*

Proof Let f be a distribution function. Then f induces a Lebesgue-Stieltje's outer measure μ_f^* with respect to which all Borel sets are measurable by Theorem 8.17. If E is a bounded Borel set then $E \subset (a, b]$ for some a, b and so by Theorem 8.16

$$\mu_f^*(E) \leq \mu_f^*((a, b]) = f(b) - f(a) < \infty$$

Hence μ_f, the restriction of μ_f^* on the class of Borel sets is a Borel measure. So, every distribution function corresponds to a Borel measure.

Let f_1 and f_2 be two distribution functions which do not differ by a constant. Then since $f_1 - f_2$ is not a constant there are two points a and b such that $f_1(a) - f_2(a) \neq f_1(b) - f_2(b)$ and so $f_1(b) - f_1(a) \neq f_2(b) - f_2(a)$. If $a < b$ then it gives by Theorem 8.16 $\mu_{f_1}((a, b]) \neq \mu_{f_2}((a, b])$ and if $b < a$ then $\mu_{f_1}((b, a]) \neq \mu_{f_2}((b, a])$. So, $\mu_{f_1} \neq \mu_{f_2}$.

Finally, let μ be any Borel measure. Let $f_\mu : \mathbb{R} \to \mathbb{R}$ be defined by $f_\mu(0) = 0$ and $f_\mu(x) = \mu((0, x])$ if $x > 0$ and $f_\mu(x) = -\mu((x, 0])$ if $x < 0$. Then it can be verified that $f_\mu(x_2) - f_\mu(x_1) = \mu((x_1, x_2])$ whenever $x_1 < x_2$ and so f_μ is non-decreasing. To show that f_μ is continuous at the right, let $\xi \in \mathbb{R}$ and let $\{x_n\}$ be any decreasing sequence such that $x_n \to \xi$ as $n \to \infty$. Then $\{(\xi, x_n]\}$ is a decreasing sequence of sets such that $\bigcap_{n=1}^{\infty}(\xi, x_n] = \emptyset$ So by Theorem 8.4

$$\lim_{n\to\infty}[f_\mu(x_n) - f_\mu(\xi)] = \lim_{n\to\infty}\mu((\xi, x_n]) = \mu(\bigcap_{n=1}^{\infty}(\xi, x_n]) = \mu(\emptyset) = 0.$$

Therefore f is continuous at the right at ξ. Since ξ is any point, f_n is a distribution function. This completes the poof. $\square$

Theorem 8.22 *Every Borel measure μ is the Lebesgue-Stieltje's measure induced by the distribution function corresponding to μ.*

Proof Let μ be any Borel measure Define $f_\mu : \mathbb{R} \to \mathbb{R}$ by $f_\mu(0) = 0$, $f_\mu(x) = \mu((0, x])$ if $x > 0$ and $f_\mu(x) = -\mu((x, 0])$ if $x < 0$. Then as in the last part of the proof of Theorem 8.21 f_μ is a distribution function which corresponds to μ and $\mu((a, b]) = f_\mu(b) - f_\mu(a)$ for all a, b, $a < b$. We show that μ is the Lebesgue-Stieltje's measure induced by f_μ. Since f_μ is non-decreasing and continuous at the right it gives a Lebesgue-Stieltje's measure v. By Theorem 1.14 there is a smallest σ algebra C containing all half-open intervals of the form $(a, b]$. Since by Theorem 8.16, $v((a, b]) = f_\mu(b) - f_\mu(a) = \mu((a, b])$ for every half-open interval $(a, b]$, $v = \mu$ on C. For $E \subset \mathbb{R}$ define

$$v^*(E) = \inf\{\sum v(A_i) : E \subset \cup A_i; A_i \in C\},$$

where the infimum is taken over all countable collection $\{A_i\} \subset C$ satisfying $E \subset \cup A_i$. Then by Theorem 8.11, v^* is an outer measure. Let M be the $\sigma-$ algebra of all v^*- measurable sets. If $\mathcal{B}$ is the Borel $\sigma-$ algebra then since all intervals of the form $(a, b]$ are Borel measurable, $C \subset \mathcal{B}$. Also since all Borel sets are measurable with respect to v, $\mathcal{B} \subset M$. Hence $C \subset \mathcal{B} \subset M$. Since μ and v are defined on $\mathcal{B}$ and $\mu = v$ on C, both μ and v are extension of each other. Since both μ and v are $\sigma-$ finite, by Theorem 8.14, $\mu = v$ on $\mathcal{B}$. Since v is the Lebesgue-Stieltje's measure induced by f_μ, the proof is complete. $\square$

8.7 Hausdorff Measure on the Real Line

Hausdorff measure is very special in the sense that it gives not only the measure of a set but also determine the strength of that set. Also it gives the Lebesgue measure as a particular case.

8.7 Hausdorff Measure on the Real Line

Definition 8.13 Choose $p > 0$ and fix it. Let $E \subset \mathbb{R}$. For $\delta > 0$ let C_δ be the collection of all countable family $\{I_k\}$ of open intervals I_k such that $|I_k| \leq \delta$ for all k and $E \subset \bigcup_{k=1}^{\infty} I_k$. Let

$$H_{p,\delta}^*(E) = \inf\{\sum_{k=1}^{\infty} |I_k|^p : \{I_k\} \in C_\delta\}$$

where $|I_k|$ denotes the length of I_k and infimum is taken over all $\{I_k\} \in C_\delta$. It is clear that if $\delta_1 < \delta_2$ then $C_{\delta_1} \subset C_{\delta_2}$ and so $H_{p,\delta_2}^*(E) \leq H_{p\delta_1}^*(E)$ and hence $\lim_{\delta \to 0} H_{p,\delta}^*(E)$ exists. The outer Hausdorff measure of E corresponding to p is defined by

$$H_p^*(E) = \lim_{\delta \to 0} H_{p,\delta}^*(E) = \sup_{\delta > 0} H_{p,\delta}^*(E)$$

$$= \sup_{\delta > 0} \inf \left\{ \sum_{k=1}^{\infty} |I_k|^p : E \subset \bigcup_{k=1}^{\infty} I_k; |I_k| \leq \delta, k = 1, 2, \ldots \right\}.$$

We show that H_p^* satisfies the condition of outer measure (see Definition 8.4).

Theorem 8.23 H_p^* is an outer measure.

Proof Clearly $H_p^*(\emptyset) = 0$ and so (i) of Definition 8.4 is satisfied. For (ii) let $E_1 \subset E_2$. Then any covering $\{I_k\}$ of E_2 is also a covering of E_1 and hence if C_δ^1 and C_δ^2 correspond to E_1 and E_2 respectively then $C_\delta^2 \subset C_\delta^1$ and so $H_{p,\delta}^*(E_1) \leq H_{p,\delta}^*(E_2)$. Letting $\delta \to 0$ $H_p^*(E_1) \leq H_p^*(E_2)$. Finally for (iii) let $E = \bigcup_{i=1}^{\infty} E_i$. We shall show that $H_p^*(E) \leq \sum_{i=1}^{\infty} H_p^*(E_i)$ which will complete the proof. Let $\epsilon > 0$ be arbitrary. Then for each i, there exists a family of intervals $\{I_{i,k}\}$ with $|I_{i,k}| \leq \delta$ such that $E_i \subset \bigcup_{k=1}^{\infty} I_{i,k}$ and

$$H_{p,\delta}^*(E_i) \geq \sum_{k=1}^{\infty} |I_{ik}|^p - \frac{\epsilon}{2^i} \tag{8.26}$$

Since $E = \bigcup_{i=1}^{\infty} E_i \subset \bigcup_{i=1}^{\infty} \bigcup_{k=1}^{\infty} I_{i,k}$, we have from (8.26)

$$H_{p,\delta}^*(E) \leq \sum_{i=1}^{\infty} \sum_{k=1}^{\infty} |I_{ik}|^p \leq \sum_{i=1}^{\infty} H_{p,\delta}^*(E_i) + \epsilon \tag{8.27}$$

Since $H_{p,\delta}^*(E_i) \leq H_p^*(E_i)$ for all $\delta > 0$, from (8.27)

$$H_{p,\delta}^*(E) \leq \sum_{i=1}^{\infty} H_p^*(E_i) + \epsilon.$$

Letting $\delta \to 0$ $H_p^*(E) \le \sum_{i=1}^{\infty} H_p^*(E_i) + \epsilon$. Since ϵ is arbitrary, the proof is complete. □

Theorem 8.24 *The outer measure of a singleton set is zero and hence an outer measure of a countable set is zero for every p.*

Proof Since $\{x\} \subset (x - \frac{\delta}{2}, x + \frac{\delta}{2})$ for every $\delta > 0$, $H_{p,\delta}^*(\{x\}) < \delta^p$ and so letting $\delta \to 0$ $H_p^*(\{x\}) = 0$. For the second part let $E = \{x_i : i = 1, 2, \ldots\}$. Then $E = \bigcup_{i=1}^{\infty} \{x_i\}$ and so by the last part of Theorem 8.23 and the first part of this theorem the result follows. □

Theorem 8.25 (Caratheodary property) *Let A and B be the sets such that* $\inf\{|x - y| : x \in A; y \in B\} > 0$. *Then*

$$H_p^*(A \cup B) = H_p^*(A) + H_p^*(B).$$

The proof is similar to that of Theorem 2.1 (vi) which is proved for Lebesgue outer measure.

Theorem 8.26 *If $p = 1$ then the Hausdorff outer measure and Lebesgue outer measure are the same.*

Proof Let E be any set. Denoting by μ^* the Lebesgue outer measure we have from the definition $\mu^*(E) \le H_{1,\delta}^*(E)$ for all $\delta > 0$ and so $\mu^*(E) \le H_1^*(E)$. We are to show that $H_1^*(E) \le \mu^*(E)$. We may suppose that $\mu^*(E) < \infty$. Let $\epsilon > 0$ be arbitrary. Then there is a countable family of open intervals $\{I_k\}$ such that $E \subset \bigcup_{k=1}^{\infty} I_k$ and

$$\mu^*(E) > \sum_{k=1}^{\infty} |I_k| - \epsilon \tag{8.28}$$

Consider a fixed I_k. We claim

$$|I_k| \ge H_1^*(I_k) - \frac{\epsilon}{2^k}. \tag{8.29}$$

For, let $\delta > 0$. Choose intervals $J_1, J_2 \ldots J_m$, such that $|J_i| \le \delta$, $I_k \subset \bigcup_{i=1}^{m} J_i$ and $\sum_{i=1}^{m} |J_i| < |I_k| + \frac{\epsilon}{2^k}$. Then $H_{1,\delta}^*(I_k) < |I_k| + \frac{\epsilon}{2^k}$. This being true for every $\delta > 0$, letting $\delta \to 0$ we have $H_1^*(I_k) \le |I_k| + \frac{\epsilon}{2^k}$, proving (8.29). From (8.28) and (8.29)

$$\mu^*(E) > \sum_{k=1}^{\infty} H_1^*(I_k) - \sum_{k=1}^{\infty} \frac{\epsilon}{2^k} - \epsilon$$

8.7 Hausdorff Measure on the Real Line

$$\geq H_1^*(\bigcup_{k=1}^\infty I_k) - 2\epsilon \geq H_1^*(E) - 2\epsilon.$$

Since ϵ is arbitrary, $\mu^*(E) \geq H_1^*(E)$. This completes the proof. □

Theorem 8.27 *Let $E \subset \mathbb{R}$ and $p > 0$. If $H_p^*(E) < \infty$ then $H_q^*(E) = 0$ for $q > p$. If $H_p^*(E) > 0$ then $H_q^*(E) = \infty$ for $0 < q < p$. If $0 < H_p^*(E) < \infty$ then $H_q^*(E) = 0$ for $q > p$ and $H_q^*(E) = \infty$ for $0 < q < p$.*

Proof Let $\delta > 0$ and let $\{I_k\}$ be a countable family of open intervals such that $|I_k| \leq \delta$ for all k and $E \subset I_k$. Then for $q > p$

$$\frac{|I_k|^q}{|I_k|^p} = |I_k|^{q-p} \leq \delta^{q-p}$$

and so $H_{q,\delta}^*(E) \leq \sum_{k=1}^\infty |I_k|^q \leq \delta^{q-p} \sum_{k=1}^\infty |I_k|^p$. Since this is true for every such collection $\{I_k\}$, taking infimum

$$H_{q,\delta}^*(E) \leq \delta^{q-p} H_{p,\delta}^*(E) \leq \delta^{q-p} H_p^*(E). \tag{8.30}$$

Letting $\delta \to 0$ the first part is proved. For the second part let $q < p$ and $H_p^*(E) > 0$. If possible suppose $H_q^*(E) < \infty$. Interchanging p and q in (8.30) $H_{p,\delta}^*(E) \leq \delta^{p-q} H_q^*(E)$. So letting $\delta \to 0$, $H_p^*(E) = 0$ which is a contradiction, proving the second part. The last part is now obvious. □

Theorem 8.28 *For any set $E \subset \mathbb{R}$ if $0 < \mu^*(E) \leq \infty$ then $H_p^*(E) = \infty$ for $0 < p < 1$, $H_p^*(E) = \mu^*(E)$ for $p = 1$ and $H_p^*(E) = 0$ for $p > 1$, where μ^* is the Lebesgue outer measure.*

Proof For $p = 1$ the result follows from Theorem 8.26. Let $0 < p < 1$. Suppose that $H_p^*(E) < \infty$. Then by the first part of Theorem 8.27, $H_1^*(E) = 0$ which is a contradiction since by Theorem 8.26 $H_1^*(E) = \mu^*(E) > 0$. Hence $H_p^*(E) = \infty$ for all p, $0 < p < 1$. Now let $p > 1$. Let $\{I_k\}$ be a countable family of open intervals of finite length such that $E \subset \bigcup_{k=1}^\infty I_k$. Then $H_p^*(E) \leq \sum_{k=1}^\infty H_p^*(I_k)$. Since by Theorem 8.26 $H_1^*(I_k) = \mu^*(I_k)$ for all k, we have $0 < H_1^*(I_k) < \infty$ for all k. So, by Theorem 8.27 $H_p^*(I_k) = 0$ for all k. Hence $H_p^*(E) = 0$ and so $H_p^*(E) = 0$ for all $p > 1$. This completes the proof. □

Corollary 8.29 *The Hausdorff measure of the real line $\mathbb{R}$ is such that $H_p^*(\mathbb{R}) = \infty$ for $0 < p \leq 1$ and $H_p^*(\mathbb{R}) = 0$ for $p > 1$.*

Proof Since $\mu^*(\mathbb{R}) = \infty$ the proof follows from Theorem 8.28. □

Theorem 8.30 *For any fixed set $E \subset \mathbb{R}$, $H_p^*(E)$ is a decreasing function of p in $(0, \infty)$.*

Proof Let $0 < \delta < 1$ and let $\{I_k\}$ be any countable collection of open intervals such that $|I_k| \leq \delta$ for all k and $E \subset \cup I_k$. Since $\delta < 1$ if $0 < p \leq q$ then $|I_k|^p \geq |I_k|^q$ and so $H^*_{p,\delta}(E) \geq H^*_{q,\delta}(E)$. Letting $\delta \to 0$ $H^*_p(E) \geq H^*_q(E)$. $\square$

Theorem 8.31 *For any set $E \subset \mathbb{R}$ exactly one of the following is true:*
(i) $H^*_p(E) = 0$ for all $p > 0$
(ii) there is $p_0, 0 < p_0 \leq 1$, such that $H^*_p(E) = \infty$ for all $p, 0 < p < p_0$ and $H^*_p(E) = 0$ for all $p, p > p_0$.

Proof Suppose that (i) is not true. Then there is $p > 0$ such that $H^*_p(E) \neq 0$. Let $p_0 = \inf\{p : H^*_p(E) = 0\}$. Then since by Corollary 8.29, $H^*_p(E) = 0$ for $p > 1$, $p_0 \leq 1$. Also $p_0 > 0$. For if $p_0 = 0$ then (i) holds which contradicts our supposition. Hence $0 < p_0 \leq 1$. By the definition of p_0, $H^*_p(E) = 0$ for all $p, p > p_0$. Also $H^*_p(E) = \infty$ for all $p, 0 < p < p_0$. For suppose that there is $p_1, 0 < p_1 < p_0$ such that $H^*_{p_1}(E) < \infty$. Choose $p_1 < p_2 < p_0$. Then by Theorem 8.27 $H^*_{p_2}(E) = 0$ and so $p_0 \leq p_2$ which is a contradiction. This completes the proof.

From Theorem 8.31 it follows that for every set $E \subset \mathbb{R}$ there is a unique $p_0 \in [0, 1]$ such that $H^*_p(E) = 0$ for $p > p_0$ and $H^*_p(E) = \infty$ for $p, 0 < p < p_0$, if $p_0 > 0$. This number p_0 is called the Hausdorff dimension of E. $\square$

Definition 8.14 The Hausdorff dimension of a set $E \subset \mathbb{R}$ is defined by

$$\dim(E) = \inf\{p : H^*_p(E) = 0\}.$$

Theorem 8.32 *Let $E \subset \mathbb{R}$ and $\alpha \in \mathbb{R}$. Define $E + \alpha = \{x + \alpha : x \in E\}$ and $\alpha E = \{\alpha x : x \in E\}$. Then for $p > 0$, (i) $H^*_p(E) = H^*_p(E + \alpha)$ and (ii) $H^*_p(\alpha E) = |\alpha|^p H^*_p(E)$.*

Proof Let $\delta > 0$ and let $\{I_k\}$ be any countable family of open intervals such that $|I_k| \leq \delta$ for all k and $E \subset \cup I_k$. Then (i) the family $\{I_k + \alpha\}$ is such that for each k, $I_k + \alpha$ is an open interval and $|I_k + \alpha| = |I_k| \leq \delta$ and $E + \alpha \subset \cup(I_k + \alpha)$ and hence

$$H^*_{p,\delta}(E + \alpha) = \inf\left\{\sum |I_k + \alpha|^p : E + \alpha \subset \cup(I_k + \alpha)\right\}$$
$$= \inf\left\{\sum |I_k|^p : E \subset \cup I_k\right\}$$
$$= H^*_{p,\delta}(E).$$

Letting $\delta \to 0$, $H^*_p(E + \alpha) = H^*_p(E)$.
(ii) The family $\{\alpha I_k\}$ is such that for each k, αI_k is an open interval and $|\alpha I_k| = |\alpha| |I_k| \leq |\alpha|\delta$ and $\alpha E \subset \cup(\alpha I_k)$ and hence

$$H^*_{p,\delta}(\alpha E) = \inf\left\{\sum |\alpha I_k|^p : \alpha E \subset \cup(\alpha I_k)\right\}$$
$$= \inf\left\{|\alpha|^p \sum |I_k|^p : E \subset \cup I_k\right\}$$

8.7 Hausdorff Measure on the Real Line

$$= |\alpha|^p H^*_{p,\delta}(E).$$

Letting $\delta \to 0$, $H^*_p(\alpha E) = |\alpha|^p H^*_p(E)$. □

Theorem 8.33 *The Hausdorff outer measure defined in Definition 8.13 remains the same if open intervals are replaced by closed or half-open intervals.*

Proof We prove that closed intervals give the same outer measure. The proof for half-open intervals is similar. Let p, E and δ be fixed. Let $H^{*c}_{p,\delta}(E)$ be obtained by using the collection of closed intervals $\{J_k\}$ just as $H^*_{p,\delta}(E)$ was obtained by using the collection of open intervals $\{I_k\}$ in Definition 8.13. Let $\overline{\{I_k\}}$ denote the closure of I_k. Then since $|I_k| = |\overline{I_k}|$, we have $\{\sum |I_k|^p : E \subset \cup I_k\} \subset \{\sum |\overline{I_k}|^p : E \subset \cup \overline{I_k}\}$ and so $H^*_{p,\delta}(E) = \inf\{\sum |I_k|^p : E \subset \cup I_k\} \geq \inf\{\sum |\overline{I_k}|^p : E \subset \cup \overline{I_k}\} \geq H^{*c}_{p,\delta}(E)$. Letting $\delta \to 0$, $H^*_p(E) \geq H^{*c}_p(E)$.

To prove the reverse inequality let $\{J_k\}$ be any countable collection of closed intervals such that $|J_k| \leq \delta$ and $E \subset \cup J_k$. For each J_k let I_k be an open interval such that $J_k \subset I_k$ and $|I_k| = |J_k|(1+\delta)$. Then since

$$\left\{\sum |J_k|^p : |J_k| \leq \delta : E \subset \cup J_k\right\} \subset \left\{(1+\delta)^{-p} \sum |I_k|^p : |I_k| \leq \delta(1+\delta); E \subset \cup I_k\right\}$$

we have

$$H^{*c}_{p,\delta}(E) = \inf\{\sum |J_k|^p : |J_k| \leq \delta; E \subset \cup J_k\}$$
$$\geq (1+\delta)^{-p} \inf\left\{\sum |I_k|^p : |I_k| \leq \delta(1+\delta); E \subset \cup I_k\right\}$$
$$= (1+\delta)^{-p} H^*_{p,\delta(1+\delta)}(E).$$

Letting $\delta \to 0$, $H^{*c}_p(E) \geq H^*_p(E)$. This completes the proof. □

The Hausdorff outer measure being defined, Hausdorff measurability of a set is defined as in Definition 8.5. As in Theorem 8.6 a set is H^*_p- measurable if and only if for any set A

$$H^*_p(A) \geq H^*_p(A \cap E) + H^*_p(A \sim E).$$

It can be shown as in Theorem 8.7 that if $\mathcal{M}$ is the class of all H^*_p- measurable set then $\mathcal{M}$ is a $\sigma-$ algebra and the restriction of the outer measure H^*_p on $\mathcal{M}$ is a measure which is the Hausdorff measure. If a set E is H^*_p- measurable, we shall write $H_p(E)$ for $H^*_p(E)$.

Theorem 8.34 *If the outer measure of a set E is zero then E is measurable.*

The proof follows from the definition of measurability.

Theorem 8.35 *A singleton set and a countable set are measurable and are of measure zero.*

The proof follows from Theorems 8.24 and 8.34.

Theorem 8.36 *If $p = 1$ then a set E is Hausdorff measurable if and only if E is Lebesgue measurable and the measures are equal.*

The proof follows from Theorem 8.26 and the definition of measurability.

Theorem 8.37 *All Borel sets are Hausdorff measurable.*

Proof Let $p > 0$ be fixed. We first show that the interval $I = (-\infty, a]$ is H_p^*-measurable for all a, $-\infty < a < \infty$. We need to show that for any set A

$$H_p^*(A) \geq H_p^*(A \cap I) + H_p^*(A \sim I). \tag{8.31}$$

We may suppose that $H_p^*(A)$ is finite. For each n let $A_n = A \cap [a + \frac{1}{n}, \infty)$. Then $A_n \subset A_{n+1}$ and $\bigcup_{n=1}^{\infty} A_n = A \sim I$. Since $A_n \subset A_{n+1} \subset A$, $H_p^*(A_n) \leq H_p^*(A_{n+1}) \leq H_p^*(A)$ for all n and so $\lim_{n \to \infty} H_p^*(A_n)$ exists and is finite. Since $\inf\{|x - y| : x \in A \cap I; y \in A_n\} \geq \frac{1}{n}$ applying Theorem 8.25 we have

$$H_p^*(A) = H_p^*((A \cap I) \cup (A \sim I)) \geq H_p^*((A \cap I) \cup A_n))$$
$$= H_p^*(A \cap I) + H_p^*(A_n) \tag{8.32}$$

Let $B_n = A_{n+1} \sim A_n$. Then

$$A \sim I = \bigcup_{k=1}^{\infty} A_k \subset A_{2n} \bigcup_{k=2n}^{\infty} B_k = A_{2n} \cup \bigcup_{k=n}^{\infty} B_{2k} \cup \bigcup_{k=n}^{\infty} B_{2k+1}$$

Hence

$$H_p^*(A \sim I) \leq H_p^*(A_{2n}) + \sum_{k=n}^{\infty} H_p^*(B_{2k}) + \sum_{k=n}^{\infty} H_p^*(B_{2k+1}) \tag{8.33}$$

Since $\bigcup_{k=1}^{n-1} B_{2k} \subset A_{2n}$ and $\inf\{|x - y| : x \in B_{2k}; y \in B_{2k+2}\} \geq \frac{1}{2k+1} - \frac{1}{2k+2} > 0$ by Theorem 8.25

$$H_p^*(A_{2n}) \geq H_p^*(\bigcup_{k=1}^{n-1} B_{2k}) = \sum_{k=1}^{n-1} H_p^*(B_{2k})$$

Hence

$$\lim_{n \to \infty} H_p^*(A_{2n}) \geq \sum_{k=1}^{\infty} H_p^*(B_{2k}) \tag{8.34}$$

8.7 Hausdorff Measure on the Real Line

Since $A_{2n} \subset A$ for all n, $H_p^*(A_{2n}) \leq H_p^*(A)$ for all n and since $H_p^*(A)$ is assumed to be finite the left-hand side of (8.34) is finite and hence from (8.34) the series $\sum_{k=1}^{\infty} H_p^*(B_{2k})$ is convergent. Similarity the series $\sum_{k=1}^{\infty} H_p^*(B_{2k+1})$ is convergent. So letting $n \to \infty$ in (8.33)

$$H_p^*(A \sim I) \leq \lim_{n \to \infty} H_p^*(A_{2n}) = \lim_{n \to \infty} H_p^*(A_n).$$

Since $A_n \subset A \sim I$, $H_p^*(A_n) \leq H_p^*(A \sim I)$ and so $\lim_{n \to \infty} H_p^*(A_n) \leq H_p^*(A \sim I)$. Hence $H_p^*(A \sim I) = \lim_{n \to \infty} H_p^*(A_n)$. So, from (8.32)

$$H_p^*(A) \geq H_p^*(A \cap I) + \lim_{n \to \infty} H_p^*(A_n) = H^*(A \cap I) + H_p^*(A \sim I),$$

which proves (8.31).

The interval $(-\infty, a]$ being proved H_p^*- measurable for all a, it is clear that (a, ∞) is measurable. So as in Theorem 2.10 and Corollary 2.11 all intervals and all open sets are H_p^*- measurable. So, if $\mathcal{M}$ is the $\sigma-$ algebra of H_p^*- measurable sets (obtained as in Theorem 8.7) then as in Corollary 2.11 $\mathcal{M}$ contains all Borel sets and hence all Borel sets are H_p^*- measurable. $\square$

By Theorem 8.28 for any set $E \subset \mathbb{R}$ if $0 < \mu^*(E) \leq \infty$ where μ^* denotes outer Lebesgue measure, then $\dim(E) = 1$ (see Definition 8.14) and so the Hausdorff dimension of $\mathbb{R}$ is 1. Also the Hausdorff dimension of any countable set is 0 by Theorem 8.24. So, Hausdorff dimension is interesting for uncountable sets of Lebesgue measure zero because it can determine the strength of that measure zero sets.

Cantor sets: Cantor sets are defined in Sect. 4.4. Let $0 < \alpha \leq 1$ and let C_α be the Cantor set corresponding to α. Then $C_\alpha \subset [0, 1]$ and $\mu(C_\alpha) = 1 - \alpha$. To construct C_α we first remove from $[0, 1]$ the open interval I_1^1 with center $\frac{1}{2}$ and of length $\frac{\alpha}{3}$, the remaining closed intervals are J_1^1 and J_2^1 each of length $\frac{1}{2}(1 - \frac{\alpha}{3})$. We again remove from each of J_1^1 and J_2^1 the central open intervals each of length $\frac{\alpha}{3^2}$ and so on. Let C_α^1 and C_α^2 be the portion of C_α which lie on J_1^1 and J_2^1 respectively. Note that J_1^1 is the closed interval $[0, \frac{1}{2}(1 - \frac{\alpha}{3})]$ and that to get C_α^1 we dissected J_1^1 in the same way that $[0, 1]$ was dissected to get C_α. Since $|J_1^1| = \frac{3-\alpha}{6}$, if we give the transformation $x \to \frac{6}{3-\alpha}x$ then $J_1^1 \to [0, 1]$ i.e. $[0, 1] = \{\frac{6}{3-\alpha}x : x \in J_1^1\} = \frac{6}{3-\alpha}J_1^1$. Therefore

$$C_\alpha = \left\{ \frac{6}{3-\alpha}x : x \in C_\alpha^1 \right\} = \frac{6}{3-\alpha}C_\alpha^1. \tag{8.35}$$

Also

$$C_\alpha^2 = \left\{ \frac{1}{2} + \frac{\alpha}{6} + x : x \in C_\alpha^1 \right\} = \frac{1}{2} + \frac{\alpha}{6} + C_\alpha^1. \tag{8.36}$$

So, applying Theorem 8.32 we get from (8.35) and (8.36) for any $p > 0$

$$H_p^*(C_\alpha) = \left(\frac{6}{3-\alpha}\right)^p H_p^*(C_\alpha^1) \text{ and } H_p^*(C_\alpha^2) = H_p^*(C_\alpha^1). \qquad (8.37)$$

Also since $C_\alpha = C_\alpha^1 \cup C_\alpha^2$ and $\inf\{|x-y| : x \in C_\alpha^1; y \in C_\alpha^2\} = \frac{\alpha}{3} > 0$ by Theorem 8.25 $H_p^*(C_\alpha) = H_p^*(C_\alpha^1) + H_p^*(C_\alpha^2)$ and so by (8.37)

$$H_p^*(C_\alpha) = 2 H_p^*(C_\alpha^1) = 2\left(\frac{3-\alpha}{6}\right)^p H_p^*(C_\alpha). \qquad (8.38)$$

So, if $H_p^*(C_\alpha) \neq 0$ and $\neq \infty$ then $2(\frac{3-\alpha}{6})^p = 1$ which gives $p = \frac{\log 2}{\log(\frac{6}{3-\alpha})}$ and hence for $\alpha = 1$, $p = \frac{\log 2}{\log 3}$. Since for $\alpha = 1$ the set C_1 is the Cantor ternary set C (see Sects. 4.4 and 4.5) we show that for this value of p, $H_p^*(C) \neq 0$ and $\neq \infty$ and prove that $p = \frac{\log 2}{\log 3}$ is the Hausdorff dimension of the Cantor ternary set C.

Theorem 8.38 *The Cantor ternary set C is such that $H_p^*(C) = 1$ if $p = \frac{\log 2}{\log 3}$, where H_p^* denotes Hausdorff outer measure corresponding to p.*

Proof Note that Cantor ternary set C is obtained by taking $\alpha = 1$ in the construction of Cantor sets C_α, $0 < \alpha \leq 1$. So at the nth step each of removed open intervals I_k^h, $1 \leq k \leq 2^{n-1}$, has length $\frac{1}{3^n}$ and each of the remaining closed intervals J_k^n, $1 \leq k \leq 2^n$, has length $\frac{1}{3^n}$ (see Sect. 4.4 and 4.5).

Now let $\delta > 0$ be arbitrary. Then taking n sufficiently large $|J_k^n| = \frac{1}{3^n} \leq \delta$. Also $C \subset \bigcup_{k=1}^{2^n} J_k^n$ and so using Theorem 8.33

$$H_{p,\delta}^*(C) \leq \sum_{k=1}^{2^n} |J_k^n|^p = \sum_{k=1}^{2^n} \left(\frac{1}{3^n}\right)^p = 2^n \frac{1}{3^{np}} = \left(\frac{2}{3^p}\right)^n.$$

Since $p = \frac{\log 2}{\log 3}$, $\frac{2}{3^p} = 1$ and hence $H_{p,\delta}^*(C) \leq 1$. Letting $\delta \to 0$, $H_p^*(C) \leq 1$.

To prove the reverse inequality let $\epsilon > 0$, $\delta > 0$ be arbitrary. Then there is a countable collection of open intervals $\{I_i\}$ such that $|I_i| \leq \delta$ for all i and $C \subset \bigcup_{i=1}^\infty I_i$ and $H_{p,\delta}^*(C) + \epsilon > \sum_{i=1}^\infty |I_i|^p$. Since $C \subset \bigcup_{i=1}^\infty I_i$ and $C = \bigcap_{n=1}^\infty \bigcup_{k=1}^{2^n} J_k^n$, and since $\bigcup_{i=1}^\infty I_i$ is open and C is closed, there exists n such that $\bigcup_{k=1}^{2^n} J_k^n \subset \bigcup_{i=1}^\infty I_i$ and so $\sum_{k=1}^{2^n} |J_k^n|^p \leq \sum_{i=1}^\infty |I_i|^p$, Hence

$$H_{p,\delta}^*(C) + \epsilon > \sum_{k=1}^{2^n} |J_k^n|^p = 2^n \left(\frac{1}{3^n}\right)^p = \left(\frac{2}{3^p}\right)^n = 1$$

Letting $\epsilon \to 0$, and $\delta \to 0$, $H_p^*(C) \geq 1$, completing the proof. $\square$

Theorem 8.39 *The Hausdorff dimension of the Cantor set C is $\frac{\log 2}{\log 3}$.*

The proof follows from Theorems 8.27 to 8.38

Theorem 8.40 (i) *If $A \subset B$ then $\dim(A) \leq \dim(B)$.*
(ii) *For any A, B $\dim(A \cup B) = \max(\dim(A), \dim(B))$.*

Proof (i) Let $A \subset B$. If $s > \dim(B)$ then $H_s^*(A) \leq H_s^*(B) = 0$ and so $\dim(A) \leq s$. Since this is true for any $s > \dim(B)$, $\dim(A) \leq \dim(B)$.
(ii) Let $s > \max(\dim(A), \dim(B))$. Then $s > \dim(A)$ and so $H_s^*(A) = 0$. Similarly $H_s^*(B) = 0$. Hence $H_s^*(A \cup B) \leq H_s^*(A) + H_s^*(B) = 0$ and so $\dim(A \cup B) \leq s$. Since this is true for any $s > \max(\dim(A), \dim(B))$, $\dim(A \cup B) \leq \max(\dim(A), \dim(B))$. Also by the first part $\dim(A \cup B) \geq \dim(A)$ and $\dim(A \cup B) \geq \dim(B)$ and hence $\dim(A \cup B) \geq \max(\dim(A), \dim(B))$, completing the proof. □

Besides Lebesgue-Stieltje's measure and Hausdorff measure which are discussed above there are other measures such as probability measure and Haar measure. We shall not discuss these. Interested readers may consult the book which are mentioned in the references.

8.8 Measurable Function and Integration

Let $(X, \mathcal{A}, \mu)$ be a measure space and let f be an extended real-valued function defined on $E \subset X$ where $E \in \mathcal{A}$. As in Definition 3.1 the function f is said to be measurable if for every $r \in \mathbb{R}$ the set $\{x \in E : f(x) > r\}$ is measurable. Most of the results which are proved in Chapter -3 are also true for general measure space and the proofs are similar. Particularly, Egorov's theorem and all convergence theorems proved in Section 3.5 are true. The proofs being similar we omit them.

In a general measure space $(X, \mathcal{A}, \mu)$ the definition of integral of a function f on a measurable set E is also defined by considering partition $\tau = \{e_i : i = 1, 2, \ldots, n\}$ of E such that $e_i \in \mathcal{A}$ for all i, $E = \bigcup_{i=1}^{n} e_i$, $e_i \cap e_j = \emptyset$ for $i \neq j$ and using Definitions 5.1 or 5.2. The properties and results on integral which are proved in Chap. 5 are also true and proofs are similar. All theorems including convergence theorems proved in Section 5.8 are true. Proving these results is just a repetition of the same argument and so we omit them.

8.9 Measure and Integration in a Product Space

We consider the measure spaces $(X, \mathcal{A}, \mu)$ and $(Y, \mathcal{B}, \nu)$ and define a $\sigma-$ algebra of subsets of the Cartesion product $X \times Y$ and then introduce a measure on this $\sigma-$ algebra which will be called product measure.

Definition 8.15 Let $(X, \mathcal{A})$ and $(Y, \mathcal{B})$ be measurable spaces. If $A \in \mathcal{A}$ and $B \in \mathcal{B}$ then $A \times B$ is called a measurable rectangle in $X \times Y$.

Since $X \in \mathcal{A}$ and $Y \in \mathcal{B}$, $X \times Y$ is a measurable rectangle. So the class of measurable rectangles in $X \times Y$ is non-void. So, by Theorem 1.14 there is a smallest $\sigma-$ algebra containing all measurable rectangles. This $\sigma-$ algebra is called the product $\sigma-$ algebra of $\mathcal{A}$ and $\mathcal{B}$ and is denoted by $\mathcal{A} \otimes \mathcal{B}$. So $(X \times Y, \mathcal{A} \otimes \mathcal{B})$ is a measurable space and this measurable space is called product measurable space of $(X, \mathcal{A})$ and $(Y, \mathcal{B})$.

Definition 8.16 If $E \subset X \times Y$ then for $x \in X$ and $y \in Y$ the $x-$ section and $y-$ section of E are defined by $E_x = \{y : (x, y) \in E\}$ and $E^y = \{x : (x, y) \in E\}$. Clearly $E_x \subset Y$ and $E^y \subset X$.

Similarly, if f is an extended real-valued function on $X \times Y$ then for $x \in X$ and $y \in Y$ the $x-$ section and $y-$ section of f are defined by $f_x(y) = f(x, y)$ and $f^y(x) = f(x, y)$ respectably.

Theorem 8.41 If $E \in \mathcal{A} \otimes \mathcal{B}$ then for each $x \in X$ and each $y \in Y$, $E_x \in \mathcal{B}$ and $E^y \in \mathcal{A}$.

Proof Let $C = \{E : E \in \mathcal{A} \otimes \mathcal{B}$ and $E_x \in \mathcal{B}$ for each $x \in X\}$. We show that C is a $\sigma-$ algebra. Let $E \in C$. Then $E \in \mathcal{A} \otimes \mathcal{B}$. Since $\mathcal{A} \otimes \mathcal{B}$ is a σ-algebra, $\widetilde{E} \in \mathcal{A} \otimes \mathcal{B}$. Since $E_x \in \mathcal{B}$, $\widetilde{E}_x \in \mathcal{B}$. But

$$\widetilde{E}_x = \{y : \widetilde{(x, y) \in E}\} = \{y : (x, y) \in \widetilde{E}\} = (\widetilde{E})_x$$

and so $(\widetilde{E})_x \in \mathcal{B}$. Therefore $\widetilde{E} \in C$. Let $E_n \in C$, $n = 1, 2, \ldots$. Then $E_n \in \mathcal{A} \otimes \mathcal{B}$ and $(E_n)_x \in \mathcal{B}$. Hence $\bigcup_{n=1}^{\infty} E_n \in \mathcal{A} \otimes \mathcal{B}$ and $\bigcup_{n=1}^{\infty} (E_n)_x \in \mathcal{B}$. So, since $\bigcup_{n=1}^{\infty} (E_n)_x = (\bigcup_{n=1}^{\infty} E_n)_x$, $(\bigcup_{n=1}^{\infty} E_n)_x \in \mathcal{B}$. Therefore $\bigcup_{n=1}^{\infty} E_n \in C$. So, C is a $\sigma-$ algebra.

Now we show that C contains all measurable rectangles. If $A \in \mathcal{A}$ and $B \in \mathcal{B}$ then $A \times B$ is a measurable rectangle and so $A \times B \in \mathcal{A} \otimes \mathcal{B}$. Also $(A \times B)_x = B$ or $\emptyset$ according as $x \in A$ or $x \in \widetilde{A}$ and so $A \times B \in C$. So, C contains all measurable rectangles. Thus C is a $\sigma-$ algebra containing all measurable rectangles. Since $\mathcal{A} \otimes \mathcal{B}$ is the smallest $\sigma-$ algebra containing all measurable rectangles $\mathcal{A} \otimes \mathcal{B} \subset C$. But $C \subset \mathcal{A} \otimes \mathcal{B}$ and so $C = \mathcal{A} \otimes \mathcal{B}$. So for every $E \in \mathcal{A} \otimes \mathcal{B}$, $E_x \in \mathcal{B}$ for each $x \in X$. Similarly for every $E \in \mathcal{A} \otimes \mathcal{B}$, $E^y \in \mathcal{A}$ for each $y \in Y$. □

Theorem 8.42 Let $(X, \mathcal{A})$ and $(Y, \mathcal{B})$ be measurable spaces and f be an extended real valued $\mathcal{A} \otimes \mathcal{B}-$ measurable function on $X \times Y$. Then for each $x \in X$ and each $y \in Y$, f_x is $\mathcal{B}$-measurable and f^y is $\mathcal{A}$-measurable.

Proof Let $r \in R$ and let $E = \{(x, y) : (x, y) \in X \times Y$ and $f(x, y) > r\}$. Then since f is $\mathcal{A} \otimes \mathcal{B}-$ measurable, $E \in \mathcal{A} \otimes \mathcal{B}$ and so $E_x = \{y : f_x(y) > r\} \in \mathcal{B}$ and $E^y = \{x : f^y(x) > r\} \in \mathcal{A}$. Hence f_x is $\mathcal{B}-$ measurable and f^y is $\mathcal{A}-$ measurable. □

8.9 Measure and Integration in a Product Space

Theorem 8.43 *Let $(X, \mathcal{A}, \mu)$ and $(Y, \mathcal{B}, \nu)$ be σ-finite measure spaces. For $V \in \mathcal{A} \otimes \mathcal{B}$ let ϕ and ψ be defined by $\phi(x) = \nu(V_x)$ for $x \in X$ and $\psi(y) = \mu(V^y)$ for $y \in Y$. Then ϕ is $\mathcal{A}$-measurable and ψ is $\mathcal{B}$-measurable and*

$$\int_X \phi \, d\mu = \int_Y \psi \, d\nu$$

Proof We first suppose that $(X, \mathcal{A}, \mu)$ and $(Y, \mathcal{B}, \nu)$ are finite measure spaces. Let $\mathcal{S}$ be the class of all sets $V \in \mathcal{A} \otimes \mathcal{B}$ for which the theorem holds. We show that $\mathcal{S}$ is a monotone class (see Definition 1.12) Let $\{V_i\}$ be any monotone increasing sequence of members of $\mathcal{S}$ and let $\phi_i(x) = \nu((V_i)_x)$ and $\psi_i(y) = \mu((V_i)^y)$. Then since $V_i \in \mathcal{S}$ the theorem is true for V_i and so ϕ_i is $\mathcal{A}$-measurable and ψ_i is $\mathcal{B}$-measurable and

$$\int_X \phi_i \, d\mu = \int_Y \psi_i \, d\nu \tag{8.39}$$

Since $V_i \subset V_{i+1}$, $(V_i)_x \subset (V_{i+1})_x$ and $(V_i)^y \subset (V_{i+1})^y$. So $\{\phi_i\}$ is a non-decreasing sequence of non-negative $\mathcal{A}$-measurable functions and $\{\psi_i\}$ is a non-decreasing sequence of non-negative $\mathcal{B}$-measurable functions. If $\lim_{i \to \infty} \phi_i = \phi$ and $\lim_{i \to \infty} \psi_i = \psi$ then we get from (8.39) by taking limit as $i \to \infty$ and using Theorem 5.28

$$\int_X \phi \, d\mu = \int_Y \psi \, d\nu. \tag{8.40}$$

Let $V = \bigcup_{i=1}^{\infty} V_i$. Then $V_x = \bigcup_{i=1}^{\infty} (V_i)_x = \lim_{i \to \infty} (V_i)_x$ and so by Theorem 2.8 $\nu(V_x) = \lim_{i \to \infty} \nu((V_i)_x) = \lim_{i \to \infty} \phi_i(x) = \phi(x)$. Similarly $\mu(V^y) = \psi(y)$. Since ϕ and ψ are respectively $\mathcal{A}$-measurable and $\mathcal{B}$-measurable this together with relation (8.40) show that $V \in \mathcal{S}$.

Next let $\{V_i\}$ be any monotone decreasing sequence of members of $\mathcal{S}$ and let $\phi_i(x) = \nu((V_i)_x)$ and $\psi_i(y) = \mu((V_i)^y)$. Then ϕ_i is $\mathcal{A}$-measurable and ψ_i is $\mathcal{B}$-measurable and (8.39) holds. Since $V_i \supset V_{i+1}$, $(V_i)_x \supset (V_{i+1})_x$ and $(V_i)^y \supset (V_{i+1})^y$. So $\{\phi_i\}$ is a non-increasing sequence of non-negative $\mathcal{A}$-measurable functions and $\{\psi_i\}$ is a non-increasing sequence of non-negative $\mathcal{B}$-measurable functions. Let $\lim_{i \to \infty} \phi_i = \phi$ and $\lim_{i \to \infty} \psi_i = \psi$. Since $(X, \mathcal{A}, \mu)$ and $(Y, \mathcal{B}, \nu)$ are finite measure spaces, $\nu(Y)$ is finite and so is integrable on X. Since $\phi_i(x) = \nu((V_i)_x) \leq \nu(Y)$, by Theorem 5.33, $\lim_{i \to \infty} \int_X \phi_i \, d\mu = \int_X \phi \, d\mu$. Similarly $\lim_{i \to \infty} \int_Y \psi_i \, d\nu = \int_Y \psi \, d\nu$. So taking limit in (8.39) we get (8.40). If $V = \bigcap_{i=1}^{\infty} V_i$ then $V_x = \bigcap_{i=1}^{\infty} (V_i)_x = \lim_{i \to \infty} (V_i)_x$ and so by Theorem 2.9, $\nu(V_x) = \lim_{i \to \infty} \nu((V_i)_x) = \lim_{i \to \infty} \phi_i(x) = \phi(x)$. Similarly $\mu(V^y) = \psi(y)$. These together with relation (8.40) show that $V \in \mathcal{S}$. So, $\mathcal{S}$ is a monotone class.

Now we show that $\mathcal{S}$ contains every measurable rectangle and the complement of every measurable rectangle. Let V be a measurable rectangle, say $A \times B$ where $A \in \mathcal{A}$ and $B \in \mathcal{B}$. Since $V_x = (A \times B)_x = B$ or $\emptyset$ according as $x \in A$ or $x \in \widetilde{A}$, we have $\nu(V_x) = \nu((A \times B)_x) = \chi_A(x)\nu(B)$ where $\chi_A(x) = 1$ or 0 accordingly as $x \in A$ or $x \in \tilde{A}$. Similarly $\mu(V^y) = \chi_B(y)\mu(A)$. Since χ_A is $\mathcal{A}-$ measurable and χ_B is $\mathcal{B}-$ measurable. ϕ is $\mathcal{A}-$ measurable and ψ is $\mathcal{B}-$ measurable. Also

$$\int_X \phi d\mu = \nu(B) \int_X \chi_A d\mu = \nu(B)\mu(A) = \mu(A) \int_Y \chi_B d\nu = \int_Y \phi d\nu.$$

So, every measurable rectangle $A \times B \in S$. If $A \times B$ is a measurable rectangle then $\tilde{A} \times Y$ and $A \times \tilde{B}$ are also measurable rectangle, and hence $\tilde{A} \times Y \in S$ and $A \times \tilde{B} \in S$. Since S is a monotone class $(\tilde{A} \times Y) \cup (A \times \tilde{B}) \in S$ and since $\widetilde{A \times B} = (\tilde{A} \times Y) \cup (A \times \tilde{B})$, $\widetilde{A \times B} \in S$. Therefore, if C is the class of all sets E such that E is either a measurable rectangle or the complement of a measurable rectangle then $C \subset S$. So, S is a monotone class containing C. Since $\mathcal{A} \otimes \mathcal{B}$ is the smallest $\sigma-$ algebra containing C, by Theorem 1.18, $\mathcal{A} \otimes \mathcal{B}$ is the smallest monotone class containing C. So, $\mathcal{A} \otimes \mathcal{B} \subset S$. But $S \subset \mathcal{A} \otimes \mathcal{B}$ and hence $S = \mathcal{A} \otimes \mathcal{B}$. This completes the proof when $(X, \mathcal{A}, \mu)$ and $(Y, \mathcal{B}, \nu)$ are finite measure spaces.

To complete the proof suppose that $(X, \mathcal{A}, \mu)$ and $(Y, \mathcal{B}, \nu)$ are $\sigma-$ finite. Then $X = \bigcup_{n=1}^{\infty} X_n$ and $Y = \bigcup_{m=1}^{\infty} Y_m$ where X_n and Y_m are respectively pairwise disjoint $\mathcal{A}-$ measurable and $\mathcal{B}-$ measurable sets such that $(X_n, \mathcal{A}_n, \mu)$ and $(Y_m, \mathcal{B}_m, \nu)$ are finite measure spaces where $\mathcal{A}_n = \{A \cap X_n : A \in \mathcal{A}\}$ and $\mathcal{B}_m = \{B \cap Y_m : B \in \mathcal{B}\}$. Let $V \in \mathcal{A} \otimes \mathcal{B}$ and write $V_{nm} = V \cap (X_n \times Y_m)$. Then $V = \bigcup_{n,m} V_{n,m}$ and so $V_x = \bigcup_{n,m} (V_{n,m})_x$. Since the theorem is true for finite measure spaces, writing $\phi_{nm}(x) = \nu((V_{n,m})_x)$ for $x \in X_n$ and $\psi_{n,m}(y) = \mu((V_{nm})^y)$ for $y \in Y_m$ the function ϕ_{nm} is $\mathcal{A}_n-$ measurable and the function $\psi_{n,m}$ is $\mathcal{B}_m-$ measurable and

$$\int_{X_n} \phi_{n,m} d\mu = \int_{Y_m} \psi_{nm} d\nu.$$

Define $\phi_{nm}(x) = 0$ if $x \in X \sim X_n$ and $\psi_{nm}(y) = 0$ if $y \in Y \sim Y_m$. Then ϕ_{nm} is $\mathcal{A}-$ measurable and ψ_{nm} is $\mathcal{B}-$ measurable and

$$\int_X \phi_{n,m} d\mu = \int_Y \psi_{nm} d\nu.$$

Since for $x \in X$

$$\phi(x) = \nu(V_x) = \nu(\bigcup_{n,m}(V_{n,m})_x) = \sum_{n,m} \nu((V_{n,m})_x) = \sum_{n,m} \phi_{n,m}(x)$$

and for $y \in Y$

$$\psi(y) = \mu(V^y) = \mu(\bigcup_{n,m}(V_{n,m})^y) = \sum_{n,m} \mu((V_{nm})^y) = \sum_{n,m} \psi_{nm} d\mu$$

we conclude that ϕ is $\mathcal{A}$— measurable and ψ is $\mathcal{B}$— measurable and

$$\int_X \phi d\mu = \sum_{n,m} \int_X \phi_{nm} d\mu = \sum_{n,m} \int_Y \psi_{nm} dv = \int_Y \psi dv.$$

This completes the proof. $\square$

Definition 8.17 Let $(X, \mathcal{A}, \mu)$ and $(Y, \mathcal{B}, v)$ be σ— finite measure spaces. Then the product measure $\mu \times v$ on $\mathcal{A} \otimes \mathcal{B}$ is defined by

$$(\mu \times v)(V) = \int_X v(V_x) d\mu = \int_Y \mu(V^y) dv \text{ for } V \in \mathcal{A} \otimes \mathcal{B}.$$

To see that $\mu \times v$ actually satisfies the condition of a measure put $\lambda = \mu \times v$. (See Definition 8.1). Clearly $\lambda(V) \geq 0$ for $V \in \mathcal{A} \otimes \mathcal{B}$ and $\lambda(\emptyset) = 0$, $\emptyset$ being the null set. Let $\{V_i\}$ be any countable family of disjoint sets in $\mathcal{A} \otimes \mathcal{B}$ then

$$\lambda(\cup V_i) = \int_X v((\cup V_i)_x) d\mu = \int_X v(\cup (V_i)_x) d\mu = \int_X \left[\sum v((V_i)_x)\right] d\mu$$
$$= \sum \int_X v((V_i)_x) d\mu = \sum \lambda(V_i)$$

(Here we have applied the analog of Corollary 5.31; Corollary 5.31 is proved for finite measure but there is no difficulty in proving it for σ— finite measure). So, λ satisfies the condition of a measure on the σ— algebra $\mathcal{A} \otimes \mathcal{B}$.

If $A \in \mathcal{A}$ and $B \in \mathcal{B}$ then $A \times B \in \mathcal{A} \otimes \mathcal{B}$ and $(A \times B)_x = B$ or $\emptyset$ according as $x \in A$ or $x \in \tilde{A}$ and so

$$\lambda(A \times B) = \int_X v((A \times B)_x) d\mu = \int_A v(B) d\mu = \mu(A)v(B).$$

Therefore $\lambda = \mu \times v$ is called the product measure on $\mathcal{A} \otimes \mathcal{B}$. Clearly $\mu \times v$ is σ— finite.

Theorem 8.44 Let $(X, \mathcal{A}, \mu)$ and $(Y, \mathcal{B}, v)$ be σ— finite measure spaces and let $f : X \times Y \to [0, \infty]$ be $\mathcal{A} \otimes \mathcal{B}$ - measurable. Let $\phi(x) = \int_Y f_x dv$ and $\psi(y) = \int_X f^y d\mu$ for $x \in X$ and $y \in Y$ respectively. Then ϕ is $\mathcal{A}$— measurable and ψ is $\mathcal{B}$— measurable and

$$\int_X \phi d\mu = \int_{X \times Y} f \, d(\mu \times v) = \int_Y \psi dv. \qquad (8.41)$$

Proof We first suppose that $(X, \mathcal{A}, \mu)$ and $(Y, \mathcal{B}, v)$ are finite measure spaces. If f is the characteristic function of an $\mathcal{A} \otimes \mathcal{B}$- measurable set then the result follows from the last theorems and from the definition of $\mu \times v$. Therefore the result is true when f is a simple function. In the general case, let $\{f_n\}$ be a non-decreasing sequence of non-negative simple functions such that $\lim_{n \to \infty} f_n = f$. Since the theorem is true for simple functions, writing $\phi_n(x) = \int_Y (f_n)_x d\nu$, ϕ_n is $\mathcal{A}-$ measurable and

$$\int_X \phi_n d\mu = \int_{X \times Y} f_n d(\mu \times \nu) \tag{8.42}$$

As $n \to \infty$, $(f_n)_x$ increases to f_x and hence by monotone converges theorem (see Theorem 5.28), as $n \to \infty$, ϕ_n increases to ϕ. Letting $n \to \infty$, we get from (8.42) by monotone converges theorem

$$\int_X \phi \, d\mu = \int_{X \times Y} f \, d(\mu \times \nu).$$

This proves the first part of (8.41). The proof of the second part is similar. Now suppose $(X, \mathcal{A}, \mu)$ and $(Y, \mathcal{B}, \nu)$ are $\sigma-$ finite. Then as in the last part of Theorem 8.43, $X = \cup X_n$ and $Y = \cup Y_m$ and $(X_n, \mathcal{A}_n, \mu)$ and $(Y_m, \mathcal{B}_m, \nu)$ are finite measure spaces and the functions $\phi_{n,m}(x) = \int_{Y_m} f_x \, d\nu$ and $\psi_{nm}(y) = \int_{X_n} f^y \, d\mu$ for $x \in X_n$ and $y \in Y_m$ are $\mathcal{A}_n-$ measurable and $\mathcal{B}_m-$ measurable respectively and

$$\int_{X_n} \phi_{n,m} \, d\mu = \int_{X_n \times Y_m} f \, d(\mu \times \nu) = \int_{Y_m} \psi_{nm} \, d\nu$$

So,

$$\int_{X_n} \{\int_{Y_m} f_x \, d\nu\} d\mu = \int_{X_n \times Y_m} f \, d(\mu \times \nu) = \int_{Y_m} \{\int_{X_n} f^y d\mu\} d\nu.$$

Hence

$$\sum_{n,m} \int_{X_n} \{\int_{Y_m} f_x \, d\nu\} d\mu = \sum_{n,m} \int_{X_n \times Y_m} f \, d(\mu \times \nu) = \sum_{n,m} \int_{Y_m} \{\int_{X_n} f^y d\mu\} d\nu,$$

which gives

$$\int_X \{\int_Y f_x \, d\nu\} d\mu = \int_{X \times Y} f \, d(\mu \times \nu) = \int_Y \{\int_X f^y d\mu\} d\nu.$$

completing the proof. $\square$

Theorem 8.45 (Fubini's theorem) *Let $(X, \mathcal{A}, \mu)$ and $(Y, \mathcal{B}, \nu)$ be $\sigma-$ finite measure spaces and let f be $\mu \times \nu-$ integrable function on $X \times Y$. Then f_x is $\nu-$ integrable for almost all $x \in X$ and f^y is $\mu-$ integrable for almost all $y \in Y$. If*

$$\phi(x) = \int_Y f_x d\nu \quad \text{and} \quad \psi(y) = \int_X f^y d\mu$$

then ϕ and ψ are respectively $\mu-$ integrable and $\nu-$ integrable and

8.9 Measure and Integration in a Product Space

$$\int_X \phi \, d\mu = \int_{X \times Y} f \, d(\mu \times \nu) = \int_Y \psi \, d\nu.$$

Proof Let $f^+ = \max[f, 0]$ and $f^- = \max[-f, 0]$. Then f^+ and f^- are non-negative and $\mu \times \nu$ integrable. Let

$$\phi_1(x) = \int_Y (f^+)_x \, d\nu \text{ and } \phi_2(x) = \int_Y (f^-)_x \, d\nu.$$

Then by Theorem 8.44

$$\int_X \phi_1 \, d\mu = \int_{X \times Y} f^+ d(\mu \times \nu) \tag{8.43}$$

and

$$\int_X \phi_2 \, d\mu = \int_{X \times Y} f^- d(\mu \times \nu) \tag{8.44}$$

So, ϕ_1 and ϕ_2 are $\mu-$ integrable and hence ϕ_1 and ϕ_2 are finite for almost all $x \in X$ and therefore for all such x, $(f^+)_x$ and $(f^-)_x$ are $\nu-$ integrable. Since $f_x = (f^+)_x - (f^-)_x$, f_x is $\nu-$ integrable for almost all $x \in X$. Also since $\phi = \phi_1 - \phi_2$ and ϕ_1 and ϕ_2 are $\mu-$ integrable, ϕ is $\mu-$ integrable. Subtracting (8.44) from (8.43) we get

$$\int_X \phi \, d\mu = \int_{X \times Y} f \, d(\mu \times \nu).$$

This proves all the statements regarding f_x and ϕ. The proof of the statements regarding f^y and ψ are similar. □

Theorem 8.46 Let $(X, \mathcal{A}, \mu)$ and $(Y, \mathcal{B}, \nu)$ be $\sigma-$ finite measure spaces and let f be $\mathcal{A} \otimes \mathcal{B}-$ measurable on $X \times Y$. Let $F(x) = \int_Y (|f|)_x d\nu$ for $x \in X$ and $G(y) = \int_X (|f|)^y d\mu$ for $y \in Y$. Then the following conditions are equivalent :
(i) F is $\mu-$ integrable on X
(ii) G is $\nu-$ integrable on Y
(iii) f is $\mu \times \nu-$ integrable on $X \times Y$.

Proof Since f is $\mathcal{A} \otimes \mathcal{B}$- measurable, $|f|$ is non-negative and $\mathcal{A} \otimes \mathcal{B}-$ measurable on $X \times Y$. So, by Theorem 8.44 F is $\mathcal{A}-$ measurable and G is $\mathcal{B}-$ measurable and

$$\int_X F \, d\mu = \int_{X \times Y} |f| \, d(\mu \times \nu) = \int_Y G \, d\nu. \tag{8.45}$$

The relation (8.45) shows that (i) and (ii) are equivalent. For (iii) suppose that (i) holds. Then by (8.45) $|f|$ is $\mu \times \nu-$ integrable on $X \times Y$ and hence f is $\mu \times \nu-$ integrable on $X \times Y$ which proves that (iii) holds. Conversely, suppose that (iii) holds. Then $|f|$ is $\mu \times \nu-$ integrable on $X \times Y$ and hence by (8.45) (i) holds. So (i) and (iii) are equivalent. □

Theorem 8.47 Let $(X, \mathcal{A}, \mu)$ and $(Y, \mathcal{B}, \nu)$ be σ-finite measure spaces. If f is μ-integrable on X and g is ν-integrable on Y then fg is $\mu \times \nu$-integrable on $X \times Y$. If $(X, \mathcal{A}, \mu)$ and $(Y, \mathcal{B}, \nu)$ are finite measure spaces then $f + g$ is $\mu \times \nu$-integrable on $X \times Y$.

Proof Let $F(x, y) = f(x)$ and $G(x, y) = g(y)$ for $(x, y) \in X \times Y$. Since f is $\mathcal{A}$-measurable and since for any $\alpha \in \mathbb{R}$, $\{(x, y) : F(x, y) > \alpha\} = \{x : f(x) > \alpha\} \times Y$, F is $\mathcal{A} \otimes \mathcal{B}$-measurable on $X \times Y$. Similarly G is $\mathcal{A} \otimes \mathcal{B}$-measurable on $X \times Y$. Hence FG and $F + G$ are $\mathcal{A} \otimes \mathcal{B}$-measurable on $X \times Y$. Since for fixed $x \in X$, $(|FG|)_x = |f(x)| \, |g|$,

$$\int_Y (|FG|)_x d\nu = |f(x)| \int_Y |g| d\nu.$$

Since f is μ-integrable on X, $|f|$ is μ-integrable on X and so $\int_Y (|FG|)_x d\nu$ is μ-integrable on X. So, by Theorem 8.46 FG is $\mu \times \nu$-integrable on $X \times Y$. Since $FG = fg$, fg is $\mu \times \nu$-integrable on $X \times Y$.

Now suppose that $(X, \mathcal{A}, \mu)$ and $(Y, \mathcal{B}, \nu)$ are finite measure spaces. Then since $(|F + G|)_x \leq (|F|)_x + (|G|)_x = |f(x)| + |g|$,

$$\int_Y (|F + G|)_x d\nu \leq \int_Y (|f(x)| + |g|) d\nu = |f(x)|\nu(Y) + \int_Y |g| d\nu.$$

So,

$$\int_X \{\int_Y (|F + G|)_x d\nu\} d\mu \leq \nu(Y) \int_X |f| d\mu + \mu(X) \int_Y |g| d\nu.$$

Since f and g are integrable and X and Y have finite measure, $\int_Y (|F + G|)_x d\nu$ is μ-integrable on X. So, by Theorem 8.46 $F + G$ is $\mu \times \nu$-integrable on $X \times Y$. Since $F + G = f + g$, $f + g$ is $\mu \times \nu$-integrable on $X \times Y$. □

Remarks

(i) For the second part of Theorem 8.47 finiteness of the measure spaces $(X, \mathcal{A}, \mu)$ and $(Y, \mathcal{B}, \nu)$ is necessary. For, consider the measure space $([1, \infty), \mathcal{M}, m)$ where $\mathcal{M}$ is the σ-algebra of all Lebesgue measurable sets in $[1, \infty)$ and m is the Lebesgue measure on $\mathcal{M}$. Clearly $([1, \infty), \mathcal{M}, m)$ is σ-finite but not finite. Let $X = Y = [1, \infty)$, $\mathcal{A}, \mathcal{B} = \mathcal{M}$ and $\mu = \nu = m$ and let $f(x) = \frac{1}{x^2}$ and $g(y) = \frac{1}{y^2}$ for $x, y \in [1, \infty)$. Then f and g are m-integrable on $[1, \infty)$ but $f + g$ is not $m \times m$ integrable on $[1, \infty) \times [1, \infty)$. For, if $f + g$ is integrable on $[1, \infty) \times [1, \infty)$ then by Theorem 8.45 $(f + g)_x$ is ν-integrable for almost all $x \in [1, \infty)$. But

$$\int_Y (\frac{1}{x^2} + \frac{1}{y^2})_x d\nu = \frac{1}{x^2}\nu(Y) + \int_Y \frac{1}{y^2} d\nu = \infty \text{ for all } x \in [1, \infty)$$

(ii) While the integrability of the product of two functions f and g on $(X, \mathcal{A}, \mu)$ and $(Y, \mathcal{B}, \nu)$ respectively is valid in the product space $(X \times Y, \mathcal{A} \otimes \mathcal{B}, \mu \times \nu)$, the integrability of the product fg is not valid in a single measure space. More precisely, if f and g are integrable in $(X, \mathcal{A}, \mu)$ then fg may not be integrable in $(X, \mathcal{A}, \mu)$. For, consider the Lebesgue measure space $([0, 1], \mathcal{M}, m)$ and $f(x) = x^{-\frac{2}{3}}, g(x) = x^{-\frac{1}{3}}$ for $x \in (0, 1]$ and $f(0) = g(0) = 0$.

Example 8.1 Let $(X, \mathcal{A}, \mu)$ and $(Y, \mathcal{B}, \nu)$ be $\sigma-$ finite measure spaces. If $E_1 \in \mathcal{A}$ and $E_2 \in \mathcal{B}$ then $E_1 \times E_2 \in \mathcal{A} \otimes \mathcal{B}$ and $(\mu \times \nu)(E_1 \times E_2) = \mu(E_1)\nu(E_2)$.

The first part is clear. For the second part, by Definition 8.17

$$(\mu \times \nu)(E_1 \times E_2) = \int_X \nu((E_1 \times E_2)_x) d\mu.$$

Since $(E_1 \times E_2)_x = E_2$ or $\emptyset$ according as $x \in E_1$ or $x \in X \sim E_1$, $\nu((E_1 \times E_2)_x) = \nu(E_2)$ or 0 according as $x \in E_1$ or $x \in X \sim E_1$ So

$$(\mu \times \nu)(E_1 \times E_2) = \int_{E_1} \nu((E_1 \times E_2)_x) d\mu + \int_{X \sim E_1} \nu((E_1 \times E_2)_x) d\mu$$
$$= \int_{E_1} \nu(E_2) d\mu = \mu(E_1)\nu(E_2).$$

8.10 Product Measure and Lebesgue Measure in $\mathbb{R}^N$

Although product measure is defined on two measurable spaces, product measure can be defined on $N-$ dimensional measurable spaces by similar method. After getting the definition of product measure it is now natural to ask why the product Lebesgue measure is not useful in $\mathbb{R}^2$. More specifically, if μ is the Lebesgue measure in $\mathbb{R}$ then why the product measure $\mu \times \mu$ will not be the Lebesgue measure in $\mathbb{R}^2$? The answer is that although the Lebesgue measure μ in $\mathbb{R}$ is complete (see Definition 8.3 and Note under Theorem 8.8) the product Lebesgue measure $\mu \times \mu$ is not complete in $\mathbb{R}^2$. To see this, suppose that $\mu \times \mu$ is complete in $\mathbb{R}^2$. Let $E_1, E_2 \subset [0, 1]$ be such that $\mu(E_1) = 0$ and E_2 is nonmeasurable. Then since $E_1 \times E_2 \subset E_1 \times [0, 1]$ and $(\mu \times \mu)(E_1 \times [0, 1]) = \mu(E_1)\mu([0, 1) = 0$ and since $\mu \times \mu$ is complete, $E_1 \times E_2$ is measurable. So, by Theorem 8.41 $(E_1 \times E_2)_x$ is measurable for each $x \in [0, 1]$. But for $x \in E_1$, $(E_1 \times E_2)_x = E_2$ which is nonmeasurable, a contradiction. From the above discussion it is clear that even when $(X, \mathcal{A}, \mu)$ and $(Y, \mathcal{B}, \nu)$ are complete measure spaces the product measure space $(X \times Y, \mathcal{A} \otimes \mathcal{B}, \mu \times \nu)$ may not be complete. However the completion of $(X \times Y, \mathcal{A} \otimes \mathcal{B}, \mu \times \nu)$ can be obtained by using Definition 8.7 and applying Theorems 8.11 and 8.8.

In fact product Lebesgue measure $m \times m$ where m is the Lebesgue measure in $\mathbb{R}$ is not effective in $\mathbb{R}^2$. To see this, consider the Lebesgue measure space $([0, 1], \mathcal{M}, m)$, where $\mathcal{M}$ is the family of all Lebesgue measurable sets in $[0, 1]$ and m is the Lebesgue measure and let $E_1, E_2 \subset [0, 1]$ be such that E_1 is of measure zero and E_2 is nonmeasurable. Let $f : [0, 1] \times [0, 1] \to \mathbb{R}$ be defined by $f(x, y) = 1$ if $(x, y) \in E_1 \times E_2$ and $f(x, y) = 0$ if $(x, y) \notin E_1 \times E_2$. Then since $E_1 \times E_2 \notin M \otimes M$, f is not $M \otimes M-$ measurable and so f is not $m \times m$ integrable. But considering Lebesgue measure μ_2 in $\mathbb{R}^2$, $E_1 \times E_2$ is of Lebesgue measure zero in $\mathbb{R}^2$ (see Theorem 7.13) and so f is μ_2- integrable.

8.11 Integration of Complex Valued Functions

Let $(X \mathcal{A}, \mu)$ be a measure space and let f be a complex-valued function on X.

Definition 8.18 The function f is said to be $\mathcal{A}-$ measurable if its real and imaginary parts are both $\mathcal{A}-$ measurable. That is, if $f = f_1 + i f_2$ then f is said to be $\mathcal{A}-$ measurable if both f_1 and f_2 are $\mathcal{A}-$ measurable.

Theorem 8.48 *If f is $\mathcal{A}-$ measurable then $|f|$ is $\mathcal{A}-$ measurable.*

Proof Let $f = f_1 + if_2$ and let f be $\mathcal{A}$- measurable. Then f_1 and f_2 are real valued $\mathcal{A}-$ measurable functions on X. Hence $f_1^2 + f_2^2$ is $\mathcal{A}-$ measurable on X. Since for $\alpha \in \mathbb{R}$, $\{x \in X : +\sqrt{f_1^2 + f_2^2} > \alpha\} = \{x \in X : f_1^2 + f_2^2 > \alpha^2\}$ if $\alpha \geq 0$ and $\{x \in X : +\sqrt{f_1^2 + f_2^2} > \alpha\} = X$ if $\alpha < 0$ and since $|f| = +\sqrt{f_1^2 + f_2^2}$, $|f|$ is $\mathcal{A}-$ measurable.

Other properties of real-valued measurable functions proved in Chapter-3 are also true for complex-valued measurable functions, and the proofs are similar. □

Definition 8.19 The function f is said to be $\mu-$ integrable if both of its real part f_1 and imaginary part f_2 are $\mu-$ integrable and in this case the integral of f is defined by

$$\int_X f \, d\mu = \int_X f_1 \, d\mu + i \int_X f_2 \, d\mu.$$

Theorem 8.49 *A complex-valued measurable function f is $\mu-$ integrable if and only if $|f|$ is $\mu-$ integrable. In either case*

$$\left| \int_X f \, d\mu \right| \leq \int_X |f| \, d\mu.$$

Proof Suppose f is integrable. Then by definition f_1 and f_2 are integrable and so $|f_1|$ and $|f_2|$ are integrable and hence $|f_1| + |f_2|$ is integrable. Since by Theorem 8.48 $|f|$ is measurable and since $|f| \leq |f_1| + |f_2|$, $|f|$ is integrable. Conversely, suppose that $|f|$ is integrable. Since $|f_1| \leq |f|$ and $|f_2| \leq |f|$, $|f_1|$ and $|f_2|$ are integrable. Since f is measurable, f_1 and f_2 are measurable and so f_1 and f_2 are integrable and hence f is integrable.

Finally, let $z = \int_X f \, d\mu$. If $z = 0$, the result follows. So, we suppose $z \neq 0$. Let $\alpha = \frac{|z|}{z}$. Then $\alpha z = |z|$ and $|\alpha| = 1$. Let u and v be the real and imaginary part of αf. Then $\alpha f = u + iv$ and

$$\int_X \alpha f \, d\mu = \int_X u \, d\mu + i \int_X v \, d\mu.$$

Hence

$$\left| \int_X f \, d\mu \right| = |z| = \alpha z = \alpha \int_X f \, d\mu = \int_X \alpha f \, d\mu = \int_X u \, d\mu + i \int_X v \, d\mu. \quad (8.46)$$

Since the left side of (8.46) is real, we must have $\int_X v \, d\mu = 0$. Also $u \leq |\alpha f| = |f|$ and hence $\int_X u \, d\mu \leq \int_X |f| \, d\mu$ and so from (8.46)

$$\left| \int_X f \, d\mu \right| = \int_X u \, d\mu \leq \int_X |f| \, d\mu$$

completing the proof.

Other properties of real-valued integrable function proved in Chapter-5 are also true for complex-valued function except those which involve inequality. Note that Dominated convergence theorem is also true for complex-valued function. □

8.12 Exercises

1. Let X be a non-void set and Let $\mathcal{A}$ be a σ− algebra of subsets of X. Let μ be a measure on $\mathcal{A}$ and μ^* be the outer measure induced by μ. If $\{E_n\}$ is a sequence of disjoint sets in $\mathcal{A}$ and if $S_n = \bigcup_{i=1}^{n} E_i$ and $S = \bigcup_{i=1}^{\infty} E_i$, prove that for any set $A \subset X$

   (i) $\mu^*(A \cap S_n) = \sum_{i=1}^{n} \mu^*(A \cap E_i)$

   (ii) $\mu^*(A \cap S) = \sum_{i=1}^{\infty} \mu^*(A \cap E_i)$

   [Hint : For (i), the result is true for $n = 1$. Supposing it is true for n, since $S_n \in \mathcal{A}$,

   $\mu^*(A \cap S_{n+1}) = \mu^*(A \cap S_{n+1} \cap S_n) + \mu^*((A \cap S_{n+1}) \sim S_n) = \sum_{i=1}^{n+1} \mu^*(A \cap E_i)$

   For (ii) $\mu^*(A \cap S) \geq \mu^*(A \cap S_n) = \sum_{i=1}^{n} \mu^*(A \cap E_i)$ by (i)

   and letting $n \to \infty$ and applying the property of μ^*, the result follows.

2. Let X be a non-void set. Define $\mu^*(\emptyset) = 0$ and $\mu^*(E) = 1$ for every set $E \subset X$, $E \neq \emptyset$. Show that μ^* is an outer measure and if X contains more than one point then μ^* is not a measure.
3. Let $X = [0, 1]$. Define μ^* on 2^X such that for $E \subset X$, $\mu^*(E) =$ number of points in E if it is a finite set and $\mu^*(E) = \infty$ otherwise. Show that μ^* is an outer measure. Determine the class of μ^*- measurable sets.
4. Let $\mathcal{B}$ and $\mathcal{S}$ be the class of Borel sets and Lebesgue measurable sets on the real line $\mathbb{R}$ then show that the measure space $(\mathbb{R}, \mathcal{B}, \mu)$ is not complete and the measure space $(\mathbb{R}, \mathcal{S}, \mu)$ is the completion of $(\mathbb{R}, \mathcal{B}, \mu)$, where μ is the Lebesgue measure.
5. Let μ^* be a regular outer measure. Show that for every non-decreasing sequence $\{A_n\}$ of sets $\mu^*(\lim A_n) = \lim \mu^*(A_n)$.
6. Let $f(x) = x$ for $x \in \mathbb{R}$. Show that f is a distribution function and that the Lebesgue-Stieltje's outer measure μ_f^* in this case is the Lebesgue outer measure μ^*.
7. Define $f : \mathbb{R} \to \mathbb{R}$ by $f(x) = 0$ if $x < 0$ and $f(x) = 1$ if $x \geq 0$. Show that f is a distribution function and the Lebesgue-Slieltje's measure μ_f satisfies $\mu_f((-1, 0)) < \mu_f((-1, 0])$.
8. Give an example of a right continuous non-decreasing function f and a pair of points $a, b, a < b$, such that

$$\mu_f((a, b)) < f(b) - f(a) < \mu_f([a, b]).$$

[Hint: See Theorem 8.18 and conclude that f must have discontinuity at a and at b.]
9. Show that, if instead of assuming right continuity of a non-decreasing function f to define Lebesgue-Stieltje's measure left continuity of f is assumed then also Lebesgue-Stieltje's measure can be defined. What is the value of $\mu_f((a, b])$ in this case?
10. Let f be the Cantor function described in Sect. 4.5. Show that f is a distribution function and find the value of $\mu_f(E)$ when (i) $E = (\frac{1}{3}, \frac{2}{3})$ and when (ii) $E = (\frac{2}{9}, \frac{1}{3})$ where μ_f is the Lebesgue-Stieltje's measure corresponding to f.
11. If $\{A_k\}$ is any countable collection of sets then prove that $\dim \left(\bigcup_{k=1}^{\infty} A_k \right) = \sup_k \dim(A_k)$.
12. Prove that dimension of any finite interval I is 1.
13. Prove that the dimension of the real line is 1.
14. Show that any countable set has dimension 0.
15. Find the Hausdorff dimension of the Lebesgue nonmeasurable set A constructed in Theorem 2.19.
16. Let E be a set whose Lebesgue outer measure is finite and positive. Find the Hausdorff dimension of E.
17. Let $(X, \mathcal{A}, \mu)$ be a $\sigma-$ finite measure space and let $(\mathbb{R}, \mathcal{M}, m)$ be the Lebesgue measure space. Let f be a real-valued non-negative $\mathcal{A}-$ measurable function on

X. Let $E = \{(x, y) : x \in X; 0 \le y \le f(x)\}$. Show that E is $\mathcal{A} \otimes \mathcal{M}$ measurable and $(\mu \times m)(E) = \int_X f d\mu$. If $G = \{(x, y) : x \in X; y = f(x)\}$ show that G is measurable and is of measure zero.
18. Show that a rectangle $A \times B$ is nonmeasurable if and only if either A is nonmeasurable and $B \ne \phi$ or B is nonmeasurable and $A \ne \phi$.
19. If $\{A_i \times B_i\}$ is a countable collection of measurable rectangles then show that $\bigcap_{i=1}^{\infty}(A_i \times B_i)$ is a measurable rectangle.
20. For any set $V \in X \times Y$ show that $(\chi_V)_x = \chi_{V_x}$ where χ_E denotes the characteristic function of E.
21. Let $(X, \mathcal{A})$ be a measurable space and let $(\mathbb{R}, \mathcal{M})$ be the Lebesgue measurable space. Let f be a non-negative real-valued function on X. Let $E(f) = \{(x, y) : x \in X; 0 \le y \le f(x)\}$. Show that f is $\mathcal{A}$– measurable if and only if $E(f)$ is $\mathcal{A} \otimes \mathcal{M}$– measurable. If $\{f_n\}$ is an increasing sequence of non-negative function on X converging to f on X then show that $\{E(f_n)\}$ is an increasing sequence of sets such that $\lim_{n \to \infty} E(f_n) = E(f)$.

Chapter 9
Function Spaces

One of the interesting contributions of the Lebesgue integral is the L^p-spaces. Since the Lebesgue integral is an absolute integral, i.e. f is Lebesgue integrable if and only if $|f|$ is Lebesgue integrable and since other integrals do not have this property, only Lebesgue integral have produced the L^p-spaces, which are useful in many branches of mathematics including Functional Analysis, Fourier Analysis and Harmonic Analysis. The L^p-spaces of functions defined on a measure space are considered. Also other function spaces, namely the space of measurable functions and the spaces of continuous functions and Riemann integrable functions are considered.

9.1 Metric Space and Linear Space

Definition 9.1 Let X be a nonempty set and let ρ be a non-negative real-valued function defined on $X \times X$ such that
(i) $\rho(x, y) = 0$ if and only if $x = y, x, y \in X$
(ii) $\rho(x, y) = \rho(y, x)$ for every $x, y \in X$
(iii) $\rho(x, z) \leq \rho(x, y) + \rho(y, z)$ for every $x, y, z \in X$.
Then ρ is called a metric (or, a distance function) on X and (X, ρ) is called a metric space.

If $x \in X$ and $\epsilon > 0$ then the set $\{y : \rho(y, x) < \epsilon\}$ is called ϵ-neighborhood of x. A set $S \subset X$ is called an open set if for every $x \in S$ there is $\epsilon > 0$ such that $\{y : \rho(y, x) < \epsilon\} \subset S$. A points $x \in X$ is called a limit point of the set $S \subset X$ if for every $\epsilon > 0$ there is $y \in S, y \neq x$ such that $\rho(y, x) < \epsilon$. That is, every ϵ neighborhood of x contains a point of S different from x. Clearly X and the null set $\emptyset$ are open sets. A set $F \subset X$ is called closed if its complement $X \sim F$ is open. If $S \subset X$ then the set $S \cup S'$ where S' is the set of all limit points of S is called the closure of S and is denoted by $\overline{S}$. A set $E \subset X$ is said to be dense in X if $\overline{E} = X$. That is, E is said to be dense in X if for every x in X and every $\epsilon > 0$ there is a member ξ in E such that

$\rho(x, \xi) < \epsilon$. A sequence $\{x_n\}$ in X is said to be a Cauchy sequence if for every $\epsilon > 0$ there is N such that $\rho(x_n, x_m) < \epsilon$ whenever $n, m \geq N$. The metric space (X, ρ) is called complete if every Cauchy sequence in X is convergent. That is, if $\{x_n\}$ is a Cauchy sequence in X then there is $x \in X$ such that for every $\epsilon > 0$ there is N such that $\rho(x_n, x) < \epsilon$ for $n \geq N$.

Definition 9.2 Let X be a nonempty set and let addition of elements of X and multiplication of elements of X by real numbers be defined in X which satisfy
(i) $x, y \in X$ imply $x + y \in X$
(ii) $x \in X$ and $\alpha \in \mathbb{R}$ imply $\alpha x \in X$.
Then X is called a linear space over the real numbers $\mathbb{R}$. If more over there is a function $\|.\|$ defined on X with valued in $\mathbb{R}$ satisfying
(iii) $\|x\| \geq 0$ for all $x \in X$ and $\|x\| = 0$ if and only if $x = 0$
(iv) $\|\alpha x\| = |\alpha| \|x\|$ for all $x \in X$ and all $\alpha \in \mathbb{R}$
(v) $\|x + y\| \leq \|x\| + \|y\|$ for all $x, y \in X$,
then X is called a normed linear space and $\|.\|$ is called its norm.

Clearly a normed linear space is a metric space, the metric being defined by $\rho(x, y) = \|x - y\|$. So, the definition of Cauchy sequence and completeness in a normed linear space is as in Definition 9.1. Other properties mentioned there are also applicable in normed linear space.

9.2 The L^p-spaces

Definition 9.3 Let $(X, \mathcal{A}, \mu)$ be a measure space and let $E \in \mathcal{A}$. Let p be a positive real number. Let $L^p(E)$ be the class of all measurable functions f on E for which $|f|^p$ is μ integrable. That is, $L^p(E) = \{f : f : E \to \mathbb{R}$ is measurable and $\int_E |f|^p d\mu < \infty\}$. For $f \in L^p(E)$ define

$$\|f\|_p = \|f\|_{p,E} = (\int_E |f|^p d\mu)^{\frac{1}{p}}.$$

For $p = \infty$, $L^\infty(E)$ is defined to be the class of all measurable functions on E for which there is $M, 0 < M < \infty$, such that $|f(x)| \leq M$ for almost all $x \in E$. For $f \in L^\infty(E)$ define

$$\|f\|_\infty = \|f\|_{\infty,E} = \operatorname*{ess.\,sup}_{E} |f|$$

where $\operatorname*{ess.\,sup}_{E} |f|$ means the essential supremum of $|f|$ on E which is defined by

$$\operatorname{ess.\,sup} |f| = \inf\{\alpha : \mu(\{x \in E; |f(x)| > \alpha\}) = 0\}.$$

Theorem 9.1 If $0 < p_1 < p_2 \leq \infty$ and E is of finite measure then $L^{p_2}(E) \subset L^{p_1}(E)$.

9.2 The L^p-spaces

Proof Let $p_2 < \infty$ and $f \in L^{p_2}(E)$. Let $E_1 = \{x \in E; |f(x)| \leq 1\}$ and $E_2 = E \sim E_1$. Then

$$\int_E |f|^{p_1} d\mu = \int_{E_1} |f|^{p_1} d\mu + \int_{E_2} |f|^{p_1} d\mu \leq \mu(E_1) + \int_{E_2} |f|^{p_2} d\mu < \infty$$

So,

$$f \in L^{p_1}(E).$$

If $p_2 = \infty$, the proof is clear. $\square$

Remarks

(i) The set E is of finite measure is necessary. For consider the Lebesgue measure space $(\mathbb{R}, \mathcal{M}, \mu)$ and $E = [1, \infty)$. Then the function $f(x) = x^{-\frac{1}{p_1}}$ is such that $f \in L^{p_2}(E)$ but $f \notin L^{p_1}(E)$.

(ii) The inclusion is proper. For, if $E = (0, 1)$ and if $f(x) = x^{-\frac{1}{p_2}}$ then $f \in L^{p_1}(E)$ but $f \notin L^{p_2}(E)$.

(iii) It is clear that if $\mu(E) < \infty$ then $L^\infty(E) \subset L^p(E)$ for all $p > 0$ and so $L^\infty(E) \subset \bigcap_{p>0} L^p(E)$. This inclusion is also proper. For, if $E = (0, 1)$ and $f(x) = \log \frac{1}{x}$ then $f \in L^p(E)$ for all $p > 0$ but $f \notin L^\infty(E)$. To see this, let $p > 0$ be fixed, choose $\alpha > 0$ such that $0 < \alpha p < 1$. Then $|\log \frac{1}{x}|^p / \frac{1}{x^{\alpha p}} = x^{\alpha p} |\log x|^p \to 0$ as $x \to 0_+$. So, for $\epsilon > 0$ there is $\delta > 0$ such that $|\log \frac{1}{x}|^p < \frac{\epsilon}{x^{\alpha p}}$ for $0 < x < \delta$. Since $0 < \alpha p < 1$, $\frac{1}{x^{\alpha p}}$ is integrable on E and so $f \in L^p(E)$. Thus $f \in \bigcap_{p>0} L^p(E)$. But f is not essentially bounded and so $f \notin L^\infty(E)$.

Theorem 9.2 *If $0 < \mu(E) < \infty$ then $\|f\|_\infty = \lim_{p \to \infty} \|f\|_p$ provided $f \in L^\infty(E)$. If $f \in \bigcap_{p>0} L^p(E) \sim L^\infty(E)$ then $\lim_{p \to \infty} \|f\|_p = \infty$.*

Proof Let $M = \|f\|_\infty$. If $M = 0$ then $f = 0$ a.e. on E and so $\|f\|_p = 0$ for all $p > 0$ and the result follows.

So suppose $M > 0$. Choose $0 < M' < M$. Let $A = \{x \in E; |f(x)| > M'\}$. Then $\mu(A) > 0$. So

$$\|f\|_p = \left(\int_E |f|^p d\mu\right)^{\frac{1}{p}} \geq \left(\int_A |f|^p d\mu\right)^{\frac{1}{p}} \geq ((M')^p \mu(A))^{\frac{1}{p}} = M'(\mu(A))^{\frac{1}{p}}.$$

Since $(\mu(A))^{\frac{1}{p}} \to 1$ as $p \to \infty$, $\liminf_{p \to \infty} \|f\|_p \geq M'$. Since M' is arbitrary, $\liminf_{p \to \infty} \|f\|_p \geq M$. Also $\|f\|_p = (\int_E |f|^p d\mu)^{\frac{1}{p}} \leq (\int_E M^p d\mu)^{\frac{1}{p}} = M(\mu(E))^{\frac{1}{p}} \to M$ as $p \to \infty$. So $\limsup_{p \to \infty} \|f\|_p \leq M$. So, the first part is proved. For the second, take $M' < \infty$ and apply the above argument. $\square$

Theorem 9.3 (Holders inequality) *If p and q are such that $1 \le p \le \infty$ and $\frac{1}{p} + \frac{1}{q} = 1$ and if $f \in L^p(E)$ and $g \in L^q(E)$ then $f.g \in L^1(E)$ and*

$$\int_E |f.g|d\mu \le \|f\|_p . \|g\|_q.$$

To prove the theorem we need

Lemma 9.1 *Let $\alpha, \beta \in [0, \infty)$ and $0 < \lambda < 1$. Then*

$$\alpha^\lambda \beta^{1-\lambda} \le \lambda\alpha + (1-\lambda)\beta.$$

The equality holds if $\alpha = \beta$.

Proof Let $\phi(t) = 1 - \lambda + \lambda t - t^\lambda, t \ge 0$. Then for $t \ne 0$

$$\phi'(t) = \lambda - \lambda t^{\lambda - 1} = \lambda(1 - t^{\lambda - 1}) = \lambda(1 - \frac{1}{t^{1-\lambda}}) = \frac{\lambda}{t^{1-\lambda}}(t^{1-\lambda} - 1).$$

Hence $\phi'(t) < 0$ if $t < 1$ and $\phi'(t) > 0$ if $t > 1$. So, for $t \ne 1$, $\phi(t) \ge \phi(1) = 0$ which gives

$$1 - \lambda + \lambda t \ge t^\lambda, \text{ the equality holds if } t = 1. \tag{9.1}$$

Now if $\beta = 0$, the proof is clear. Suppose $\beta \ne 0$. Then putting $t = \frac{\alpha}{\beta}$ in (9.1)

$$1 - \lambda + \lambda\frac{\alpha}{\beta} \ge (\frac{\alpha}{\beta})^\lambda$$

So,

$$(1-\lambda)\beta + \lambda\alpha \ge \alpha^\lambda \beta^{1-\lambda},$$

the equality holds if $\alpha = \beta$. $\square$

Proof of the theorem Let $p = 1$. Then $q = \infty$. In this case $f \in L^1(E)$ and $g \in L^\infty(E)$. Let $M = \text{ess sup } |g|$. Then $|fg| \le |f|M$ a.e. on E. So

$$\int_E |fg|d\mu \le M\int_E |f|d\mu = M\|f\|_1 = \|f\|_1 . \|g\|_\infty$$

So, the result is proved in this case. If $p = \infty$ then $q = 1$ and so the result is proved as above.

Now suppose $1 < p < \infty$. Then $1 < q < \infty$. We further suppose, as a special case, that $\|f\|_p = \|g\|_q = 1$. For fixed t writing $\alpha = |f(t)|^p, \beta = |g(t)|^q, \lambda = \frac{1}{p}, 1 - \lambda = \frac{1}{q}$ and applying the above Lemma

$$|f(t)g(t)| \le \frac{1}{p}|f(t)|^p + \frac{1}{q}|g(t)|^q.$$

9.2 The L^p-spaces

Since this is true for all $t \in E$

$$\int_E |fg|d\mu \leq \frac{1}{p}\int_E |f|^p d\mu + \frac{1}{q}\int_E |g|^q d\mu$$
$$= \frac{1}{p}(\|f\|_p)^p + \frac{1}{q}(\|g\|_q)^q = \frac{1}{p} + \frac{1}{q} = 1 = \|f\|_p \|g\|_q$$

So, the result is proved in this special case. To prove the general case note that if $\|f\|_p = 0$ or $\|g\|_q = 0$ then the theorem is trivially true. Therefore suppose that $\|f\|_p \neq 0$ and $\|g\|_q \neq 0$. Then the functions $\frac{f}{\|f\|_p}$ and $\frac{g}{\|g\|_q}$ satisfy the condition of the special case, and so by the special case

$$\frac{1}{\|f\|_p} \cdot \frac{1}{\|g\|_q} \int_E |fg|d\mu \leq \|\frac{f}{\|f\|_p}\|_p \cdot \|\frac{g}{\|g\|_q}\|_q = 1$$

That is

$$\int_E |fg|d\mu \leq \|f\|_p \cdot \|g\|_q$$

Theorem 9.4 (Minkowski's inequality) *If $1 \leq p \leq \infty$ and $f, g \in L^p(E)$ then $f + g \in L^p(E)$ and $\|f + g\|_p \leq \|f\|_p + \|g\|_p$*

Proof If $p = 1$ then since $|f + g| \leq |f| + |g|$,

$$\int_E |f + g|d\mu \leq \int_E |f|d\mu + \int_E |g|d\mu$$

and so the result follows. If $p = \infty$ then $|f| \leq \|f\|_\infty$ a.e. in E and $|g| \leq \|g\|_\infty$ a.e. in E and so $|f + g| \leq \|f\|_\infty + \|g\|_\infty$ a.e. in E and hence $\|f + g\|_\infty \leq \|f\|_\infty + \|g\|_\infty$. Therefore the result follows in this case also. So, we suppose $1 < p < \infty$. Since for fixed t,

$$|f(t) + g(t)| \leq 2|f(t)| \text{ if } |f(t)| \geq |g(t)|$$
$$\leq 2|g(t)| \text{ if } |f(t)| \leq |g(t)|$$

we have

$$|f(t) + g(t)|^p \leq 2^p(|f(t)|^p + |g(t)|^p) \text{ for all } t \in E.$$

Therefore

$$\int_E |f + g|^p d\mu \leq 2^p \left(\int_E |f|^p d\mu + \int_E |g|^p d\mu \right) < \infty$$

and so $f + g \in L^p(E)$. If $\|f + g\|_p = 0$, the proof is complete.
So suppose that $\|f + g\|_p \neq 0$. Then

$$\int_E |f+g|^p d\mu = \int_E |f+g|^{p-1}|f+g|d\mu \le \int_E |f+g|^{p-1}(|f|+|g|)d\mu$$
$$= \int_E |f+g|^{p-1}|f|d\mu + \int_E |f+g|^{p-1}|g|d\mu \qquad (9.2)$$

Let $q = \frac{p}{p-1}$. Then $\frac{1}{p} + \frac{1}{q} = 1$. Hence $\int_E |f+g|^{(p-1)q} d\mu = \int_E |f+g|^p d\mu < \infty$ and so $|f+g|^{p-1} \in L^q(E)$. Therefore, by Theorem 9.3

$$\int_E |f+g|^{p-1}|f|d\mu \le \|f\|_p \||f+g|^{p-1}\|_q$$

and

$$\int_E |f+g|^{p-1}|g|d\mu \le \|g\|_p \||f+g|^{p-1}\|_q.$$

So by (9.2)

$$\int_E |f+g|^p d\mu \le (\|f\|_p + \|g\|_p)\||f+g|^{p-1}\|_q \qquad (9.3)$$

Since $(p-1)q = p$, we have

$$\||f+g|^{p-1}\|_q = \left(\int_E |f+g|^{(p-1)q} d\mu\right)^{\frac{1}{q}} = \left(\int_E |f+g|^p d\mu\right)^{\frac{1}{q}} = (\|f+g\|_p)^{\frac{p}{q}}$$

Hence from (9.3)

$$\int_E |f+g|^p d\mu \le (\|f\|_p + \|g\|_p)(\|f+g\|_p)^{\frac{p}{q}}$$

That is

$$(\|f+g\|_p)^p \le (\|f\|_p + \|g\|_p)(\|f+g\|_p)^{\frac{p}{q}}$$

$p - \frac{p}{q} = 1$, this gives

$$\|f+g\|_p \le \|f\|_p + \|g\|_p.$$

completing the proof. $\square$

Remark Theorem 9.4 is not true if $0 < p < 1$. For consider the Lebesgue measure space $(\mathbb{R}, \mathcal{M}, \mu)$ and take $E = (0,1)$, $f(x) = 1$ if $0 < x < \frac{1}{2}$ and $f(x) = 0$ if $\frac{1}{2} \le x < 1$ and $g(x) = 0$ if $0 < x < \frac{1}{2}$ and $g(x) = 1$ if $\frac{1}{2} \le x < 1$. Then $\|f+g\|_p = 1$ while $\|f\|_p + \|g\|_p = (\frac{1}{2})^{\frac{1}{p}} + (\frac{1}{2})^{\frac{1}{p}} = 2^{1-\frac{1}{p}} < 1$.

Now consider the space $L^p(E)$, $1 \le p \le \infty$. We show that $L^p(E)$ is a normed linear space if we agree to consider the elements f and g of $L^p(E)$ to be equal if the functions f and g are equal *a.e.* on E. Under this agreement $\|f\|_p = 0$ if and only if $f = 0$ and so the condition (iii) of Definition 9.2 is satisfied. By Minkowski's inequality the conditions (i) and (v) of Definition 9.2 are satisfied. If $f \in L^p(E)$

and $\alpha \in \mathbb{R}$ then since

$$\left(\int_E |\alpha f|^p\right)^{\frac{1}{p}} = |\alpha| \left(\int_E |f|^p\right)^{\frac{1}{p}} < \infty$$

which gives $\|\alpha f\|_p = |\alpha| \|f\|_p$ the conditions (ii) and (iv) of Definition 9.2 are satisfied. Therefore $L^p(E)$ is a normed linear space for $1 \leq p \leq \infty$. We shall show that $L^p(E)$ is complete for $1 \leq p \leq \infty$.

Definition 9.4 A sequence of functions $\{f_n\} \subset L^p(E), 1 \leq p < \infty$, is said to converge to $f \in L^p(E)$ in the L^p—mean if $\|f_n - f\|_p \to 0$ as $n \to \infty$ and we write $f_n \to f [L^p - \text{mean}]$ on E.

9.3 Counting Measure and Application

Let $\{a_i\}$ and $\{b_i\}$ be sequences of real numbers. The Holder and Minkowski inequality for $\{a_i\}$ and $\{b_i\}$ are respectively

$$\sum_{i=1}^{\infty} |a_i b_i| \leq \left(\sum_{i=1}^{\infty} |a_i|^p\right)^{\frac{1}{p}} \left(\sum_{i=1}^{\infty} |b_i|^q\right)^{\frac{1}{q}} \text{ for } 1 \leq p < \infty, \ \frac{1}{p} + \frac{1}{q} = 1$$

and $\left(\sum_{i=1}^{\infty} |a_i + b_i|^p\right)^{\frac{1}{p}} \leq \left(\sum_{i=1}^{\infty} |a_i|^p\right)^{\frac{1}{p}} + \left(\sum_{i=1}^{\infty} |b_i|^p\right)^{\frac{1}{p}}$ for $1 \leq p < \infty$.

These inequalities can be proved directly. But using the above theorems these can be proved by considering counting measures.

Let $\mathbb{N}$ be the set of positive integers and let $\mathcal{A}$ be the collection of all subsets of $\mathbb{N}$. Then $\mathcal{A}$ is a σ-algebra and $(\mathbb{N}, \mathcal{A})$ is a measurable space. Define $\mu : \mathcal{A} \to (0, \infty)$ by $\mu(E) = $ number of elements of E, for $E \in \mathcal{A}$.

Then $\mu(\emptyset) = 0$ where $\emptyset$ is the null set. If $\{E_i\}$ is a countable collection of disjoint subsets of $\mathbb{N}$ then $\mu(\cup E_i) = \sum \mu(E_i)$. So μ is a measure on $\mathcal{A}$. This measure is called counting measure and the measure space $(\mathbb{N}, \mathcal{A}, \mu)$ is called counting measure space.

By definition μ is a complete measure space. For any sequence $\{a_i\}$, define a functions $f : \mathbb{N} \to \mathbb{R}$ such that $f(i) = a_i$. Then f is $\mathcal{A}$-measurable.

If $E = \{1, 2, ..., M\}$ then writing $E_i = \{i\}$ we have $E = \bigcup_{i=1}^{M} E_i$ and therefore since $\mu(E_i) = 1$, the integral of f on E is

$$\int_E f d\mu = \sum_{i=1}^{M} \int_{E_i} f d\mu = \sum_{i=1}^{M} a_i$$

If we consider the whole set $\mathbb{N}$ then $\int_{\mathbb{N}} f d\mu = \sum_{i=1}^{\infty} a_i$. Now we prove Holder and Minkowski inequality.

For any two sequences $\{a_i\}$ and $\{b_i\}$ define two functions $f : \mathbb{N} \to \mathbb{R}$ and $g : \mathbb{N} \to \mathbb{R}$ such that $f(i) = a_i$ and $g(i) = b_i$, $i = 1, 2, \ldots$. Then applying Theorem 9.3 and Theorem 9.4 we have if $1 \leq p < \infty$ and $\frac{1}{p} + \frac{1}{q} = 1$ then

$$\text{Holder inequality}: \sum_{i=1}^{\infty} |a_i b_i| \leq \left(\sum_{i=1}^{\infty} |a_i|^p\right)^{\frac{1}{p}} \left(\sum_{i=1}^{\infty} |b_i|^q\right)^{\frac{1}{q}} \quad \text{and}$$

$$\text{Minkowski inequality}: \left(\sum_{i=1}^{\infty} |a_i + b_i|^p\right)^{\frac{1}{p}} \leq \left(\sum_{i=1}^{\infty} |a_i|^p\right)^{\frac{1}{p}} + \left(\sum_{i=1}^{\infty} |b_i|^p\right)^{\frac{1}{p}}.$$

9.4 Completeness of the L^p-spaces

Theorem 9.5 *The spaces $L^p(E)$, $1 \leq p \leq \infty$ are complete.*

Proof Let $p = \infty$. Let $\{f_n\}$ be a Cauchy sequence in $L^\infty(E)$. For each pair of members f_m, f_n of $\{f_n\}$ let $Z_{m,n}$ denote the set of measure zero outside which $|f_m(x) - f_n(x)| \leq \|f_m - f_n\|_\infty$ holds. Let $Z = \bigcup_{m=1}^{\infty} \bigcup_{n=1}^{\infty} Z_{m,n}$. Then Z is of measure zero and $|f_m(x) - f_n(x)| \leq \|f_m - f_n\|_\infty$ outside Z for each pair m, n. Let $\epsilon > 0$ be arbitrary. Since $\{f_n\}$ is Cauchy in $L^\infty(E)$ there is N such that $\|f_m - f_n\|_\infty < \frac{\epsilon}{2}$ for $m, n \geq N$ which gives

$$|f_m(x) - f_n(x)| < \frac{\epsilon}{2} \text{ for all } m, n \geq N \text{ and all } x \notin Z \quad (9.4)$$

So, $\{f_n(x)\}$ converges uniformly outside Z. Let $f(x)$ be the limit of $\{f_n(x)\}$. Then letting $m \to \infty$ and keeping n fixed in (9.4)

$$|f_n(x) - f(x)| \leq \frac{\epsilon}{2} \text{ for } n \geq N \text{ and all } x \notin Z. \quad (9.5)$$

Since Z is of measure zero, $f_n - f \in L^\infty(E)$ for all $n \geq N$. Since $f_n \in L^\infty(E)$, $f \in L^\infty(E)$. Also from (9.5) $\|f_n - f\|_\infty \leq \frac{\epsilon}{2} < \epsilon$ for all $n \geq N$ and hence $\{f_n\}$ converges to f in the L^∞-mean

Let $1 \leq p < \infty$. We first prove that if $\{f_n\}$ is a sequence in $L^p(E)$ and $\sum_{n=1}^{\infty} \|f_n\|_p \leq M < \infty$ for some M then there is $S \in L^p(E)$ such that $\sum_{n=1}^{\infty} f_n$ converges to S in the L^p- mean.

Let $g_n(x) = \sum_{k=1}^{n} |f_k(x)|$. Then by Minkowski's inequality

9.4 Completeness of the L^p-spaces

$$\|g_n\|_p \le \sum_{k=p}^{n} \|f_k\|_p \le M \text{ i.e. } \int_E g_n^p d\mu \le M^p. \tag{9.6}$$

Since $\{g_n(x)\}$ is a non-decreasing sequence of non-negative extended real-valued functions on E, it converges to a function g on E. This being true for the sequence $\{[g_n(x)]^p\}$, by Theorem 5.28 (Monotone convergence theorem) and (9.6) above

$$\int_E g^p d\mu \le M^p \tag{9.7}$$

So, g^p is integrable on E and therefore g is finite $a.e.$ on E. Hence $\sum_{k=1}^{\infty} f_k(x)$ converges $a.e.$ on E. to a function say $S(x)$. Writing $S_n(x) = \sum_{k=1}^{n} f_k(x)$, since $|S_n(x)| \le g_n(x) \le g(x)$ we have $|S(x)| \le g(x)$ $a.e.$ on E. Hence by (9.7) $S \in L^p(E)$. Also $|S_n(x) - S(x)| \le 2g(x)$ $a.e.$ on E and hence $|S_n(x) - S(x)|^p \le 2^p[g(x)]^p$ $a.e.$ on E. Since by (9.7) g^p is integrable on E and $|S_n(x) - S(x)|^p$ converges to 0 $a.e.$ on E, by Theorem 5.33 (Lebesgue dominated convergence theorem)

$$\int_E |S_n - S|^p \to 0 \text{ as } n \to \infty$$

and so $\sum_{k=1}^{\infty} f_k$ has the sum S in L^p-mean which proves our assertion.

Now let $\{f_n\}$ be a Cauchy sequence in $L^p(E)$. So, for each k there is n_k such that

$$\|f_n - f_m\|_p < \frac{1}{2^k} \text{ whenever } n, m \ge n_k.$$

Choose $n_1 < n_2 < \cdots < n_k < ..$ and define $g_1 = f_{n_1}$ and $g_k = f_{n_k} - f_{n_{k-1}}$ for $k > 1$. Then $\|g_k\|_p \le 2^{-k+1}$ for $k > 1$. Hence

$$\sum_{k=1}^{\infty} \|g_k\|_p \le \|g_1\|_p + \sum_{k=2}^{\infty} 2^{-k+1} = \|g_1\|_p + 1.$$

So by the above assertion, there is $f \in L^p(E)$ such that $\sum_{k=1}^{\infty} g_k$ converges to $f \in L^p(E)$ in the L^p-mean. But the kth partial sum of $\sum g_k$ is f_{n_k} and hence $\{f_{n_k}\}$ converges to f in the L^p mean. Let $\epsilon > 0$ be arbitrary. Then there is k_0 such that

$$\|f_{n_k} - f\|_p < \frac{\epsilon}{2} \text{ for } k \ge k_0.$$

Since $\{f_n\}$ is a Cauchy sequence there is N such that $\|f_n - f_m\|_p < \frac{\epsilon}{2}$ whenever $n, m \geq N$. Let k be such that $k > k_0$ and $n_k > N$. Then

$$\|f_n - f\|_p \leq \|f_n - f_{n_k}\|_p + \|f_{n_k} - f\|_p < \frac{\epsilon}{2} + \frac{\epsilon}{2} = \epsilon \text{ whenever } n > N.$$

So, $\{f_n\}$ converges to f in the L^p-mean. $\square$

Remark A complete normed linear space is called a Banach space. So, if the functions in $L^p(E)$ are considered to be equal if they differ only on a set of measure zero, then the spaces $L^p(E)$, $1 \leq p \leq \infty$ are Banach spaces.

Theorem 9.6 *If E is a measurable set of finite measure and $0 < p < q \leq \infty$ then $L^q(E) \subset L^p(E)$ and*

$$\|f\|_p \leq \|f\|_q (\mu(E))^{\frac{1}{p}-\frac{1}{q}}$$

for all $f \in L_q(E)$.

Proof The first part is proved in Theorem 9.1. Here we give a complete proof. Let $q < \infty$.

Let $r = \frac{q}{p}$. Then $r > 1$. Let $f \in L_q(E)$. Then

$$\int_E |f|^{pr} d\mu = \int_E |f|^q d\mu < \infty.$$

So,

$$\left(\int_E |f|^{pr} d\mu\right)^{\frac{1}{r}} = \left(\int_E |f|^q d\mu\right)^{\frac{1}{r}} < \infty.$$

Hence $|f|^p \in L_r(E)$. Let $\frac{1}{r} + \frac{1}{s} = 1$. Then since $1 \in L_s(E)$ we have by Holder's inequality

$$\int_E |f|^p d\mu = \int_E |f|^p 1 d\mu \leq \left(\int_E |f|^{pr} d\mu\right)^{\frac{1}{r}} \left(\int_E 1^s d\mu\right)^{\frac{1}{s}}$$

$$= \left(\int_E |f|^q d\mu\right)^{\frac{p}{q}} (\mu(E))^{\frac{q-p}{q}} < \infty$$

Hence $f \in L_p(E)$ and

$$\|f\|_p \leq \|f\|_q (\mu(E))^{\frac{1}{p}-\frac{1}{q}}.$$

Let $q = \infty$. If $f \in L^\infty(E)$ then f is essentially bounded. Let $M = \text{ess sup}\,|f|$. Then

$$\left(\int_E |f|^p d\mu\right)^{\frac{1}{p}} \leq \left(\int_E M^p d\mu\right)^{\frac{1}{p}} = M(\mu(E))^{\frac{1}{p}} < \infty$$

and so $f \in L^p(E)$. Also since $\|f\|_\infty = \text{ess sup}|f| = M$, this gives $\|f\|_p = (\int_E |f|^p d\mu)^{\frac{1}{p}} \leq \|f\|_\infty (\mu(E))^{\frac{1}{p}}$, completing the proof. $\square$

9.5 Other Properties of L^p-spaces

All properties of L^p-spaces proved above are considered in arbitrary measure space $(X, \mathcal{A}, \mu)$. There are other properties for which X needs to be topological space and the proofs are not simple. So, we consider in this Section the Lebesgue measure space $(\mathbb{R}, \mathcal{M}, \mu)$. We shall use the following well-known theorem.

Theorem 9.7 *If f is continuous in a closed interval $[a, b]$ then for every $\epsilon > 0$ there is a polynomial P such that*

$$|f(x) - P(x)| < \epsilon \text{ for all } x \in [a, b].$$

For proof see any book of Mathematical Analysis.

Theorem 9.8 *Let $\mu(E) < \infty$. Then the class of continuous functions is everywhere dense in $L^p(E)$ for $1 \leq p < \infty$.*

Proof Let $f \in L^p(E)$ and let $\epsilon > 0$ be arbitrary. Since $|f|^p$ is integrable on E, by Theorem 5.25 (vii) there exists $\delta > 0$ such that

$$\int_e |f|^p < (\frac{\epsilon}{2})^p \text{ for every measurable set } e \subset E \text{ for which } \mu(e) < \delta... \quad (9.8)$$

By Luzins theorem *i.e* Theorem 3.24 there is a closed set $F \subset E$ such that $\mu(E \sim F) < \delta$ and the restriction f/F of f on F is continuous. Let $g(x) = f(x)$ for $x \in F$ and $g(x) = 0$ for $x \in E \sim F$. Then g is measurable and bounded. Since $\mu(E \sim F) < \delta$ by (9.8)

$$\|f - g\|_p = \left(\int_E |f - g|^p\right)^{\frac{1}{p}} = \left(\int_{E \sim F} |f|^p\right)^{\frac{1}{p}} < \frac{\epsilon}{2}. \quad (9.9)$$

Let $k = \sup |g(x)|$. By Luzin's theorem there is a closed set $H \subset E$ such that $\mu(E \sim H) < (\frac{\epsilon}{4k})^p$ and g/H is continuous. Since H is closed its complement $\mathbb{R} \sim H$ is open and so it is a countable union of disjoint open intervals. We shall define a continuous function ϕ on $\mathbb{R}$. Let $\phi(x) = g(x)$ for $x \in H$. To define ϕ on $\mathbb{R} \sim H$ we consider the following cases: (i) H is bounded, (ii) inf $H = -\infty$, sup $H < \infty$, (iii) sup $H = \infty$, inf $H > -\infty$, (iv) inf $H = -\infty$, sup $H = \infty$. For (i) let $m = \inf H$, $M = \sup H$. Then $m, M \in H$. Define $\phi(x) = g(m)$ for $x \in (-\infty, m)$ and $\phi(x) = g(M)$ for $x \in (M, \infty)$. Since $(m, M) \sim H$ is open, $(m, M) \sim H = \cup(a_i, b_i)$ where (a_i, b_i) are disjoint open intervals. Define

$$\phi(x) = \frac{g(b_i) - g(a_i)}{b_i - a_i}(x - a_i) + g(a_i) \text{ for } x \in (a_i, b_i). \tag{9.10}$$

So, ϕ is defined everywhere and is continuous. For (ii) let $M = \sup H$ and define $\phi(x) = g(M)$ for $x \in (M, \infty)$. Since $(-\infty, M) \sim H$ is open, $(-\infty, M) \sim H = \cup(a_i, b_i)$ where (a_i, b_i) are disjoint open intervals and define ϕ as in (9.10). The case (iii) is similar to (ii). For (iv) $\mathbb{R} \sim H = \cup(a_i, b_i)$ and define ϕ as in (9.10). Thus in any case ϕ is a continuous function on $\mathbb{R}$ and $\phi = g$ on H, and from the definition of ϕ it follows that $|\phi(x)| \leq k$ for all $x \in \mathbb{R}$. Also taking the restriction ϕ_E of ϕ on E

$$\|g - \phi_E\|_p = \left(\int_E |g - \phi_E|^p\right)^{\frac{1}{p}} = \left(\int_{E \sim H} |g - \phi_E|^p\right)^{\frac{1}{p}}$$
$$\leq \{(2k)^p \mu(E \sim H)\}^{\frac{1}{p}} = 2k\frac{\epsilon}{4k} = \frac{\epsilon}{2}. \tag{9.11}$$

From (9.9) and (9.11) $\|f - \phi_E\|_p < \epsilon$, completing the proof. $\square$

Theorem 9.9 *Let E be bounded and measurable. Then the class of step functions is everywhere dense in $L^p(E)$, $1 \leq p < \infty$.*

Proof Let $f \in L^p(E)$ and let $\epsilon > 0$ be arbitrary. Then by Theorem 9.8 there is a continuous function ϕ on $\mathbb{R}$ such that $\|f - \phi_E\|_p < \frac{\epsilon}{2}$. Let $a = \inf E$, $b = \sup E$. Then ϕ is continuous in the closed interval $[a, b]$. So, we can divide the interval $[a, b]$ by points $a = c_0 < c_1 < \cdots < c_n = b$ such that the oscillation ϕ in each sub-interval $[c_k, c_{k+1}]$ is less that $\frac{\epsilon}{2}(\frac{1}{b-a})^{\frac{1}{p}}$. Define

$$S(x) = \phi(c_k) \text{ if } c_k \leq x < c_{k+1}, k = 0, 1, 2, ..., n-2$$
$$= \phi(c_{n-1}) \text{ if } c_{n-1} \leq x \leq b.$$

Then S is a step function on $[a, b]$. Also $|S(x) - \phi(x)| < \frac{\epsilon}{2}\left(\frac{1}{b-a}\right)^{\frac{1}{p}}$ everywhere in $[a, b]$. Therefore if S_E and ϕ_E are the restrictions of S and ϕ on E

$$\|S_E - \phi_E\|_p = \left(\int_E |S_E - \phi_E|^p\right)^{\frac{1}{p}} \leq \left(\left(\frac{\epsilon}{2}\right)^p \frac{1}{b-a}\mu(E)\right)^{\frac{1}{p}} \leq \frac{\epsilon}{2}.$$

Hence
$$\|f - S_E\|_p \leq \|f - \phi_E\|_p + \|\phi_E - S_E\|_p < \epsilon,$$

completing the proof. $\square$

Theorem 9.10 *Let E be bounded and measurable. Then the class of polynomials is everywhere danse in $L^p(E)$, $1 \leq p < \infty$.*

Proof Let $f \in L^p(E)$ and let $\epsilon > 0$ be arbitrary. Then by Theorem 9.8 there is a continuous function ϕ on $\mathbb{R}$ such that $\|f - \phi_E\|_p < \frac{\epsilon}{2}$. Let $a = \inf E$, $b = \sup E$. Then by Theorem 9.7 there is a polynomial P such that

$$|\phi(x) - P(x)| < \frac{\epsilon}{2}(\frac{1}{b-a})^{\frac{1}{p}} \text{ for all } x \in [a,b], \text{ Hence}$$

$$\|\phi_E - P_E\|_p = (\int_E |\phi_E - P_E|^p)^{\frac{1}{p}} \leq \frac{\epsilon}{2}. \text{ Hence}$$

$$\|f - P_E\|_p \leq \|f - \phi_E\|_p + \|\phi_E - P_E\|_p < \epsilon,$$

completing the proof. □

Theorem 9.11 *Let E be bounded and measurable. Then the space $L^p(E)$ is separable for $1 \leq p < \infty$. That is $L^p(E)$ has a countable dense subset.*

Proof Let $f \in L^p(E)$ and let $\epsilon > 0$ be arbitrary. By Theorem 9.10 there exists a polynomial P such that its restriction P_E on E satisfies $\|f - P_E\|_p < \frac{\epsilon}{2}$. Let $P(x) = \sum_{k=0}^{n} a_k x^k$ and let $a = \inf E$, $b = \sup E$. Then for $x \in [a,b]$, $|x| \leq |a| + |b|$. Let $\sum_{k=0}^{n}(|a| + |b|)^k = A$. Choose rational numbers b_k, $k = 0, 1, ..., n$, such that $|a_k - b_k| < \frac{\epsilon}{2A}(\frac{1}{\mu(E)})^{\frac{1}{p}}$. Then the polynomial Q where $Q(x) = \sum_{k=0}^{n} b_k x^k$ is such that for all $x \in [a,b]$

$$|P(x) - Q(x)| \leq \sum_{k=0}^{n} |a_k - b_k| |x^k| \leq \frac{\epsilon}{2A}\left(\frac{1}{\mu(E)}\right)^{\frac{1}{p}} \sum_{k=0}^{n} |x|^k$$

$$\leq \frac{\epsilon}{2A}\left(\frac{1}{\mu(E)}\right)^{\frac{1}{p}} \sum_{k=0}^{n}(|a| + |b|)^k = \frac{\epsilon}{2}\left(\frac{1}{\mu(E)}\right)^{\frac{1}{p}}.$$

So,

$$\|P_E - Q_E\|_p = (\int_E |P_E - Q_E|^p)^{\frac{1}{p}} \leq \frac{\epsilon}{2}$$

Hence

$$\|f - Q_E\|_p \leq \|f - P_E\|_p + \|P_E - Q_E\|_p < \epsilon.$$

So the class of all polynomials whose coefficients are rational numbers is everywhere dense in $L^p(E)$. The theorem will be proved if we can show that the class of all polynomials whose coefficients are rational numbers is countable. To do this let T be the set of all rational numbers and let for fixed positive integer n, $S = T^n = T \times T \times ...n$ times. Then S is the set of all points $(t_1, t_2, ..., t_n)$ where $t_r \in T$ for $r = 1, 2, ...n$. Since T is countable, its cardinal number is $\underline{a}$ and so the cardinal number of S is $\underline{a}^n = \underline{a}$ (see Theorem 4.9) and so S is countable. Let $\mathbb{P}^n$ be the class of all polynomials of degree n and whose coefficients are rational numbers. Then $\mathbb{P}^n$ and S are in one-one correspondence and so the cardinal number of $\mathbb{P}^n$ is $\underline{a}$. If $\mathbb{P}$ is the set of all polynomials whose coefficients are rational numbers then $\mathbb{P} = \mathbb{P}^o \cup \bigcup_{n=1}^{\infty} \mathbb{P}^n$ where $\mathbb{P}^o$ is the class of all polynomials with rational coefficients of degree zero *i.e* T. Hence the cardinal number of $\mathbb{P}$ is $\underline{a} + \underline{a} + \underline{a} + \ldots = \underline{a}$ (see Theorem 4.9) and so $\mathbb{P}$ is countable. This completes the proof. □

Example 9.1 Let $f(x) = e^{\frac{1}{x}}$ for $x \in (0, 1)$. Then $f \notin L^p((0, 1))$ for any $p, 0 < p \leq \infty$.

Let $0 < p < \infty$. Since $(e^{\frac{1}{x}})^p = e^{\frac{p}{x}} > \frac{p}{x}$ for all $x > 0$, $\int_\epsilon^1 (e^{\frac{1}{x}})^p > p \int_\epsilon^1 \frac{1}{x} \to \infty$ as $\epsilon \to 0$.

So f^p is not improper Riemann integrable and hence by Theorem 5.39 $f \notin L^p(0, 1)$. If $p = \infty$, then since f is not essentially bounded $f \notin L^\infty(0, 1)$.

Example 9.2 Let $f(x) = x^{-\frac{1}{2}}(1 + |\log x|)^{-1}$ for $x \in (0, \infty)$. Then $f \in L^p(0, \infty)$ for $p = 2$ but $f \notin L^p(0, \infty)$ for $p \neq 2$.

Let $p = 2$. Then $(f(x))^p = x^{-1}(1 + |\log x|)^{-2}$ and so for $0 < \epsilon < 1$

$$\int_\epsilon^1 f^p = \int_\epsilon^1 x^{-1}(1 + |\log x|)^{-2} = \int_\epsilon^1 \frac{d(\log x)}{(1 - \log x)^2} = \int_{\log \epsilon}^0 \frac{dz}{(1 - z)^2} \to 1 \text{ as } \epsilon \to 0$$

and for $1 < A < \infty$

$$\int_1^A f^p = \int_1^A x^{-1}(1 + |\log x|)^{-2} = \int_1^A \frac{d(\log x)}{(1 + \log x)^2} = \int_0^{\log A} \frac{dz}{(1 + z)^2} \to 1 \text{ as } A \to \infty$$

So, f^p is improper Riemann integrable in $(0, \infty)$. Also f is non-negative and therefore by Theorem 5.38 $f \in L^p(0, \infty)$.

Now let $p < 2$ then $\frac{p}{2} < 1$. Let $g(x) = x^{-1}$ for $x \in (1, \infty)$. Then for

$$x \in [I, \infty), \quad \frac{(f(x))^p}{g(x)} = x^{-\frac{p}{2}}(1 + \log x)^{-p} x$$

$$= x^{1-\frac{p}{2}}(1 + \log x)^{-p} \to \infty \text{ as } x \to \infty$$

Since $\int_0^1 g$ diverges, by limit test $\int_0^1 f^p$ diverges and therefore $\int_0^\infty f^p$ diverges. Hence by Theorem 5.38 $f \notin L^p(0, \infty)$. Again let $p > 2$. Then $\frac{p}{2} > 1$. Let $\alpha > 0$ be such that $\frac{p}{2} - \alpha > 1$. Let $g(x) = x^{-\frac{p}{2}+\alpha}$ for $x \in (0, 1)$. Then for $x \in (0, 1)$

$$\frac{(f(x))^p}{g(x)} = x^{-\frac{p}{2}}(1 - \log x)^{-p} \cdot x^{\frac{p}{2}-\alpha}$$

$$= x^{-\alpha}(1 - \log x)^{-p} \to \infty \text{ as } x \to 0$$

Since $\int_0^1 g$ diverges, $\int_0^1 f^p$ diverges and therefore $\int_0^\infty f^p$ diverges. Hence by Theorem 5.38 $f \notin L^p(0, \infty)$.

Remark In theorem 9.1 it is shown that if $0 < p_1 < p_2 \leq \infty$ and if E is of finite measure then $L^{p_2}(E) \subset L^{p_1}(E)$. Example 9.2 above shows that in Theorem 9.1 the finiteness of $\mu(E)$ is necessary.

9.6 Space of Measurable Functions

After considering L^p-spaces with its various properties it is now natural to consider the space of measurable functions. We need the following theorem.

Theorem 9.12 Let $(X, \mathcal{A}, \mu)$ be a measure space and let $E \in \mathcal{A}$, $\mu(E) < \infty$. Let $f, f_n, n = 1, 2, \ldots$ be measurable functions defined on E which are finite a.e. Then the sequence $\{f_n\}$ converges to f in measure on E if and only if

$$\lim_{n \to \infty} \int_E \frac{|f_n - f|}{1 + |f_n - f|} dx = 0 \tag{9.12}$$

Proof Let $\{f_n\}$ converge to f in measure on E. Let $\sigma > 0$ be arbitrary and let $E_n(\sigma) = \{x \in E; |f_n(x) - f(x)| \geq \sigma\}$. Then $\mu(E_n(\sigma)) \to 0$ as $n \to \infty$. So, writing $E_n = E_n(\sigma)$ and $F_n = E \sim E_n$, we have

$$0 \leq \int_E \frac{|f_n - f|}{1 + |f_n - f|} = \int_{E_n} \frac{|f_n - f|}{1 + |f_n - f|} + \int_{F_n} \frac{|f_n - f|}{1 + |f_n - f|} \leq \mu(E_n) + \sigma\mu(F_n)$$

Hence

$$0 \leq \lim_{n \to \infty} \int_E \frac{|f_n - f|}{1 + |f_n - f|} \leq \sigma \lim_{n \to \infty} \mu(F_n) \leq \sigma\mu(E)$$

Since σ is arbitrary, (9.12) holds.

Conversely let (9.12) hold. Let $0 < \sigma < 1$. Then taking E_n as above we have $E_n \subset E$ and so from (9.12)

$$\lim_{n \to \infty} \int_{E_n} \frac{|f_n - f|}{1 + |f_n - f|} = 0. \tag{9.13}$$

Since for points on E_n $|f_n - f| \geq \sigma$, we have $2|f_n - f| > |f_n - f| + \sigma|f_n - f| \geq \sigma + \sigma|f_n - f| = \sigma(1 + |f_n - f|)$ which gives

$$\int_{E_n} \frac{|f_n - f|}{1 + |f_n - f|} \geq \frac{\sigma}{2} \int_{E_n} 1 = \frac{\sigma}{2} \mu(E_n).$$

So, by (9.13) $\lim_{n \to \infty} \frac{\sigma}{2} \mu(E_n) = 0$. Since $\sigma \neq 0$, the result follows. □

Theorem 9.13 Let $(X, \mathcal{A}, \mu)$ be a measure space and let $E \in \mathcal{A}$. Let $\mathcal{F}$ be the family of all $\mathcal{A}$ - measurable functions defined on E. Define

$$\rho(f, g) = \int_E \frac{|f - g|}{1 + |f - g|} d\mu \text{ for } f, g \in \mathcal{F}.$$

Then ρ is a metric on $\mathcal{F}$, if we agree to consider the elements f and g of $\mathcal{F}$ be equal if $f = g$ a.e. on E. Hence $(\mathcal{F}, \rho)$ is a metric space.

Proof Clearly (i) and (ii) of the property of a metric space (see Definition 9.1) are satisfied, To prove (iii), note that if a, b and c are non-negative real numbers and $a + b \geq c$ then

$$\frac{a}{1+a} + \frac{b}{1+b} \geq \frac{c}{1+c}.$$

Now let $f, g, h \in \mathcal{F}$. Then putting $a = |f - g|$, $b = |g - h|$ and $c = |f - h|$, since $a + b = |f - g| + |g - h| \geq |f - h|$ we have from above

$$\frac{|f - g|}{1 + |f - g|} + \frac{|g - h|}{1 + |g - h|} \geq \frac{|f - h|}{1 + |f - h|}.$$

So, integrating we have $\rho(f, g) + \rho(g, h) \geq \rho(f, h)$, completing the proof. □

Theorem 9.14 *The metric space $(\mathcal{F}, \rho)$ in Theorem 9.13 is complete.*

Proof Let $\{f_n\}$ be any Cauchy sequence in $(\mathcal{F}, \rho)$ and let $\epsilon > 0$ be arbitrary. Then there is N such that $\rho(f_n, f_m) < \epsilon$ whenever $n, m \geq N$. So

$$\int_E \frac{|f_n - f_m|}{1 + |f_n - f_m|} d\mu < \epsilon \text{ whenever } n, m \geq N. \tag{9.14}$$

Let $0 < \sigma < 1$. Then writing $E_{n,m} = \{x \in E; |f_n(x) - f_m(x)| \geq \sigma\}$ we have from (9.14)

$$\int_{E_{n,m}} \frac{|f_n - f_m|}{1 + |f_n - f_m|} d\mu < \epsilon \text{ whenever } n, m \geq N. \tag{9.15}$$

For points on $E_{n,m}$, $|f_n - f_m| \geq \sigma$ and so $2|f_n - f_m| \geq |f_n - f_m| + \sigma|f_n - f_m| \geq \sigma + \sigma|f_n - f_m| = \sigma(1 + |f_n - f_m|)$ and hence

$$\int_{E_{n,m}} \frac{|f_n - f_m|}{1 + |f_n - f_m|} d\mu \geq \frac{\sigma}{2} \mu(E_{nm}).$$

So, from (9.15)

$$\mu(E_{nm}) < \frac{2}{\sigma} \epsilon \text{ whenever } n, m \geq N.$$

So, $\{f_n\}$ is a Cauchy sequence in measure (see Definition 3.9). Therefore by Theorem 3.34 $\{f_n\}$ converges in measure. Let f be the function to which $\{f_n\}$ converges in measure. Clearly $f \in \mathcal{F}$. So, by Theorem 9.12

$$\lim_{n \to \infty} \int_E \frac{|f_n - f|}{1 + |f_n - f|} d\mu = 0 \text{ i.e } \lim_{n \to \infty} \rho(f_n, f) = 0.$$

So, $(\mathcal{F}, \rho)$ is complete. □

9.7 Spaces of Continuous Functions and Riemann Integrable Functions

We shall consider the space of continuous functions and the space of Riemann integrable functions with metric similar to L_1-metric and show that these spaces are not complete with respect to this metric.

Let $[a, b]$ be any finite interval and let $C[a, b]$ and $\mathcal{R}[a, b]$ be respectively the classes of all continuous functions and Riemann integrable functions on $[a, b]$. Define

$$\rho_1(f, g) = \int_a^b |f - g| dx \text{ for } f, g \in C[a, b], \text{ and}$$

$$\rho_2(f, g) = \int_a^b |f - g| dx \text{ for } f, g \in \mathcal{R}[a, b].$$

Then it can be verified that ρ_1 satisfies the property (i), (ii) and (iii) of a metric on $C[a, b]$ and so $(C[a, b], \rho_1)$ is a metric space and ρ_2 satisfies (ii) and (iii) but for (i) $\rho_2(f, g) = 0$ may not imply $f = g$ and so ρ_2 is not a metric but it is what is called a pseudo metric and hence $(\mathcal{R}[a, b], \rho_2)$ is a pseudo metric space. We show that none of these spaces is complete. We first consider $(C[a, b], \rho_1)$. We take $C[0, 2]$. For each n let

$$f_n(x) = 0 \text{ if } 0 \le x \le \frac{1}{2}$$
$$= n(x - \frac{1}{2}) \text{ if } \frac{1}{2} \le x \le \frac{1}{2} + \frac{1}{n}$$
$$= 1 \text{ if } \frac{1}{2} + \frac{1}{n} \le x \le 2.$$

Then f_n is continuous in $[0, 2]$ and so $f_n \in C[0, 2]$ for each n. To prove that $\{f_n\}$ is a Cauchy sequence let $\epsilon > 0$ be arbitrary. Choose $n > m > \frac{2}{\epsilon}$. Then $\frac{1}{2} + \frac{1}{n} < \frac{1}{2} + \frac{1}{m}$ and hence $f_n = f_m$ in $[0, \frac{1}{2}] \cup [\frac{1}{2} + \frac{1}{m}, 2]$ and $|f_m - f_n| \le 2$ for all $x \in [0, 2]$. So

$$\int_0^2 |f_n - f_m| dx = \int_{\frac{1}{2}}^{\frac{1}{2} + \frac{1}{m}} |f_n - f_m| dx \le 2\frac{1}{m} < \epsilon$$

and so $\{f_n\}$ is a Cauchy sequence in $C[0, 2]$. If possible suppose that there is a continuous function f on $[0, 2]$ such that $\rho_1(f_n, f) \to 0$ as $n \to \infty$. Then

$$\int_0^{\frac{1}{2}} |f| dx = \int_0^{\frac{1}{2}} |f_n - f| dx \le \int_0^2 |f_n - f| dx = \rho_1(f_n, f) \to 0 \text{ as } n \to \infty$$

and so $\int_0^{\frac{1}{2}} |f| dx = 0$. Since f is continuous, $f = 0$ in $[0, \frac{1}{2}]$. Let $\frac{1}{2} < \alpha \le 2$. Choose n_0 such that $\frac{1}{2} + \frac{1}{n_0} < \alpha$. Then for $n \ge n_0$

$$\int_\alpha^2 |1-f|dx = \int_\alpha^2 |f_n - f|dx \leq \int_0^2 |f_n - f|dx = \rho_1(f_n, f) \to 0 \text{ as } n \to \infty$$

and so $\int_\alpha^2 |1-f|dx = 0$. Since $1-f$ is continuous, $f = 1$ in $[\alpha, 2]$. Since α is arbitrary, $f = 1$ in $(\frac{1}{2}, 2]$. So, f is not continuous in $[0, 2]$, which is a contradiction. Thus $C[0, 2]$ is not complete.

Now we show that $\mathcal{R}[a, b]$ is not complete. We take $\mathcal{R}[0, 1]$. Let C_α be the Cantor set in $[0, 1]$ of measure $1 - \alpha$, $0 < \alpha < 1$ (see Sect. 4.4). If $\{I_k^n : k = 1, 2, \ldots, 2^{n-1}\}$ is the collection of all open intervals removed from $[0, 1]$ at the nth step then $\sum_{n=1}^\infty \sum_{k=1}^{2^{n-1}} |I_k^n| = \alpha$. Define for each positive integer r, $f_r(x) = 1$ for $x \in \bigcup_{n=1}^r \bigcup_{k=1}^{2^{n-1}} I_k^n$ and $f_r(x) = 0$ for $x \in [0, 1] \sim \bigcup_{n=1}^r \bigcup_{k=1}^{2^{n-1}} I_k^n$. Let $\epsilon > 0$ be arbitrary. Since $\sum_{n=1}^\infty \sum_{k=1}^{2^{n-1}} |I_k^n| = \alpha$, there is N such that $\sum_{n=m}^\infty \sum_{k=1}^{2^{n-1}} |I_k^n| < \epsilon$ for $m \geq N$. Hence $(\mathcal{R}) \int_0^1 |f_{m+p} - f_m|dx = \sum_{n=m+1}^{m+p} \sum_{k=1}^{2^{n-1}} |I_k^n| < \epsilon$ for $m \geq N$ and p is any positive integer. So $\{f_r\}$ is a Candly sequence in $\mathcal{R}[0, 1]$. Suppose that there is $f \in \mathcal{R}[0, 1]$ such that $\rho_2(f_r, f) \to 0$ as $r \to \infty$. We shall get a contradiction. Since $\{f_r\}$ is a Cauchy sequence in $\mathcal{R}[0, 1]$, $\{f_r\}$ is also a Cauchy sequence in $L[0, 1]$ with L^1-metric (where $L[0, 1]$ is the family of all Lebesgue integrable function on $[0, 1]$ and L^1-metric in $L[0, 1]$ is defined as $\|f - g\|_1 = (L) \int_0^1 |f - g|d\mu$) and since $L[0, 1]$ is complete by Theorem 9.5 there is $g \in L[0, 1]$ such that $\|f_r - g\|_1 \to 0$ as $r \to \infty$. Since $\mathcal{R}[0, 1] \subset L[0, 1]$, $f \in L[0, 1]$ and $f_r \in L[0, 1]$ and so

$$(L) \int_0^1 |f - g|d\mu \leq (L) \int_0^1 |f_r - f|d\mu + (L) \int_0^1 |f_r - g|d\mu$$
$$= \rho_2(f_r, f) + \|f_r - g\|_1 \to 0 \text{ as } r \to \infty.$$

So $f = g$ a.e. on $[0, 1]$. Since

$$(L) \int_0^1 \frac{|f_r - g|}{1 + |f_r - g|}d\mu \leq (L) \int_0^1 |f_r - g|d\mu \to 0 \text{ as } r \to \infty,$$

by Theorem 9.12 $\{f_r\}$ converges to g in measure in $[0, 1]$. Hence by Theorem 3.32 there is a subsequence $\{f_{r_i}\}$ of $\{f_r\}$ which converges to g a.e. in $[0, 1]$. But $\{f_r\}$ converges to a function ψ on $[0, 1]$ such that $\psi(x) = 1$ for $x \in [0, 1] \sim C_\alpha$ and $\psi(x) = 0$ for $x \in C_\alpha$. Hence $g = \psi$ a.e. on $[0, 1]$. Since $f = g$ a.e. on $[0, 1]$, $f = 0$ a.e. on C_α and $f = 1$ a.e. on $[0, 1] \sim C_\alpha$. Since C_α is nowhere danse, f is discontinuous a.e. on C_α. Since C_α is of positive measure, f cannot be Riemann integrable which is a contradiction. Thus $\mathcal{R}[0, 1]$ is not complete.

We state this as a theorem.

Theorem 9.15 *The space of continuous functions and the space of Riemann integrable functions on a closed interval $[a, b]$ with metrics similar to L^1-metric are not complete.*

9.8 Convergence Theorems (2)

Convergence theorems (1) are given in Sect. 3.5. While uniform convergence is well known, convergence almost everywhere, almost uniform convergence and convergence in measure are discussed there and examples are given. Here we shall complete the remaining portion.

Theorem 9.16 *Let $\mu(E) < \infty$ and $1 < q \leq \infty$. If $\{f_n\} \subset L^q(E)$ converges to $f \in L^q(E)$ in the L^q-mean then $\{f_n\}$ converges to f in the L^p-mean for all p, $1 \leq p < q$.*

Proof Let $\{f_n\} \subset L^q(E)$ converges to $f \in L^q(E)$ in the L^q-mean and let $1 \leq p < q$. Then $f_n - f \in L^q(E)$ and so by Theorem 9.6 $f_n - f \in L^p(E)$ and

$$\|f_n - f\|_p \leq \|f_n - f\|_q (\mu(E))^{\frac{1}{p}-\frac{1}{q}}.$$

Applying Definition 9.4 the proof is clear. □

Theorem 9.17 *Let $1 \leq p \leq \infty$. If a sequence $\{f_n\} \subset L^p(E)$ converges to $f \in L^p(E)$ in the L^p-mean then $\{f_n\}$ converges to f in measure on E, where $\mu(E) < \infty$*

Proof Let $p = 1$. Since

$$\int_E \frac{|f_n - f|}{1 + |f_n - f|} d\mu \leq \int_E |f_n - f| d\mu = \|f_n - f\|_1,$$

if $\{f_n\} \subset L^1(E)$ converges to $f \in L^1(E)$ in the L^1-mean then $\|f_n - f\|_1 \to 0$ as $n \to \infty$ and so by Theorem 9.12 $\{f_n\}$ converges to f in measure and so the theorem is true for $p = 1$.

Let $1 < p \leq \infty$ and let $\{f_n\} \subset L^p(E)$ converge to $f \in L^p(E)$ in the L^p-mean. Then by Theorem 9.16 $\{f_n\}$ converges to f in the L^1-mean and so by the first part, $\{f_n\}$ converges to f in measure. □

Corollary 9.18 *Let $\mu(E) < \infty$. If a sequence $\{f_n\} \subset L^p(E)$ converges to $f \in L^p(E)$ in the L^p-mean then there is a subsequence of $\{f_n\}$ which converges to f a.e. on E.*

Proof By Theorem 9.17, $\{f_n\}$ converges to f in measure on E. So, by Theorem 3.32 there is a subsequence of $\{f_n\}$ which converges to f a.e. on E. □

Remark Convergence in L^p-mean of a sequence $\{f_n\} \subset L^p(E)$ to $f \in L^p(E)$ does not imply that $\{f_n\}$ converges to f a.e. on E. For, in Theorem 3.31 a sequence $\{f_n\}$ is constructed which converges in measure but $\{f_n\}$ does not converge at any point. It can be verified that the sequence $\{f_n\}$ constructed in Theorem 3.31 also converges to 0 in the L^p-mean for any p, $1 \le p < \infty$.

Theorem 9.19 *Let $\mu(E) < \infty$. If $\{f_n\}$ is a sequence of functions in $L^p(E)$, $1 \le p \le \infty$ which converges uniformly to f on E then $f \in L^p(E)$ and $\{f_n\}$ converges to f in the L^p-mean.*

Proof Let $1 \le p < \infty$. Let $\epsilon > 0$ be arbitrary. Then there is n_ϵ such that

$$|f_n(x) - f(x)| < \epsilon \text{ for } n \ge n_\epsilon \text{ and for all } x \in E. \tag{9.16}$$

Hence

$$\int_E |f_n - f|^p d\mu \le \epsilon^p \mu(E) \text{ for } n \ge n_\epsilon.$$

So, $f_n - f \in L^p(E)$ for $n \ge n_\epsilon$, which shows that $f \in L^p(E)$. Also the above inequality gives $\|f_n - f\|_p \le \epsilon(\mu(E))^{\frac{1}{p}}$ for $n \ge n_\epsilon$ showing that $\{f_n\}$ converges to f in the L^p-mean.

If $p = \infty$ then from (9.16) $\|f_n - f\|_\infty \le \epsilon$ for $n \ge n_\epsilon$ and hence $\{f_n\}$ converges to f in the L^∞-mean. $\square$

Theorem 9.20 *Let $\mu(E) < \infty$. If $\{f_n\}$ is a sequence of measurable functions on E which converges uniformly to f on E and if $f \in L^p(E)$, $1 \le p \le \infty$, then there is N such that $f_n \in L^p(E)$ for all $n \ge N$ and $\{f_n\}_{n \ge N}$ converges to f in the L^p-mean.*

Proof If $\{f_n\}$ converges uniformly to f on E then there is N such that

$$|f_n(x) - f(x)| < 1 \text{ for all } n \ge N \text{ and for all } x \in E. \tag{9.17}$$

So,

$$\int_E |f_n - f|^p d\mu \le \mu(E) \text{ for } n \ge N, \text{ and } 1 \le p < \infty,$$

and hence $f_n - f \in L^p(E)$. Since $f \in L^p(E)$, $f_n \in L^p(E)$ for $n \ge N$, and so by Theorem 9.19 the proof is complete when $1 \le p < \infty$.

If $p = \infty$ then from (9.17) $f_n - f \in L^\infty(E)$ for all $n \ge N$ and since $f \in L^\infty(E)$, $f_n \in L^\infty(E)$ for all $n \ge N$. Let $\epsilon > 0$ be arbitrary. Then there is n_ϵ such that

$$|f_n(x) - f(x)| < \epsilon \text{ for } n \ge n_\epsilon \text{ and for all } x \in E$$

and so $\|f_n - f\|_\infty \le \epsilon$ for all $n \ge \max[n_\epsilon, N]$ and hence $\{f_n\}_{n \ge N}$ converges to f in the L^∞-mean. $\square$

Remarks

(a) The condition $\mu(E) < \infty$ in Theorem 9.19 and Theorem 9.20 cannot be omitted. For, consider the sequence $\{f_n\}$ on $[0, \infty)$ defined by $f_n(x) = n^{-\frac{1}{p}}$ for $0 \le x \le n$ and $f_n(x) = 0$ for $x > n$, where $1 \le p < \infty$. Then $\{f_n\}$ converges uniformly to 0 in $[0, \infty)$ but $\{f_n\}$ does not converge to 0 in the L^p-mean.

(b) Convergence everywhere does not imply convergence in the L^p-mean for $1 \le p \le \infty$.

For consider $\{f_n\}$ where

$$f_n(x) = n^2 \text{ for } 0 < x < \frac{1}{n},$$
$$= 0 \text{ for } \frac{1}{n} \le x \le 1$$
$$= 0 \text{ for } x = 0$$

Then $\{f_n\}$ converges to 0 everywhere in $[0, 1]$. But for $1 \le p < \infty$

$$\int_0^1 |f_n|^p d\mu = \int_0^{\frac{1}{n}} n^{2p} d\mu = n^{2p-1}$$

and so $\{f_n\} \subset L^p([0, 1)]$ but $\{f_n\}$ cannot converge to 0 in the L^p mean. For $p = \infty$, $\|f_n\|_\infty = n^2$ and so this is clear.

(c) Convergence in the L^p-mean does not imply convergence $a.e.$
This is discussed in a remark under Corollary 9.18.

(d) Convergence in measure does not imply convergence in the L^p-mean. Since convergence everywhere implies convergence in measure, (d) follows from (b).

(e) Almost uniform convergence does not imply convergence in the L^p-mean.

The example in (b) shows that the sequence $\{f_n\}$ converges almost uniformly to f on $[0, 1]$ (see Theorem 3.27) and so the result follows from (b).

For a full discussion see Munroe $[p206 - 208]$

9.9 Exercises

1. Show that the function $f : [0, \infty) \to \mathbb{R}$ defined by

$$f(x) = \frac{(-1)^n}{n} \text{ for } n - 1 \le x < n, n = 1, 2, 3...$$

is improper Riemann integrable but not Lebesgue integrable in $[0, \infty)$.

2. Show that $f(x) = \frac{\cos x}{\sqrt{1+x^3}}$ is Lebesgue integrable in $[0, \infty)$.
3. Examine the Lebesgue integrability of the functions
   (i) $f(x) = e^{-a^2 x^2} \cos bx$ in $[0, \infty)$ (ii) $g(x) = \frac{\sin x}{\sqrt{x}}$ in $[0, \infty)$.
4. Let $f(x) = x^{-\frac{1}{p}} (\log \frac{1}{x})^{-\frac{2}{p}}$ for $x \in (0, 1), 0 < p < \infty$. Examine whether $f \in L^q((0, 1))$ for different values of q, $0 < q < 1$.
5. Examine whether the function f defined by $f(x) = \log \frac{1}{x}, x \in (0, 1)$ is in $L^p((0, 1))$ for $0 < p < \infty$.
6. Let $f(x) = x^{-\frac{1}{p}}, 0 < p \leq \infty$. Show that $f \notin L^p((0, 1))$ but $f \in L^q((0, 1))$ if $0 < q < p$.
7. If $f, g \in L^p(E)$ then prove that $(f^2 + g^2)^{\frac{1}{2}} \in L^p(E)$ for $1 \leq p \leq \infty$. If $f, g \in L^1(E)$ and if $p, q \in (0, 1), p + q = 1$ then show that $|f|^p |g|^q \in L^1(E)$.
8. Show that for each of the five kinds of convergence considered in this Section and in Section 3.5, if a sequence $\{f_n\}$ is convergent then $\{f_n\}$ is a Cauchy sequence.
9. Let $\{f_n\}$ be a sequence of functions in $L^1(E)$, and let $g \in L^\infty(E)$. If $\lim_{n \to \infty} f_n = f$ a.e. in E and if there is $F \in L^1(E)$ such that $|f_n| \leq F$ for all n then show that $fg \in L^1(E)$ and $\{f_n g\}$ converges to fg in the L^1-mean.
10. Let $f_n(x) = n^{\frac{1}{p}} e^{-nx}, p > 0$, for $0 \leq x \leq 1$. Show that $\{f_n\} \in L^p([0, 1])$ and $\{f_n\}$ converges to 0 almost uniformly on $[0, 1]$ but $\{f_n\}$ does not converge in L^p-mean.
11. Let $\{f_n\}$ be a sequence of measurable functions defined on a measurable set E and let $|f_n| \leq g$ where $g \in L^1(E)$. Then proved that if $\{f_n\}$ converges to f a.e. on E then $\{f_n\}$ converges almost uniformly to f.
12. Let $\{f_n\} \in L^2([a, b])$ converge to $f \in L^2([a, b])$ in the L^2-mean prove that

$$\lim_{n \to \infty} \int_a^b f_n^2 = \int_a^b f^2.$$

13. If $f \in L^p(E)$ and $g \in L^q(E)$ when $p > 0, q > 0$, then show that

$$|fg|^\lambda \in L^1(E) \text{ where } \lambda = \frac{p+q}{pq}.$$

Chapter 10
Signed Measure and Complex Measure

10.1 Signed Measure

Signed measure is a generalization of measure by allowing it to take negative values. It can be verified that if μ_1 and μ_2 are measures then $\mu_1 + \mu_2$ is also a measure, but $\mu_1 - \mu_2$ is not necessarily a measure. For $\mu_1 - \mu_2$ may not be non-negative. Moreover if for a set E, $\mu_1(E) = \infty = \mu_2(E)$ then $\mu_1 - \mu_2$ is undefined on E. To overcome this difficulty signed measure is introduced :

Definition 10.1 Let $(X, \mathcal{A})$ be a measurable space and let μ be an extended real-valued set function defined on $\mathcal{A}$ satisfying

(i) $\mu(\emptyset) = 0$
(ii) μ assumes at most one of the values ∞ and $-\infty$.
(iii) For every sequence $\{E_n\}$ of disjoint measurable sets, $\mu(\cup E_n) = \sum \mu(E_n)$, the series on the right being absolutely convergent if $\mu(\cup E_n)$ is finite and properly divergent if $\mu(\cup E_n)$ is infinite.
Then μ is called a signed measure on $\mathcal{A}$.

Clearly every measure is also a signed measure.

Theorem 10.1 *If μ_1 and μ_2 are signed measures on the same measurable space $(X, \mathcal{A})$ at least one of which is finite and if α and β are real numbers then $\alpha \mu_1 + \beta \mu_2$ is also signed measure on $(X, \mathcal{A})$.*

Proof Let $\mu = \alpha \mu_1 + \beta \mu_2$. Clearly $\mu(\emptyset) = 0$ and since at least one of μ_1 and μ_2 is finite, μ assumes at most one of the values ∞ and $-\infty$ and so (i) and (ii) are satisfied. To prove (iii), let $\{E_n\}$ be any sequence of disjoint measurable sets. Since at least one of μ_1 and μ_2 is finite, we suppose that μ_2 is finite. (The proof is similar if μ_1 is finite). Therefore $\mu(\cup E_n)$ is finite if and only if $\mu_1(\cup E_n)$ is finite. Let $\mu(\cup E_n)$ be finite. Then $\mu_1(\cup E_n)$ and $\mu_2(\cup E_n)$ being finite, the series $\sum \mu_1(E_n)$ and $\sum \mu_2(E_n)$ are absolutely convergent. So,

$$\mu(\cup E_n) = \alpha\mu_1(\cup E_n) + \beta\mu_2(\cup E_n) = \alpha\sum\mu_1(E_n) + \beta\sum\mu_2(E_n)$$
$$= \sum(\alpha\mu_1(E_n) + \beta\mu_2(E_n)) = \sum\mu(E_n)$$

and

$$\sum|\mu(E_n)| \leq \sum|\alpha\mu_1(E_n) + \beta\mu_2(E_n)|$$
$$\leq |\alpha|\sum|\mu_1(E_n)| + |\beta|\sum|\mu_2(E_n)|.$$

showing that $\sum\mu(E_n)$ is absolutely convergent, proving (iii) in this case. Next, let $\mu(\cup E_n)$ be infinite. Then $\mu_1(\cup E_n)$ is infinite. So, $\sum\mu_1(E_n)$ is properly divergent. Since $\sum\mu(E_n) = \sum(\alpha\mu_1(E_n) + \beta\mu_2(E_n))$ and since $\sum\beta\mu_2(E_n)$ is absolutely convergent, $\sum\mu(E_n)$ is properly divergent, showing that (iii) is satisfied, so μ is a signed measure. □

Corollary 10.2 *The class of all finite signed measures defined on a measurable space is a linear space.*

10.2 Hahn and Jordan Decomposition

Definition 10.2 Let μ be a signed measure on a measurable space $(X, \mathcal{A})$. Then a set $E \in \mathcal{A}$ is said to be positive set with respect to μ if $\mu(F) \geq 0$ whenever $F \subset E$ and $F \in \mathcal{A}$ and a set $E \in \mathcal{A}$ is said to be negative set with respect to μ if $\mu(F) \leq 0$ whenever $F \subset E$ and $F \in \mathcal{A}$.

Theorem 10.3 *The union of a countable collection of positive sets is a positive set and the union of a countable collection of negative sets is a negative set.*

Proof Let μ be a signed measure on a measurable space $(X, \mathcal{A})$. Let $\{E_n\}$ be a countable collection of positive sets and let $E = \cup E_n$. Let $F \subset E$ and $F \in \mathcal{A}$. Let $F_1 = F \cap E_1$, $F_n = F \cap E_n \cap \widetilde{E}_{n-1} \cap \ldots \cap \widetilde{E}_1$ for $n \geq 2$, where $\widetilde{E}_i$ denotes complement of E_i. Then $F_n \subset E_n$, $F_n \in \mathcal{A}$ and so $\mu(F_n) \geq 0$ for all n. Clearly the sets F_n are disjoint and $F = \cup F_n$ and so $\mu(F) = \sum\mu(F_n) \geq 0$. Hence E is a positive set. For negative sets the proof is similar. □

Theorem 10.4 *Let μ be a signed measure on a measurable space $(X, \mathcal{A})$. Let $E \in \mathcal{A}$ be such that $0 < \mu(E) < \infty$. Then there is a positive set $A \subset E$ such that $\mu(A) > 0$.*

Proof If E is a positive set take $A = E$ and the result follows. Suppose that E is not a positive set. So E contains sets of negative measure. Let n_1 be the smallest positive integer such that there is a measurable set $E_1 \subset E$ with $\mu(E_1) < -\frac{1}{n_1}$. Since $\mu(E) > 0$, $E \sim E_1 \neq \emptyset$. If $E \sim E_1$ is positive set take $A = E \sim E_1$ and the result follows. So suppose $E \sim E_1$ is not a positive set. Hence $E \sim E_1$ contains a set of negative measure. Let n_2 be the smallest positive integer such that there

10.2 Hahn and Jordan Decomposition

is a measurable set $E_2 \subset E \sim E_1$ with $\mu(E_2) < -\frac{1}{n_2}$. Clearly $E \sim (E_1 \cup E_2) \neq \emptyset$. If it is a positive set take $A = E \sim (E_1 \cup E_2)$ and the proof follows. Suppose $E \sim (E_1 \cup E_2)$ contains sets of negative measure. Proceeding inductively, let n_k be the smallest positive integer such that there is a measurable set $E_k \subset E \sim \bigcup_{i=1}^{k-1} E_i$ with $\mu(E_k) < -\frac{1}{n_k}$. Clearly $E_k \neq E \sim \bigcup_{i=1}^{k-1} E_i$. If $E \sim \bigcup_{i=1}^{k} E_i$ is a positive set take $A = E \sim \bigcup_{i=1}^{k} E_i$ and the proof follows. So suppose $E \sim \bigcup_{i=1}^{k} E_i$ is not positive set. If the process does not end, put $A = E \sim \bigcup_{i=1}^{\infty} E_i$. Then $E = A \cup \bigcup_{i=1}^{\infty} E_i$ and since E_i's are disjoint, $\mu(E) = \mu(A) + \sum_{i=1}^{\infty} \mu(E_i)$. Now $\mu(E_i) < 0$, for all i and $\mu(E) > 0$, and so $\mu(A) > 0$. We are to show that A is a positive set. Let ϵ be arbitrary, since $\mu(E_k) < -\frac{1}{n_k}$ for all k, we have $|\mu(E_k)| > \frac{1}{n_k}$ for all k. Since $\mu(E) = \mu(A) + \sum_{k=1}^{\infty} \mu(E_k)$ and since $\mu(E)$ is finite, the series $\sum_{k=1}^{\infty} |\mu(E_k)|$ is convergent, and so $\sum_{k=1}^{\infty} \frac{1}{n_k}$ is convergent. So $n_k \to \infty$ as $k \to \infty$. Hence there is k_0 such that $\frac{1}{n_{k_0}-1} < \epsilon$. Since $A \subset E - \bigcup_{i=1}^{k_0-1} E_i$, A can contain no measurable subsets of measure less than $-\frac{1}{n_{k_0}-1}$ since n_{k_0} is the smallest positive integer having this property. Hence A can contain no measurable subset of measure less then $-\epsilon$. Since ϵ is arbitrary, A can contain no measurable set of negative measure and hence A is a positive set. $\square$

With the help of above two theorems, we prove an important theorem in signed measure.

Theorem 10.5 (Hahn decomposition theorem) *Let μ be a signed measure on a measurable space $(X, \mathcal{A})$. Then there is a positive set A and a negative set B such that $A \cup B = X$, $A \cap B = \emptyset$.*

Proof Suppose that μ does not assume ∞, the proof for $-\infty$ is similar. Let $\lambda = \sup\{\mu(A) : A \text{ is a positive set}\}$. Since $\emptyset$ is positive set $\lambda \geq 0$. Let $\{A_i\}$ be a sequence of positive sets such that $\lambda = \lim_{i \to \infty} \mu(A_i)$. By Theorem 10.3, $A = \bigcup_{i=1}^{\infty} A_i$ is a positive set, so $\lambda \geq \mu(A)$. Since $A - A_i \subset A$, $\mu(A - A_i) \geq 0$ and so

$$\mu(A) = \mu(A_i) + \mu(A - A_i) \geq \mu(A_i) \text{ for all } i.$$

Hence $\mu(A) \geq \lim_{i \to \infty} \mu(A_i) = \lambda$. Hence $\lambda = \mu(A)$ and so $0 \leq \lambda < \infty$. Let $B = X - A$. We claim that B is a negative set. For if possible let $E \subset B$, $E \in \mathcal{A}$ and $\mu(E) > 0$. Then $0 < \mu(E) < \infty$. So by Theorem 10.4 there is a positive set $E_1 \subset E$ such that $\mu(E_1) > 0$. Now for disjoint positive sets A and E_1,

$$\lambda \geq \mu(A \bigcup E_1) = \mu(A) + \mu(E_1) = \lambda + \mu(E_1).$$

So $\mu(E_1) \leq 0$, a contradiction. Hence B is a negative set and $X = A \bigcup B$. □

The decomposition of X into a positive set A and a negative set B is called Hahn decomposition of X with respect to μ and we write (A, B).

Hahn decomposition of X is not unique. For if E be both positive and negative set, that is, a set with μ-measure zero, then $(A \cup E, B - E)$ and $(A - E, B \cup E)$ are also Hahn decomposition. Hence Hahn decomposition is unique up to sets of μ-measure zero.

Definition 10.3 Let μ_1 and μ_2 be any two measures on the same measurable space $(X, \mathcal{A})$, then μ_1 and μ_2 are said to be mutually singular if there are measurable sets A and B such that $A \cup B = X$, $A \cap B = \emptyset$ and $\mu_1(A) = 0 = \mu_2(B)$.

If μ_1 and μ_2 mutually singular then we write $\mu_1 \perp \mu_2$.

Theorem 10.6 *If μ_1 and μ_2 are mutually singular measures on $(X, \mathcal{A})$ then the pair (μ_1, μ_2) determines a Hahn decomposition of X with respect to the signed measure $\mu = \mu_1 - \mu_2$.*

Proof Since μ_1 and μ_2 are mutually singular there are measurable sets A and B such that $A \cup B = X$, $A \cap B = \phi$ and $\mu_1(A) = 0 = \mu_2(B)$. Let $E \subset A$. Since μ_1 and μ_2 are measures and since $\mu_1(A) = 0$, $\mu_1(E) = 0$ and so $\mu(E) = \mu_1(E) - \mu_2(E) = -\mu_2(E) \leq 0$. Since E is arbitrary, A is a negative set with respect to μ. Similarly if $E \subset B$ then $\mu_2(E) = 0$ and so $\mu(E) = \mu_1(E) - \mu_2(E) = \mu_1(E) \geq 0$ and hence B is a positive set with respect to μ. So, (B, A) is a Hahn decomposition of X with respect to μ. □

In Theorem 10.1, it is seen that if μ_1 and μ_2 are two measures then $\mu = \mu_1 - \mu_2$ is a signed measure. Now the question is : do all signed measures arise in this way? The next theorem answers this question.

Theorem 10.7 (Jordan decomposition theorem) *Let μ be a signed measure on a measurable space $(X, \mathcal{A})$. Then there are two mutually singular measures μ^+ and μ^- on $(X, \mathcal{A})$ at least one of which is finite such that $\mu = \mu^+ - \mu^-$.*

Proof Let (A, B) be a Hahn decomposition with respect to μ. For $E \in \mathcal{A}$ define $\mu^+(E) = \mu(A \cap E)$ and $\mu^-(E) = -\mu(B \cap E)$. It is easy to verify that μ^+ and μ^- are measures. Clearly $\mu^+(B) = \mu(A \cap B) = 0$ and $\mu^-(A) = -\mu(B \cap A) = 0$. Since $A \cup B = X$ and $A \cap B = \emptyset$, $\mu^+ \perp \mu^-$. If possible suppose μ^+ and μ^- be both infinite, then

$$\infty = \mu^+(X) = \mu^+(A) + \mu^+(B) = \mu^+(A) = \mu(A) \text{ and}$$

$$\infty = \mu^-(X) = \mu^-(A) + \mu^-(B) = \mu^-(B) = -\mu(B)$$

which is a contradiction since μ can assume at most one of $+\infty$ and $-\infty$.

10.2 Hahn and Jordan Decomposition

Finally let $E \in \mathcal{A}$, so

$$\mu^+(E) - \mu^-(E) = \mu(A \cap E) + \mu(B \cap E) = \mu(E)$$

Hence $\mu = \mu^+ - \mu^-$. □

The decomposition of μ into two measures μ^+ and μ^- is called Jordan decomposition of μ. It is denoted by (μ^+, μ^-).

Theorem 10.8 *The Jordan decomposition of a signed measure is unique.*

Proof Let μ be a signed measure on a measurable space $(X, \mathcal{A})$. Let (μ_1, μ_2) and (λ_1, λ_2) be two Jordan decomposition of μ. Since $\mu_1 \perp \mu_2$ and $\lambda_1 \perp \lambda_2$ there are measurable sets A, B and C, D such that

$$A \cup B = X, \quad A \cap B = \emptyset, \quad \mu_1(B) = 0 = \mu_2(A) \text{ and}$$

$$C \cup D = X, \quad C \cap D = \emptyset, \quad \lambda_1(D) = 0 = \lambda_2(C).$$

For each $E \in \mathcal{A}$, we have

$$\mu(E \cap A) = \mu_1(E \cap A) - \mu_2(E \cap A) = \mu_1(E \cap A) \geq 0.$$

So A is positive set. Similarly B is negative set. So (A, B) is a Hahn decomposition. Similarly (C, D) is also a Hahn decomposition. Let $E \in \mathcal{A}$, then since

$$\mu_1(E) = \mu_1(A \cap E) + \mu_1(B \cap E)$$
$$= \mu_1(A \cap E)$$
$$= \mu_1(A \cap E) - \mu_2(A \cap E)$$
$$= \mu(A \cap E)$$

we have

$$\mu_1(E) = \mu(A \cap E) = \lambda_1(A \cap E) - \lambda_2(A \cap E) \leq \lambda_1(A \cap E) \leq \lambda_1(E)$$

Similarly

$$\lambda_1(E) = \mu(C \cap E) = \mu_1(C \cap E) - \mu_2(C \cap E) \leq \mu_1(C \cap E) \leq \mu_1(E).$$

Hence $\lambda_1 = \mu_1$. Similarly $\lambda_2 = \mu_2$. □

Definition 10.4 If (μ^+, μ^-) is the Jordan decomposition of μ then μ^+ is called positive part or upper variation and μ^- is called negative part or lower variation of μ. The measure $\mu^+ + \mu^-$ is called the absolute value or total variation of μ and is denoted by $|\mu|$.

Since the sum of two measures is a measure, $|\mu|$ is a measure.

Theorem 10.9 *Let μ be a signed measure on a measurable space $(X, \mathcal{A})$. Then for every $E \in \mathcal{A}$*

$$|\mu|(E) = \sup\left\{\sum_{i=1}^{n}|\mu(E_i)| : E_1, ..., E_n \text{ is a measurable dissection of } E\right\} \quad (10.1)$$

where 'sup' is taken over all measurable dissection of E.

Proof Let S denote the right side of (10.1). Let $E \in \mathcal{A}$ be fixed and let $E_1, E_2, ..E_n$ be any measurable dissection of E. (That is $E_i's$ are disjoint and $E = \bigcup_{i=1}^{n} E_i$). If (μ^+, μ^-) is the Jordan decomposition of μ then since $|\mu|$ is a measure we have

$$\sum_{i=1}^{n}|\mu(E_i)| = \sum_{i=1}^{n}|\mu^+(E_i) - \mu^-(E_i)| \leq \sum_{i=1}^{n}|\mu^+(E_i) + \mu^-(E_i)| = \sum_{i=1}^{n}|\mu|(E_i) = |\mu|(E).$$

Since this is true for all measurable dissection, taking 'sup' we have $S \leq |\mu|(E)$. To prove the reverse inequality, consider the measurable dissection $E \cap A$, $E \cap B$ of E where (A, B) is a Hahn decomposition of X for μ. Since μ^+ and μ^- are measures on $(X, \mathcal{A})$, $\mu^+(E) = \mu^+(E \cap A) + \mu^+(E \cap B) = \mu^+(E \cap A) = \mu(E \cap A)$ and similarly $\mu^-(E) = \mu(E \cap B)$. So,

$$S \geq |\mu(E \cap A)| + |\mu(E \cap B)| = \mu^+(E) + \mu^-(E) = |\mu|(E).$$

So $S = |\mu|(E)$ which completes the proof. $\square$

10.3 Integration with Respect to Signed Measure

Definition 10.5 Let $f : X \to \mathbb{R}$ be a measurable function on a measurable space $(X, \mathcal{A})$ and μ be a signed measure on $\mathcal{A}$. Let (μ^+, μ^-) be the Jordan decomposition of μ. The integral of f on X with respect to μ is defined to be

$$\int_X f d\mu = \int_X f d\mu^+ - \int_X f d\mu^-;$$

provided the right-hand side exists. If $E \in \mathcal{A}$ define $\int_E f d\mu = \int_X f \chi_E d\mu$, χ_E being the characteristic function of E.

The function f is said to be integrable if f is integrable with respect to both of μ^+ and μ^-.

Remark Just as signed measure does not have all the properties of usual measure, integral with respect to a signed measure also does not have all the properties of the

10.3 Integration with Respect to Signed Measure

usual integral. For instance, like usual integral this integral is linear *i.e* if f and g are integrable in X and α and β are real numbers then $\alpha f + \beta g$ is also integrable and

$$\int_X (\alpha f + \beta g) d\mu = \alpha \int_X f d\mu + \beta \int_X g d\mu,$$

but if f and g are integrable in X and $f \geq g$ on X then it is not necessary that $\int_X f d\mu \geq \int_X g d\mu$. For if E is a negative set with respect to μ and if $\mu(E) < 0$ then the characteristic function χ_E of E is such that $\chi_E \geq 0$ everywhere but $\int_E \chi_E d\mu = \mu(E) < 0 = \int_E 0 d\mu$.

Example 10.1 Let $(X, \mathcal{A}, \nu)$ be a measure space and let $f : X \to \mathbb{R}$ be such that $\int_X f d\nu$ exists. Let $\mu(E) = \int_E f d\nu$ for every $E \in \mathcal{A}$. Then

1. μ is a signed measure on $(X, \mathcal{A})$.
2. What are the Hahn and Jordan decomposition of μ.

Solution

1. Since $\int_X f d\nu$ exists either $\int_X f^+ d\nu$ or $\int_X f^- d\nu$ is finite.
   Thus μ cannot assume both ∞ and $-\infty$.
   Obviously $\mu(\emptyset) = 0$. Finally let $\{E_i\}$ be any sequence of disjoint measurable sets. If $\mu(\bigcup E_i)$ is finite then $\int_{\bigcup E_i} f^+ d\nu$ and $\int_{\bigcup E_i} f^- d\nu$ are finite, and hence $\sum \int_{E_i} f^+ d\nu$ and $\sum \int_{E_i} f^- d\nu$ are finite, and so

$$\sum |\mu(E_i)| = \sum |\int_{E_i} f d\nu|$$

$$\leq \sum |\int_{E_i} f^+ d\nu - \int_{E_i} f^- d\nu|$$

$$\leq \sum \int_{E_i} f^+ d\nu + \sum \int_{E_i} f^- d\nu$$

$$< \infty$$

So $\sum \mu(E_i)$ is absolutely convergent and in that case

$$\mu(\bigcup E_i) = \int_{\bigcup E_i} f d\nu = \sum \int_{E_i} f d\nu = \sum \mu(E_i).$$

If $\mu(\bigcup E_i) = \infty$ then $\sum \int_{E_i} f^+ dv$ diverges to ∞ and $\sum \int_{E_i} f^- dv$ is finite. So $\sum \mu(E_i) = \sum \int_{E_i} f^+ dv - \sum \int_{E_i} f^- dv = \infty$. Similar is the case if $\mu(\bigcup E_i) = -\infty$. So, μ is a signed measure on $(X, \mathcal{A})$.

2. Let $A = \{x \in X : f(x) \geq 0\}$ and $B = \{x \in X : f(x) < 0\}$. So $A \bigcup B = X$, $A \bigcap B = \emptyset$. If $E \subset A$, then $\mu(E) = \int f dv \geq 0$. Hence A is a positive set. Similarly B is a negative set. Therefore (A, B) is a Hahn decomposition of X with respect to μ.

Let $\mu^+(E) = \int_E f^+ dv$ and $\mu^-(E) = \int_E f^- dv$. Since f^+ and f^- are non-negative, μ^+ and μ^- are measures on $(X, \mathcal{A})$. Since $f^+ = 0$ on B and $f^- = 0$ on A, we get $\mu^+(B) = 0 = \mu^-(A)$ and so $\mu^+ \perp \mu^-$. Since either $\int_X f^+ dv$ is finite or $\int_X f^- dv$ is finite, one of μ^+ and μ^- must be finite. Finally for any $E \in \mathcal{A}$, we have

$$\mu^+(E) - \mu^-(E) = \int_E f^+ dv - \int_E f^- dv = \int_E f dv = \mu(E).$$

Hence (μ^+, μ^-) is the Jordan decomposition of μ.

Example 10.2 If μ_1 and μ_2 are signed measures on $(X, \mathcal{A})$ then

$$|\mu_1 + \mu_2| \leq |\mu_1| + |\mu_2|$$

This is an easy application of Theorem 10.9.

Example 10.3 Suppose (A, B) is Hahn decomposition of X with respect to a signed measure μ and E be a measurable set. Let $f = \chi_{E \cap A} - \chi_{E \cap B}$. Show that $\int_E f d\mu = |\mu|(E)$.

Solution : Clearly $f^+ = \chi_{E \cap A}$ and $f^- = \chi_{E \cap B}$. If (μ^+, μ^-) is Jordan decomposition of μ then we find

$$\int_E f d\mu = \int_E (f^+ - f^-) d\mu^+ - \int_E (f^+ - f^-) d\mu^-$$

$$= \int_E (\chi_{E \cap A} - \chi_{E \cap B}) d\mu^+ - \int_E (\chi_{E \cap A} - \chi_{E \cap B}) d\mu^-$$

$$= \mu^+(E \cap A) - \mu^+(E \cap B) - \mu^-(E \cap A) + \mu^-(E \cap B)$$

$$= \mu^+(E) - 0 - 0 + \mu^-(E)$$

$$= |\mu|(E).$$

10.3 Integration with Respect to Signed Measure

Theorem 10.10 *Let $(X, \mathcal{A})$ be a measurable space and let μ and ν be measures on $\mathcal{A}$ and let $\lambda = \mu + \nu$. Then λ is a measure on $\mathcal{A}$. If f is integrable on a set $E \in \mathcal{A}$ with respect to μ and ν then f is integrable on E with respect to λ and*

$$\int_E f d\lambda = \int_E d\mu + \int_E f d\nu.$$

Proof The proof that λ is a measure is easy. We prove the second part. Note that the definition of integral for Lebesgue measurable function discussed in Chap. 5 is the same for all measures and so it is applicable to λ, μ and ν also. Let f be integrable on E with respect to μ and ν. First suppose that f is bounded. Consider any measurable partition $\tau = \{e_1, e_2, \ldots, e_n\}$ of E and form the lower sums $s(\tau, f, \lambda)$, $s(\tau, f, \mu)$ and $s(\tau, f, \nu)$ corresponding to λ, μ and ν respectively. Then since $\lambda(e_i) = \mu(e_i) + \nu(e_i)$ for $i = 1, 2, \ldots, n$, we have $s(\tau, f, \lambda) = s(\tau, f, \mu) + s(\tau, f, \nu)$. Since f is integrable with respect to μ and ν using Definition 5.1, $s(\tau, f, \lambda) \leq \int_E f d\mu + \int_E f d\nu$ and hence

$$\underline{\int_E} f d\lambda \leq \int_E f d\mu + \int_E f d\nu. \tag{10.2}$$

Similarly considering the upper sums $S(\tau, f, \lambda)$, $S(\tau, f, \mu)$ and $S(\tau, f, \nu)$ and using Definition 5.1 we get

$$\overline{\int_E} f d\lambda \geq \int_E f d\mu + \int_E f d\nu \tag{10.3}$$

From (10.2) and (10.3), the proof is complete when f is bounded. When f is unbounded and non-negative then considering the truncated function f_n as in Section 5.5 we have from above

$$\int_E f_n d\lambda = \int_E f_n d\mu + \int_E f_n d\nu$$

Since f is integrable with respect to μ and ν, the right-hand side tends to $\int_E f d\mu + \int_E f d\nu$ as $n \to \infty$ and so letting $n \to \infty$, the proof is complete in this case.

In the general case applying Definition 5.5

$$\int_E f d\mu = \int_E f^+ d\mu - \int_E f^- d\mu \text{ and } \int_E f d\nu = \int_E f^+ d\nu - \int_E f^- d\nu$$

Since the theorem is true for non-negative function

$$\int_E f^+ d\lambda = \int_E f^+ d\mu + \int_E f^+ d\nu \text{ and } \int_E f^- d\lambda = \int_E f^- d\mu + \int_E f^- d\nu$$

and so
$$\int_E f^+ d\lambda - \int_E f^- d\lambda = \int_E f d\mu + \int_E f dv$$
and hence
$$\int_E f d\lambda = \int_E f d\mu + \int_E f dv$$
completing the proof. □

Theorem 10.11 *If f is integrable with respect to a signed measure μ on a measurable space $(X, \mathcal{A})$ then $|f|$ is integrable with respect to $|\mu|$ and*
$$\left| \int_E f d\mu \right| \leq \int_E |f| d|\mu| \text{ for every } E \in \mathcal{A}.$$

Proof Since f is integrable with respect to μ, f is integrable with respect to μ^+ and μ^- and so $|f|$ is integrable with respect to μ^+ and μ^- and so by Theorem 10.10 $|f|$ is integrable with respect to $\mu^+ + \mu^- = |\mu|$ and

$$\left| \int_E f d\mu \right| = \left| \int_E f d\mu^+ - \int_E f d\mu^- \right| \leq \left| \int_E f d\mu^+ \right| + \left| \int_E f d\mu^- \right|$$
$$\leq \int_E |f| d\mu^+ + \int_E |f| d\mu^-$$
$$= \int_E |f| d(\mu^+ + \mu^-) = \int_E |f| d|\mu|.$$
□

Theorem 10.12 *If μ and v are measures on a measurable space $(X, \mathcal{A})$ such that $\mu(E) \leq v(E)$ for all $E \in \mathcal{A}$ and if f is a non-negative measurable function on X then*
$$\int_E f d\mu \leq \int_E f dv \text{ for all } E \subset \mathcal{A}.$$

Hence if f is non-negative and integrable with respect to v then f is integrable with respect to μ.

Proof The first part can be proved as in Theorem 10.10 by considering lower sums The second part is clear. □

Theorem 10.13 *If f is integrable with respect to a signed measure μ on a measurable space $(X, \mathcal{A})$ then $|f|$ is integrable with respect to μ. Also if g is integrable with respect to μ and $|f| \leq g$ on X then g is integrable with respect to $|\mu|$ and*
$$\left| \int_E |f| d\mu \right| \leq \int_E g d|\mu| \text{ for all } E \in \mathcal{A}.$$

Proof By Theorem 10.11 $|f|$ is integrable with respect to $|\mu|$. Since $\mu^+ \leq |\mu|$, by Theorem 10.12 $|f|$ is integrable with respect to μ^+. Similarly $|f|$ is integrable with respect to μ^-. So $|f|$ is integrable with respect to μ.

For the second part, by Theorem 10.11, g is integrable with respect to $|\mu|$ and so by Theorem 10.10

$$\left| \int_E |f| d\mu \right| = \left| \int_E |f| d\mu^+ - \int_E |f| d\mu^- \right| \leq \int_E |f| d\mu^+ + \int_E |f| d\mu^-$$
$$= \int_E |f| d(\mu^+ + \mu^-) = \int_E |f| d|\mu| \leq \int_E g \, d|\mu|.$$

completing the proof. □

Example 10.4 If $|f| \leq M$ on a measurable set E where f is integrable with respect to a signed measure μ then $\left| \int_E f \, d\mu \right| \leq M|\mu|(E)$.

Solution Since $|f| \leq M$ on a measurable set E we get

$$\left| \int_E f \, d\mu \right| \leq \int_E |f| \, d\mu^+ + \int_E |f| \, d\mu^- \leq M\{\mu^+(E) + \mu^-(E)\} = M|\mu|(E).$$

10.4 Absolute Continuity of Measures

Definition 10.6 Let μ and ν be signed measures on the measurable space $(X, \mathcal{A})$. Then ν is said to be absolutely continuous with respect to μ if $\nu(E) = 0$ whenever $|\mu|(E) = 0$ for $E \in \mathcal{A}$, and in this case we write $\nu << \mu$.

Theorem 10.14 Let μ and ν be signed measures on the measurable space $(X, \mathcal{A})$ Then the following conditions are equivalent
(i) $\nu << \mu$, (ii) $|\nu| << |\mu|$, (iii) $\nu^+ << \mu$ and $\nu^- << \mu$

Proof From the definition of $\nu << \mu$, it follows that $\nu << \mu$ if and only if $\nu << |\mu|$. So, we may suppose that $\mu \geq 0$. Let (i) hold. Let (A, B) be a Hahn decomposition of X with respect to ν. Suppose $|\mu|(E) = 0$, $E \in \mathcal{A}$. then

$$0 \leq |\mu|(E \cap A) \leq |\mu|(E) = 0 \text{ and } 0 \leq |\mu|(E \cap B) \leq |\mu|(E) = 0$$

So, $\nu^+(E) = \nu(E \cap A) = 0$ and $\nu^-(E) = \nu(E \cap B) = 0$. This proves (ii) and (iii). Suppose (ii) holds and let $\mu(E) = 0$. Then $|\nu|(E) = 0$ and hence (i) holds and since $|\nu|(E) = \nu^+(E) + \nu^-(E)$, (iii) holds. Now suppose (iii) holds and let $\mu(E) = 0$. Then $\nu^+(E) = 0$ and $\nu^-(E) = 0$ and so $\nu^+(E) + \nu^-(E) = |\nu|(E) = 0$. So, (ii) holds. Also if (iii) holds then clearly (i) holds. This completes the proof. □

Theorem 10.15 *Let ν and μ be signed measures on a measurable space $(X, \mathcal{A})$ of which ν is finite. Then $\nu << \mu$ if and only if for every $\epsilon > 0$ there is $\delta > 0$ such that $|\nu|(E) < \epsilon$ for every $E \in \mathcal{A}$ for which $|\mu|(E) < \delta$.*

Proof Let $\nu << \mu$. Suppose, if possible that there exists $\epsilon_0 > 0$ for which the assertion is not true. Then for each n there is $E_n \in \mathcal{A}$ such that $|\mu|(E_n) < \frac{1}{2^n}$ and $|\nu|(E_n) \geq \epsilon_0$. Let $F_k = \bigcup_{n=k}^{\infty} E_n$. Then $\limsup E_n = \bigcap_{k=1}^{\infty} F_k$. So for each k,

$$|\mu|(\limsup E_n) \leq |\mu|(F_k) \leq \sum_{n=k}^{\infty} |\mu|(E_n) \leq \sum_{n=k}^{\infty} \frac{1}{2^n} = \frac{1}{2^{k-1}}.$$

Since this is true for all k, $|\mu|(\limsup E_n) = 0$. But for each k, $|\nu|(F_k) \geq |\nu|(E_k) \geq \epsilon_0$ and hence, since ν is finite, $|\nu|(\limsup E_n) = |\nu|(\lim F_k) = \lim |\nu|(F_k) \geq \epsilon_0$. This is a contradiction since $\nu << \mu$.

Conversely suppose that the condition holds. Let $A \in \mathcal{A}$ and let $|\mu|(A) = 0$. Then, taking $\epsilon = \frac{1}{n}$, we have by the given condition for every n there is $\delta_n > 0$ such that $|\mu|(E) < \delta_n$ implies $|\nu|(E) < \frac{1}{n}$ for every $E \in \mathcal{A}$. Since $|\mu|(A) < \delta_n$ for all n, $|\nu|(A) < \frac{1}{n}$ for all n and hence $|\nu|(A) = 0$, giving $\nu(A) = 0$ which shows that $\nu << \mu$. $\square$

Theorem 10.16 *Let $(X, \mathcal{A}, \mu)$ be a measure space and let $f : X \to \mathbb{R}$ be μ-integrable. Then the set function $\phi(E) = \int_E f d\mu$, $E \in \mathcal{A}$ is a signed measure and ϕ is absolutely continuous with respect to μ.*

Proof The first part is proved in Example 10.1. We prove the second part. If f is bounded then there is $M > 0$ such that $|f| \leq M$ on X and so $|\int_E f d\mu| \leq \int_E |f| d\mu \leq M\mu(E) < \epsilon$ if $\mu(E) < \frac{\epsilon}{M}$ and so putting $\delta = \frac{\epsilon}{M}$ the theorem is proved by Theorem 10.15. Suppose that f is unbounded. Since f is integrable, f^+ and f^- are integrable and so $|f| = f^+ + f^-$ is integrable. Hence there is a sequence $\{f_n\}$ of bounded measurable function on X such that $0 \leq f_n \leq |f|$ and $\lim_{n \to \infty} \int_X f_n d\mu = \int_X |f| d\mu$. Let $\epsilon > 0$ be arbitrary. Then there is f_n such that $|\int_X |f| d\mu - \int_X f_n d\mu| < \frac{\epsilon}{2}$. Since the theorem is true for bounded functions, applying the theorem on f_n, there is $\delta > 0$ such that $\int_E f_n d\mu < \frac{\epsilon}{2}$ whenever $\mu(E) < \delta$ and so

$$\left|\int_E f d\mu\right| \leq \int_E |f| d\mu = \int_E (|f| - f_n) d\mu + \int_E f_n d\mu$$

$$\leq \int_X (|f| - f_n) d\mu + \int_E f_n d\mu < \epsilon.$$

whenever $\mu(E) < \delta$. So, the theorem is proved by Theorem 10.15. $\square$

10.4 Absolute Continuity of Measures

Signed measure possesses some of the properties of measure but not all. Consider the following

Example 10.5 Let $(X, \mathcal{A})$ be a measurable space and let ν be a finite signed measure on $\mathcal{A}$. Then

(i) if $A, B \in \mathcal{A}$ and $B \subset A$ then $\nu(B) \leq \nu(A)$ may not hold
(ii) if $A, B \in \mathcal{A}$ and $B \subset A$ then $\nu(A \sim B) = \nu(A) - \nu(B)$ holds
(iii) if $\{E_n\}$ is an increasing sequence of sets in $\mathcal{A}$ then

$$\nu(\bigcup_{n=1}^{\infty} E_n) = \lim_{n \to \infty} \nu(E_n)$$

(iv) if $\{E_n\}$ is a decreasing sequence of sets in $\mathcal{A}$ and $\nu(E_1)$ is finite then

$$\nu(\cap E_n) = \lim_{n \to \infty} \nu(E_1).$$

Solution

(i) For a measure μ this property holds. For, since $A = B \cup (A \sim B)$, $\mu(A) = \mu(B) + \mu(A \sim B)$ by definition of μ and so since $\mu(A \sim B) \geq 0$, $\mu(B) \leq \mu(A)$. But for signed measure ν, since $\nu(A \sim B)$ may be < 0 the argument fails. Consider $X = [-2, 1]$ and let $\mathcal{A}$ be the class of all Lebesgue measurable sets in X and let μ be the Lebesgue measure on $\mathcal{A}$. Let $f(x) = x$ for $x \in [-2, 1]$ and $\nu(E) = \int_E f d\mu$ for $E \in \mathcal{A}$. Then by Theorem 10.16 ν is a signed measure and if $A = [-2, 1]$ and $B = [0, 1]$ then $B \subset A$. But $\nu(A) = \int_A x d\mu = -\frac{3}{2}$ and $\nu(B) = \int_B x d\mu = \frac{1}{2}$ and so $\nu(B) \leq \nu(A)$ does not hold.

(ii) Since $A = B \cup (A \sim B)$, $\nu(A) = \nu(B) + \nu(A \sim B)$ by definition of ν giving $\nu(A \sim B)(= \nu(A) - \nu(B))$.

(iii) Since a signed measure is a completely additive set function (iii) and (iv) follow from Theorem 1.9 and Theorem 1.10 respectively. However, we give independent proof. Since $\bigcup_{n=1}^{\infty} E_n = E_1 \cup (\bigcup_{n=2}^{\infty} (E_n \sim E_{n-1}))$, using (ii) we have

$$\nu\left(\bigcup_{n=1}^{\infty} E_n\right) = \nu(E_1) + \sum_{n=2}^{\infty} \nu(E_n \sim E_{n-1}) = \nu(E_1) + \lim_{m \to \infty} \sum_{n=2}^{m} \nu(E_n \sim E_{n-1})$$

$$= \lim_{m \to \infty} [\nu(E_1) + \sum_{n=2}^{m} \nu(E_n \sim E_{n-1})]$$

$$= \lim_{m \to \infty} [\nu(E_1) + \sum_{n=2}^{m} (\nu(E_n) - \nu(E_{n-1}))]$$

$$= \lim_{m \to \infty} \nu(E_m)$$

(iv) Since $E_1 = \left(\bigcap_{n=1}^{\infty} E_n\right) \cup \left(\bigcup_{n=1}^{\infty}(E_1 \sim E_n)\right)$ and since $\{E_1 \sim E_n\}$ is an increasing sequence, using (iii) and (ii)

$$\nu(E_1) = \nu\left(\bigcap_{n=1}^{\infty} E_n\right) + \nu\left(\bigcup_{n=1}^{\infty}(E_1 \sim E_n)\right) = \nu\left(\bigcap_{n=1}^{\infty} E_n\right)$$

$$+ \lim_{n \to \infty} \nu(E_1 \sim E_n) = \nu\left(\bigcap_{n=1}^{\infty} E_n\right) + \lim_{n \to \infty}(\nu(E_1) - \nu(E_n))$$

$$= \nu\left(\bigcap_{n=1}^{\infty} E_n\right) + \nu(E_1) - \lim_{n \to \infty} \nu(E_n).$$

Hence

$$\nu\left(\bigcap_{n=1}^{\infty} E_n\right) = \lim_{n \to \infty} \nu(E_n).$$

10.5 Radon-Nikodym Theorems

Radon-Nikodym Theorem gives the converse of Theorem 10.16. To prove Radon-Nikodym theorem we need the following lemma.

Lemma 10.17 *Let $(X, \mathcal{B}, \mu)$ be a measure space and let Q be a countable set of non-negative real numbers such that for each $\alpha \in Q$ there exists $B_\alpha \in \mathcal{B}$ satisfying $\mu(B_\alpha \sim B_\beta) = 0$ whenever $\alpha < \beta, \beta \in Q$. Then there exists a non-negative measurable function f on X such that $f(x) \leq \alpha$ a.e. on B_α and $f(x) \geq \alpha$ a.e. on $X \sim B_\alpha$ for $\alpha \in Q$.*

Proof Let $C = \bigcup\{B_\alpha \sim B_\beta : \alpha < \beta; \alpha, \beta \in \mathbb{Q}\}$. Then $\mu(C) = 0$. Let $D_\alpha = B_\alpha \cup C$. Then $D_\alpha \sim D_\beta = B_\alpha \sim B_\beta \sim C = \emptyset$ whenever $\alpha < \beta$. So $D_\alpha \subset D_\beta$ for $\alpha < \beta$. For each $x \in X$, define

$$f(x) = \inf\{\alpha \in Q : x \in D_\alpha\} \text{ if } \{\alpha \in Q : x \in D_\alpha\} \neq \emptyset$$

and $f(x) = \infty$ if $\{\alpha \in Q : x \in D_\alpha\} = \emptyset$. Clearly f is non-negative and if $x \in D_\alpha$ then $f(x) \leq \alpha$. Also if $x \in X \sim D_\alpha$ then $x \notin D_\alpha$ and hence $x \notin D_\gamma$ for all $\gamma \in Q$ such that $\gamma < \alpha$. So, $f(x) \geq \alpha$. Hence $f(x) \leq \alpha$ for $x \in B_\alpha$ and $f(x) \geq \alpha$ for $x \in X \sim B_\alpha \sim C$ i.e. $f(x) \geq \alpha$ a.e. on $X \sim B_\beta$ since $\mu(C) = 0$. We show that f is measurable. For any constant k, we have

$$\{x \in X : f(x) < k\} = \bigcup_\alpha \{D_\alpha : \alpha < k; \alpha \in Q\} \tag{10.4}$$

10.5 Radon-Nikodym Theorems

For, if $f(x) < k$ then there is $\alpha < k$ such that $\alpha \in Q$ and $x \in D_\alpha$. On the other hand if $x \in D_\alpha$ for some $\alpha \in Q$ and $\alpha < k$ then $f(x) \leq \alpha < k$. So the above equality holds. Since the right side of (10.4) is a countable union of members of $\mathcal{B}$, the union is in $\mathcal{B}$ and hence the left side of (10.4) is in $\mathcal{B}$ showing that f is measurable, completing the proof. □

Theorem 10.18 (Radon-Nikodym Theorem) *Let $(X, \mathcal{B}, \mu)$ be a measure space where μ is finite and let ν be a measure on $\mathcal{B}$ which is absolutely continuous with respect to μ. Then there is a non-negative $\mathcal{B}$-measurable function f on X such that*

$$\nu(E) = \int_E f\, d\mu \text{ for } E \in \mathcal{B}.$$

If ν is finite then f is integrable and the function f is unique if we ignore sets of μ-measure zero.

Proof Let Q be the set of all non-negative rational numbers. Then Q is a countable set of non-negative real numbers. For each $\alpha \in Q$, $\nu - \alpha\mu$ is a signed measure on $(X, \mathcal{B})$ by Theorem 10.1. Let $\alpha < \beta$, $\alpha, \beta \in Q$. Let (A_α, B_α) and (A_β, B_β) be Hahn decomposition for $\nu - \alpha\mu$ and $\nu - \beta\mu$ respectively. Then since $B_\alpha \sim B_\beta = B_\alpha \cap A_\beta$ we have $(\nu - \alpha\mu)(B_\alpha \sim B_\beta) \leq 0$ and $(\nu - \beta\mu)(B_\alpha \sim B_\beta) \geq 0$. Hence $\nu(B_\alpha \sim B_\beta) \leq \alpha\mu(B_\alpha \sim B_\beta)$ and $\nu(B_\alpha \sim B_\beta) \geq \beta\mu(B_\alpha \sim B_\beta)$. So, $\beta\mu(B_\alpha \sim B_\beta) \leq \alpha\mu(B_\alpha \sim B_\beta)$. Since $\alpha < \beta$ this gives $\mu(B_\alpha \sim B_\beta) = 0$. Hence by Lemma 10.17 there is a non-negative measurable function f on X such that for each $\alpha \in Q$ $f(x) \leq \alpha$ a.e. $[\mu]$ on B_α and $f(x) \geq \alpha$ a.e. $[\mu]$ on $X \sim B_\alpha = A_\alpha$. This property will be used in the following.

Let $E \in \mathcal{B}$. Let N be a fixed positive integer. For each non-negative integer k let $\alpha_k = \frac{k}{N}$. Then $\alpha_k \in Q$ for each k. Let

$$E_k = E \cap (B_{\alpha_{k+1}} \sim \bigcup_{i=0}^{k} B_{\alpha_i}) \text{ and } E_\infty = E \sim \bigcup_{k=0}^{\infty} B_{\alpha_k}$$

Then the sets $E_0, E_1, ..., E_\infty$ are disjoint and

$$E = E_\infty \cup \left(\bigcup_{k=0}^{\infty} E_k\right) \quad (10.5)$$

Hence

$$\nu(E) = \nu(E_\infty) + \sum_{k=0}^{\infty} \nu(E_k) \quad (10.6)$$

Since $E_k \subset B_{\alpha_{k+1}} \cap A_{\alpha_k}$ we have $\alpha_k \leq f \leq \alpha_{k+1}$ a.e.$[\mu]$ on E_k and so

$$\alpha_k \mu(E_k) \leq \int_{E_k} f\, d\mu \leq \alpha_{k+1} \mu(E_k) \quad (10.7)$$

Again since $E_k \subset B_{\alpha_{k+1}} \cap A_{\alpha_k}$ we have

$$(\nu - \alpha_{k+1}\mu)(E_k) \leq 0 \text{ and } (\nu - \alpha_k\mu)(E_k) \geq 0$$

and so

$$\alpha_k\mu(E_k) \leq \nu(E_k) \leq \alpha_{k+1}\mu(E_k) \tag{10.8}$$

From (10.7) and (10.8)

$$|\nu(E_k) - \int_{E_k} f\,d\mu| \leq (\alpha_{k+1} - \alpha_k)\mu(E_k) = \frac{1}{N}\mu(E_k).$$

Hence

$$\nu(E_k) - \frac{1}{N}\mu(E_k) \leq \int_{E_k} f\,d\mu \leq \nu(E_k) + \frac{1}{N}\mu(E_k) \text{ for } k = 0, 1, 2, \ldots \tag{10.9}$$

Regarding the set E_∞, if $\mu(E_\infty) = 0$ then since ν is absolutely continuous with respect to μ, $\nu(E_\infty) = 0$ and so

$$\nu(E_\infty) = \int_{E_\infty} f\,d\mu \tag{10.10}$$

If $\mu(E_\infty) > 0$ then since $E_\infty \subset A_{\alpha_k}$ for all $\alpha_k \in Q$, $(\nu - \alpha_k\mu)(E_\infty) \geq 0$ i.e. $\nu(E_\infty) \geq \alpha_k\mu(E_\infty)$ for all $\alpha_k \in Q$ showing that $\nu(E_\infty) = \infty$. Also since $E_\infty \subset A_{\alpha_k}$ for all $\alpha_k \in Q$, $f \geq \alpha_k$ a.e. $[\mu]$ on E_∞ for all $\alpha_k \in Q$ and hence $\int_{E_\infty} f\,d\mu \geq \alpha_k\mu(E_\infty)$ for all $\alpha_k \in Q$ and so $\int_{E_\infty} f\,d\mu = \infty$. Thus in any case (10.10) holds. Adding all the inequalities in (10.9) and (10.10) we have from (10.5) and (10.6)

$$\nu(E) - \frac{1}{N}\mu(E) \leq \int_E f\,d\mu \leq \nu(E) + \frac{1}{N}\mu(E).$$

Since N is arbitrary and $\mu(E)$ is finite, letting $N \to \infty$

$$\nu(E) = \int_E f\,d\mu$$

Since E is any set in $\mathcal{B}$ the first part is proved. The second part is clear. For the last part, suppose that there is another function g with this property. Then $\int_E (f - g)\,d\mu = 0$ for all $E \in \mathcal{B}$. Let $F = f - g$. Then $\int_E F\,d\mu = 0$ for all $E \in \mathcal{B}$. So if $E^+ = \{x \in E : F(x) \geq 0\}$ then since $E^+ \in \mathcal{B}$, $\int_{E^+} F\,d\mu = 0$ and so $F = 0$ a.e. $[\mu]$

10.5 Radon-Nikodym Theorems

on E^+ (see Theorem 5.25 (xi)). Similarly $F = 0$ a.e. $[\mu]$ on $E^- = \{x \in E : F(x) \leq 0\}$. Since $E^+ \cup E^- = E$, $F = 0$ a.e. $[\mu]$ on E i.e $f = g$ a.e. $[\mu]$ on E, completing the proof. $\square$

The following two theorems show that Radon-Nikodym theorem is also true when μ and ν are signed measures.

Theorem 10.19 (Radon-Nikodym Theorem) *Let $(X, \mathcal{B}, \mu)$ be a measure space where μ is finite and ν be a signed measure on $\mathcal{B}$ which is absolutely continuous with respect to μ. Then there exists a $\mathcal{B}$-measurable function f on X such that*

$$\nu(E) = \int_E f d\mu \text{ for } E \in \mathcal{B}.$$

If ν is finite then f is integrable and the function f is unique if we ignore sets of μ—measure zero.

Proof By Theorem 10.7 there are two mutually singular measures ν^+ and ν^- on $(X, \mathcal{B})$ at least one of which is finite such that $\nu = \nu^+ - \nu^-$. By Theorem 10.14 (iii) ν^+ and ν^- are absolutely continuous with respect to μ. So, by Theorem 10.18 there are non-negative $\mathcal{B}$-measurable functions f_1 and f_2 such that

$$\nu^+(E) = \int_E f_1 d\mu \text{ and } \nu^-(E) = \int_E f_2 d\mu \text{ for } E \in \mathcal{B}.$$

Since at least one of ν^+ and ν^- is finite, at least one of f_1 and f_2 is integrable. Let $f = f_1 - f_2$. Then $\int_E f d\mu$ is defined for each $E \in \mathcal{B}$. Also since $\nu = \nu^+ - \nu^-$,

$$\nu(E) = \nu^+(E) - \nu^-(E) = \int_E f_1 d\mu - \int_E f_2 d\mu = \int_E f d\mu \text{ for } E \in \mathcal{B}.$$

So the first part is proved. For the second part if ν is finite then f is integrable. The rest is as in Theorem 10.19. $\square$

Theorem 10.20 (Radon-Nikodym Theorem) *Let $(X, \mathcal{B})$ be a measurable space and let μ and ν be signed measures on $\mathcal{B}$ such that μ is finite and ν is absolutely continuous with respect to μ. Then there exists a $\mathcal{B}$ - measurable function f on X such that*

$$\nu(E) = \int_E f d\mu \text{ for } E \in \mathcal{B}.$$

If ν is finite then f is integrable with respect to μ and the function f is unique if we ignore sets of $|\mu|$ - measure zero.

Proof Since μ is a signed measure on $(X, \mathcal{B})$, by Hahn decomposition theorem (Theorem 10.5) there are measurable sets A and B such that $A \cup B = X$ and $A \cap B = \emptyset$ where A is a positive set and B is a negative set with respect to μ. Let $\mathcal{B}_1 = \{E \cap A : E \in \mathcal{B}\}$ and $\mathcal{B}_2 = \{E \cap B; E \in \mathcal{B}\}$. Then $\mathcal{B}_1$ and $\mathcal{B}_2$ are σ-algebras on $X \cap A$ and $X \cap B$ respectively. Let (μ^+, μ^-) be the Jordan decomposition of μ defined by $\mu^+(E) = \mu(E \cap A)$ and $\mu^-(E) = -\mu(E \cap B)$. Then μ^+ and μ^- are measures on the measurable spaces $(X \cap A, \mathcal{B}_1)$ and $(X \cap B, \mathcal{B}_2)$ respectively. Also ν is a signed measure on $\mathcal{B}_1$ as well as on $\mathcal{B}_2$ and ν satisfies $\nu << \mu^+$ on $\mathcal{B}_1$ and $\nu << \mu^-$ on $\mathcal{B}_2$ and so by Theorem 10.19 there exists a $\mathcal{B}_1$-measurable function f_1 on $X \cap A$ and a $\mathcal{B}_2$-measurable function f_2 on $X \cap B$ such that

$$\nu(E \cap A) = \int_{E \cap A} f_1 d\mu^+ \text{ and } \nu(E \cap B) = \int_{E \cap B} f_2 d\mu^- \text{ for } E \in \mathcal{B}.$$

Let $f = f_1$ on A and $f = -f_2$ on B. Then f is a $\mathcal{B}$-measurable function on X and

$$\nu(E) = \nu(E \cap A) + \nu(E \cap B) = \int_{E \cap A} f_1 d\mu^+ + \int_{E \cap B} f_2 d\mu^- = \int_{E \cap A} f d\mu^+ - \int_{E \cap B} f d\mu^-.$$

Since ν is a signed measure the right-hand side cannot be of the form $\infty - \infty$ and so $\nu(E)$ is well defined and

$$\nu(E) = \int_{E \cap A} f d\mu + \int_{E \cap B} f d\mu = \int_E f d\mu \text{ for } E \in \mathcal{B}$$

So, if ν is finite then f is integrable. The rest is an easy exercise. □

Remark Radon-Nikodym Theorems considered in Theorem 10.18, 10.19 and 10.20 are also true when μ is σ-finite. For, suppose that Theorem 10.18 is proved considering μ finite. Now suppose that μ is σ—finite. Then there exists a countable collection $\{X_n\}$ of sets such that $X = \bigcup_{n=1}^{\infty} X_n$, $X_n \in \mathcal{B}$ and $\mu(X_n) < \infty$ for all n. We may suppose that the sets X_n are disjoint. Let $\mathcal{B}_n = \{E \cap X_n : E \in \mathcal{B}\}$. Then $\mathcal{B}_n$ is a σ-algebra and $(X_n, \mathcal{B}_n, \mu)$ is a measure space where μ is finite. Then by Theorem 10.18 applied on $(X_n, \mathcal{B}_n, \mu)$ there is a non-negative $\mathcal{B}_n$-measurable function on X_n such that

$$\nu(E) = \int_E f_n d\mu \text{ for } E \in \mathcal{B}_n \text{ and for each } n.$$

Let $E \in \mathcal{B}$. Then $E = \bigcup_{n=1}^{\infty} E_n$ where $E_n \in \mathcal{B}_n$ for each n. Let f be defined on X by $f = f_n$ on X_n. Then f is a non-negative $\mathcal{B}$-measurable function on X and

$$\nu(E) = \sum_{n=1}^{\infty} \nu(E_n) = \sum_{n=1}^{\infty} \int_{E_n} f_n d\mu = \int_E f d\mu$$

Thus Theorem 10.18 is true when μ is σ—finite. Similarly Theorem 10.19 and Theorem 10.20 are also true when μ is σ—finite.

10.6 Application of Radon-Nikodym Theorem

Definition 10.7 Let $(X, \mathcal{A})$ be a measurable space and let ψ be a set function defined on $\mathcal{A}$ and let μ be a signed measure on $\mathcal{A}$. Then ψ is called absolutely continuous with respect to μ if for every $\epsilon > 0$ there is $\delta > 0$ such that $|\psi(E)| < \epsilon$ for every $E \in \mathcal{A}$ for which $\mu(E) < \delta$.

Absolute continually is an intrinsic property of indefinite integral. It is proved in Theorem 6.41 that a point function F is absolutely continuous if and only if F is an indefinite integral. It is natural to ask whether this property of absolute continually is also true for set function.

Note that if a set function ψ is an indefinite integral then by Theorem 10.16 ψ is a signed measure. But there are set functions ψ which are not signed measure. Consider the measurable space $(X, \mathcal{A})$ where $X = [2, 6]$ and $\mathcal{A}$ is the family of all Lebesgue measurable sets in X. Applying Axiom of Choice choose exactly one point from each $E \in \mathcal{A}$ and call it ξ_E. Define $\psi : \mathcal{A} \to \mathbb{R}$ by $\psi(E) = \xi_E$ for $E \in \mathcal{A}$. Let $E_1 = (2, 3)$, $E_2 = (4, 5)$. Then $\psi(E_1) = \xi_{E_1} \in (2, 3)$ and $\psi(E_2) = \xi_{E_2} \in (4, 5)$ and so $\psi(E_1) + \psi(E_2) \geq 6 > 5 \geq \psi(E_1 \cup E_2)$ and so ψ cannot be a signed measure since it does not satisfy condition (iii) of Definition 10.1. Thus the analog of Theorem 6.41 is not true for all set functions. However, if the set function is a signed measure then this is true.

Definition 10.8 Let $(X, \mathcal{A}, \mu)$ be a measure space. Then a signed measure ν on $\mathcal{A}$ is called an indefinite μ - integral if there exists a μ—integrable function f on X such that

$$\nu(E) = \int_E f d\mu \quad \text{for all } E \in \mathcal{A}.$$

The following theorem is an analog of Theorem 6.41.

Theorem 10.21 *Let $(X, \mathcal{A}, \mu)$ be a measure space where μ is finite and let ν be a finite signed measure on $\mathcal{A}$. Then ν is absolutely continuous with respect to μ if and only if ν is an indefinite μ - integral of some $\mathcal{A}$ - measurable function on X.*

Proof Let ν be absolutely continuous with respect to μ. Then by Theorem 10.19 there exists a μ integrable function f on X such that

$$\nu(E) = \int_E f d\mu \quad \text{for all } E \in \mathcal{A}$$

So ν is an indefinite μ-integral of f. Conversely suppose that ν is an indefinite μ-integral of some $\mathcal{A}$-measurable function f on X. Then

$$\nu(E) = \int_E f d\mu \text{ for all } E \in \mathcal{A}.$$

Since ν is finite f is μ-integrable and so f^+ and f^- are integrable and

$$\int_E f d\mu = \int_E f^+ d\mu - \int_E f^- d\mu$$

So if $\nu_1(E) = \int_E f^+ d\mu$ and $\nu_2(E) = \int_E f^- d\mu$ then $\nu(E) = \nu_1(E) - \nu_2(E)$.

Let $\epsilon > 0$ be arbitrary. Then by Theorem 5.25 (vii) there exists $\delta > 0$ such that if $\mu(E) < \delta$ then $\nu_1(E) < \frac{\epsilon}{2}$ and $\nu_2(E) < \frac{\epsilon}{2}$. Since $\nu = \nu_1 - \nu_2$ and since Jordan decomposition is unique $\nu_1 = \nu^+$, $\nu_2 = \nu^-$ and so

$$|\nu|(E) = \nu^+(E) + \nu^-(E) < \epsilon \text{ whenever } \mu(E) < \delta.$$

So, by Theorem 10.15 the signed measure ν is absolutely continuous with respect to μ. □

Remark It is known that a point function F is absolutely continuous if and only if F is an indefinite Lebesgue integral (See Theorem 6.41). Theorem 10.21 gives an analog for set functions, more precisely for measurable functions.

Theorem 10.22 (Lebesgue Decomposition Theorem) *Let $(X, \mathcal{B}, \mu)$ be a σ-finite measure space and let ν be a σ-finite measure on $\mathcal{B}$. Then $\nu = \nu_0 + \nu_1$ where ν_0 and ν_1 are measures on $\mathcal{B}$ such that $\nu_0 \perp \mu$ and $\nu_1 \ll \mu$.*

Proof To prove the theorem we shall use Theorem 10.18 which, in view of the Remark after Theorem 10.20 is true when μ and ν are σ-finite. Let $\lambda = \mu + \nu$. Then λ is a σ-finite measure on $\mathcal{B}$. Also μ and ν are absolutely continuous with respect to λ. So, by Theorem 10.18 there is a non-negative $\mathcal{B}$-measurable function f on X such that

$$\mu(E) = \int_E f d\lambda \text{ for all } E \in \mathcal{B}.$$

Let $A = \{x \in X : f(x) > 0\}$ and $B = \{x \in X : f(x) = 0\}$. Then $A \cup B = X$ and $A \cap B = \emptyset$. Define ν_0 and ν on $\mathcal{B}$ by $\nu_0(E) = \nu(E \cap B)$ and $\nu_1(E) = \nu(E \cap A)$ for $E \in \mathcal{B}$. Clearly ν_0 and ν_1 are measures on $\mathcal{B}$ and

$$\nu_0(E) + \nu_1(E) = \nu(E \cap B) + \nu(E \cap A) = \nu(E) \text{ for all } E \in \mathcal{B}.$$

Also $\nu_0(A) = \nu(A \cap B) = 0$ and $\mu(B) = \int_B f d\lambda = 0$. Hence $\nu_0 \perp \mu$. Finally let $\mu(E) = 0$ for some $E \in \mathcal{B}$. Then $\int_E f d\lambda = 0$. Since f is non-negative, $f = 0$

almost everywhere with respect to λ on E. Since $f > 0$ on $A \cap E$ we must have $\lambda(A \cap E) = 0$. Since $\nu \ll \lambda$, $\nu(A \cap E) = 0$ and hence $\nu_1(E) = 0$. So $\nu_1 \ll \mu$. This complete the proof. □

The decomposition of ν into two measures ν_0 and ν_1 is called the Lebesgue decomposition of ν with respect to μ and is defined by (ν_0, ν_1). Like Jordan decomposition Lebesgue decomposition is unique.

Theorem 10.23 *Let $(X, \mathcal{B}, \mu)$ be a σ-finite measure space and let ν be a σ-finite measure on $\mathcal{B}$. Then the Lebesgue decomposition of ν with respect to μ is unique.*

Proof Let (ν_0, ν_1) and (ν_0', ν_1') be two Lebesgue decomposition of ν with respect to μ. Then
$$\nu = \nu_0 + \nu_1 = \nu_0' + \nu_1', \ \nu_0 \perp \mu, \ \nu_0' \perp \mu, \ \nu_1 \ll \mu, \ \nu_1' \ll \mu.$$

So there are sets A, B, A', B' such that $X = A \cup B = A' \cup B'$, $A \cap B = \emptyset = A' \cap B'$ and $\nu_0(B) = \mu(A) = \nu_0'(B') = \mu(A') = 0$. Let $E \in \mathcal{B}$. Then
$$E = (E \cap B \cap B') \cup (E \cap A' \cap B) \cup (E \cap A \cap A') \cup (E \cap A \cap B').$$

Clearly μ is zero for the last three sets in this union and since ν_1 and ν_1' are absolutely continuous with respect to μ, ν_1 and ν_1' are also zero for the last three sets in this union. Therefore since $\nu_1' - \nu_1 = \nu_0 - \nu_0'$ we have
$$(\nu_1' - \nu_1)(E) = (\nu_1' - \nu_1)(E \cap B \cap B') = (\nu_0 - \nu_0')(E \cap B \cap B') = 0.$$

Since E is any set in $\mathcal{B}$, this gives $\nu_1' - \nu_1 = 0$ and hence $\nu_0 - \nu_0' = 0$. So, $\nu_0 = \nu_0'$ and $\nu_1 = \nu_1'$. □

10.7 Radon Nikodym Derivative

Definition 10.9 Let $(X, \mathcal{B})$ be a measurable space and let μ and ν be signed measures on $\mathcal{B}$ such that μ is σ-finite and ν is absolutely continuous with respect to μ. Then the function f given by Radon-Nikodym Theorem satisfying
$$\nu(E) = \int_E f \, d\mu \ \text{ for } E \in \mathcal{B}$$

is called the Radon-Nikodym derivative of ν with respect to μ and is denoted by $\frac{d\nu}{d\mu}$.

The Radon-Nikodym Theorem used in the above definition may be any one of Theorems 10.18, 10.19 and 10.20 (together with Remark after Theorem 10.20) whichever is suitable for μ and ν which are chosen. The Radon-Nikodym derivative is unique if two functions are considered equal when they are equal almost everywhere.

Theorem 10.24 Let $(X, \mathcal{B})$ be a measurable space and let v_1, v_2 and μ be finite signed measures on $\mathcal{B}$ such that v_1 and v_2 are absolutely continuous with respect to μ, Then

$$\frac{d(v_1 + v_2)}{d\mu} = \frac{dv_1}{d\mu} + \frac{dv_2}{d\mu}.$$

Proof Since v_1 and v_2 are finite signed measures, $v_1 + v_2$ is also a finite signed measure on $\mathcal{B}$. Also since v_1 and v_2 are absolutely continuous with respect to μ the sum $v_1 + v_2$ is absolutely continuous with respect to μ. So applying Theorem 10.20

$$(v_1 + v_2)(E) = v_1(E) + v_2(E) = \int_E \frac{dv_1}{d\mu} d\mu + \int_E \frac{dv_2}{d\mu} d\mu$$

$$= \int_E \left(\frac{dv_1}{d\mu} + \frac{dv_2}{d\mu} \right) d\mu$$

and so

$$\frac{d(v_1 + v_2)}{d\mu} = \frac{dv_1}{d\mu} + \frac{dv_2}{d\mu}.$$

□

Remark The above theorem is also true when v_1, v_2 and μ are σ-finite, provided $v_1 + v_2$ is signed measure. For if v_1 and v_2 are σ-finite then $v_1 + v_2$ may not be a signed measure, since in this case $v_1 + v_2$ may assume both the values ∞ and $-\infty$ contradicting condition (ii) of Definition 10.1.

Corollary 10.25 Under the hypothesis of Theorem 10.24

$$\frac{d(v_1 - v_2)}{d\mu} = \frac{dv_1}{d\mu} - \frac{dv_2}{d\mu}.$$

Proof Applying Theorem 10.24

$$\frac{dv_1}{d\mu} = \frac{d((v_1 - v_2) + v_2)}{d\mu} = \frac{d(v_1 - v_2)}{d\mu} + \frac{dv_2}{d\mu}$$

and hence

$$\frac{d(v_1 - v_2)}{d\mu} = \frac{dv_1}{d\mu} - \frac{dv_2}{d\mu}.$$

□

Theorem 10.26 Let v be a σ-finite signed measure and let μ and λ be σ-finite measures on a measurable space $(X, \mathcal{B})$ such that $v \ll \mu \ll \lambda$. Then

$$\frac{dv}{d\lambda} = \frac{dv}{d\mu} \frac{d\mu}{d\lambda}$$

10.7 Radon Nikodym Derivative

Proof We first suppose that ν is a measure. So, by Theorem 10.18 there are non-negative measurable functions f and g on X such that

$$\nu(E) = \int_E f d\mu \text{ and } \mu(E) = \int_E g d\lambda \text{ for all } E \in \mathcal{B}. \tag{10.11}$$

Let ψ be a non-negative measurable simple function on X defined by $\psi = \sum_{i=1}^{n} a_i \chi_{E_i}$, where $E_1, E_2, \ldots, E_n$ are disjoint measurable sets such that $\bigcup_{i=1}^{n} E_i = X$ and χ_{E_i} is the characteristic function of E_i. Then for $E \in \mathcal{B}$

$$\int_E \psi d\mu = \sum_{i=1}^{n} a_i \mu(E_i \cap E) = \sum_{i=1}^{n} a_i \int_{E_i \cap E} g d\lambda = \int_E \psi g d\lambda. \tag{10.12}$$

Let $\{\psi_n\}$ be a non-decreasing sequence of non-negative measurable simple functions such that ψ_n converges to f. Then $\{\psi_n g\}$ is a non-decreasing sequence of non-negative measurable functions such that $\psi_n g$ converges to fg. Since (10.12) is true if ψ is replaced by ψ_n we have by Monotone convergence Theorem

$$\nu(E) = \int_E f d\mu = \lim_{n \to \infty} \int_E \psi_n d\mu = \lim_{n \to \infty} \int_E \psi_n g d\lambda = \int_E f g d\lambda.$$

So, by (10.11)

$$\frac{d\nu}{d\lambda} = fg = \frac{d\nu}{d\mu} \frac{d\mu}{d\lambda}.$$

Suppose that ν is a signed measure. By Jordan decomposition Theorem (Theorem 10.7) $\nu = \nu^+ - \nu^-$ where ν^+ and ν^- are measures on $(X, \mathcal{B})$. So by the above

$$\frac{d\nu^+}{d\lambda} = \frac{d\nu^+}{d\mu} \frac{d\mu}{d\lambda} \text{ and } \frac{d\nu^-}{d\lambda} = \frac{d\nu^-}{d\mu} \frac{d\mu}{d\lambda}.$$

Hence using Corollary 10.25

$$\frac{d\nu}{d\lambda} = \frac{d(\nu^+ - \nu^-)}{d\lambda} = \frac{d\nu^+}{d\lambda} - \frac{d\nu^-}{d\lambda} = \frac{d\nu^+}{d\mu} \frac{d\mu}{d\lambda} - \frac{d\nu^-}{d\mu} \frac{d\mu}{d\lambda}$$

$$= \left(\frac{d\nu^+}{d\mu} - \frac{d\nu^-}{d\mu} \right) \frac{d\mu}{d\lambda} = \frac{d\nu}{d\mu} \frac{d\mu}{d\lambda}$$

completing the proof. $\square$

Theorem 10.27 *Let $(X, \mathcal{B})$ be a measurable space and let g be any non-negative $\mathcal{B}$ - measurable function on X. If μ and ν are σ finite measures on $\mathcal{B}$ such that $\nu \ll \mu$ then*

$$\int_E g\,dv = \int_E g\frac{dv}{d\mu}d\mu \quad \text{for } E \in \mathcal{B}.$$

If g is v integrable then the non-negative condition on g is not necessary.

Proof we first suppose that g is simple and let $g = \sum_{i=1}^{n} c_i \chi_{E_i}$ where c_i are non-negative constants and $E_i's$ are disjoint $\mathcal{B}$ - measurable sets with $X = \bigcup_{i=1}^{n} E_i$. Then using the Radon-Nikodym derivative we have for any $E \in \mathcal{B}$

$$\int_E g\,dv = \sum_{i=1}^{n} c_i v(E \cap E_i) = \sum_{i=1}^{n} c_i \int_{E \cap E_i} \frac{dv}{d\mu}d\mu = \sum_{i=1}^{n} \int_{E \cap E_i} c_i \frac{dv}{d\mu}d\mu = \int_E g\frac{dv}{d\mu}d\mu.$$

So the result is true when g is simple. Let g be any non-negative $\mathcal{B}$ measurable function. Let $\{\phi_n\}$ be a non-decreasing sequence of non-negative $\mathcal{B}$ - measurable simple functions such that $\lim_{n \to \infty} \phi_n = g$. Since the result is true for simple functions

$$\int_E \phi_n dv = \int_E \phi_n \frac{dv}{d\mu}d\mu \quad \text{for all n, and } E \in \mathcal{B}.$$

Taking limit as $n \to \infty$ we have by Monotone convergence Theorem

$$\int_E g\,dv = \int_E g\frac{dv}{d\mu}d\mu$$

proving the first part.

For the second part consider the positive part g^+ and negative part g^- separately and we get from above

$$\int_E g^+ dv = \int_E g^+ \frac{dv}{d\mu}d\mu \quad \text{and} \quad \int_E g^- dv = \int_E g^- \frac{dv}{d\mu}d\mu.$$

Since g is v - integrable, these integrals are finite and so

$$\int_E g\,dv = \int_E g^+ dv - \int_E g^- dv = \int_E (g^+ - g^-)\frac{dv}{d\mu}d\mu = \int_E g\frac{dv}{d\mu}d\mu$$

completing the proof. □

10.8 Complex Measure and Integration

Definition 10.10 Let $(X, \mathcal{A})$ be a measurable space. A complex-valued set function μ defined on $\mathcal{A}$ is called a complex measure if μ satisfies the following conditions:

(i) $\mu(\emptyset) = 0$

(ii) $\mu(\bigcup_{n=1}^{\infty} E_n) = \sum_{n=1}^{\infty} \mu(E_n)$ for any countable collection $\{E_n\}$ of disjoint members of $\mathcal{A}$, the series on the right being absolutely convergent,

If μ is a complex measure, we write $\mu_1(E) = Re \, \mu(E)$ and $\mu_2(E) = Im \, \mu(E)$, where $Re \, \mu(E)$ and $Im \, \mu(E)$ denote real and imaginary parts of $\mu(E)$ respectively. Since $|x| \leq |z|$ and $|y| \leq |z|$ for every complex number $z = x + iy$, the set functions μ_1 and μ_2 satisfies (i) and (ii) and so $\mu = \mu_1 + i\mu_2$ where μ_1 and μ_2 are finite signed measures on the measurable space $(X, \mathcal{A})$. The total variation $|\mu|$ (also called absolute measure) is defined to be

$$|\mu|(E) = \sup\{\sum_{i=1}^{n} |\mu(E_i)| : E_i \in \mathcal{A}_i; E_i \cap E_j = \emptyset; E = \bigcup_{i=1}^{n} E_i\}$$

The sup is taken over all such measurable dissection of E.

Theorem 10.28 Let μ be a complex measure on the measurable space $(X, \mathcal{A})$ and let μ_1 and μ_2 be the real and imaginary parts of μ. Then

(i) μ_1 and μ_2 are finite signed measures on $(X, \mathcal{A})$.
(ii) $\mu = \mu_1^+ - \mu_1^- + i(\mu_2^+ - \mu_2^-)$ where $\mu_1^+ - \mu_1^-$ and $\mu_2^+ - \mu_2^-$ are Jordan decomposition of μ_1 and μ_2 respectively
(iii) $|\mu| \leq \mu_1^+ + \mu_1^- + \mu_2^+ + \mu_2^-$
(iv) $\mu_i^+ \leq |\mu|$ and $\mu_i^- \leq |\mu|$ for $i = 1, 2$

Proof The proofs of (i) and (ii) are easy. To prove (iii) note that for any set $E \in \mathcal{A}$

$$|\mu(E)| = |\mu_1(E) + i\mu_2(E)| = |\mu_1^+(E) - \mu_1^-(E) + i\mu_2^+(E) - i\mu_2^-(E)|$$
$$\leq \mu_1^+(E) + \mu_1^-(E) + \mu_2^+(E) + \mu_2^-(E)$$

and so for any measurable dissection $E_1, E_2, ..., E_n$ of E

$$\sum_{i=1}^{n} |\mu(E_i)| \leq \sum_{i=1}^{n} [\mu_1^+(E_i) + \mu_1^-(E_i) + \mu_2^+(E_i) + \mu_2^-(E_i)]$$
$$= \mu_1^+(E) + \mu_1^-(E) + \mu_2^+(E) + \mu_2^-(E).$$

Taking 'sup' over all measurable dissection of E we have $|\mu|(E) \leq \mu_1^+(E) + \mu_1^-(E) + \mu_2^+(E) + \mu_2^-(E)$ proving (iii).

To prove (iv) let $(A_1 B_1)$ be the Hahn decomposition of X with respect to μ_1. Then since for a complex number $z = x + iy$, $|x| \leq |z|$ and $|y| \leq |z|$, we have for any set $E \in \mathcal{A}$

$$\mu_1^+(E) = \mu_1(E \cap A_1) = |\mu_1(E \cap A_1)| \leq |\mu(E \cap A_1)| \leq |\mu(E \cap A_1)| + |\mu(E \cap B_1)| \leq |\mu|(E)$$

and

$$\mu_1^-(E) = -\mu_1(E \cap B_1) = |\mu_1(E \cap B_1)| \leq |\mu(E \cap B_1)| \leq |\mu(E \cap A_1)| + |\mu(E \cap B_1)| \leq |\mu|(E)$$

So, $\mu_1^+ \leq |\mu|$ and $\mu_1^- \leq |\mu|$.
Similarly, considering the Hahn decomposition (A_2, B_2) of X with respect to μ_2 we get $\mu_2^+ \leq |\mu|$ and $\mu_2^- \leq |\mu|$ completing the proof. □

The decomposition $\mu_1^+ - \mu_1^- + i(\mu_2^+ - \mu_2^-)$ of the complex measure μ in (ii) above is called the Jordan decomposition of μ.

Theorem 10.29 *Let μ be a complex measure on the measurable space $(X, \mathcal{A})$. Then the set function $|\mu|$ is a measure on $(X, \mathcal{A})$.*

Proof Clearly $|\mu|(E) = 0$ and $|\mu|$ is non-negative. We are to show that $|\mu|$ is countably additive. Let $\{A_i\}$ be a countable collection of sets in $\mathcal{A}$ such that $A_i \cap A_j = \emptyset$ for $i \neq j$ and let $A = \bigcup_{i=1}^{\infty} A_i$. Let $\{E_1, E_2, \ldots, E_n\}$ be a measurable dissection of A.

Since $E_k = \bigcup_{i=1}^{\infty} (E_k \cap A_i)$ for $k = 1, 2, \ldots, n$, and since the sets $E_k \cap A_i$ are disjoint for fixed k,

$$\mu(E_k) = \sum_{i=1}^{\infty} \mu(E_k \cap A_i) \text{ and so}$$

$$|\mu(E_k)| \leq \sum_{i=1}^{\infty} |\mu(E_k \cap A_i)|$$

and hence

$$\sum_{k=1}^{n} |\mu(E_k)| \leq \sum_{k=1}^{n} \sum_{i=1}^{\infty} |\mu(E_k \cap A_i)| = \sum_{i=1}^{\infty} \sum_{k=1}^{n} |\mu(E_k \cap A_i)|$$

$$\leq \sum_{i=1}^{\infty} |\mu|(A_i) \text{ (by Definition of } |\mu|).$$

Hence

$$|\mu|(A) \leq \sum_{i=1}^{\infty} |\mu|(A_i). \tag{10.13}$$

To prove the reverse inequality note that by Theorem 10.28 μ_1 and μ_2 are finite signed measures and so $\mu_1^+, \mu_1^-, \mu_2^+$ and μ_2^- are all finite measures and since $|\mu| \leq \mu_1^+ +$

10.8 Complex Measure and Integration

$\mu_1^- + \mu_1^+ + \mu_2^-$, $|\mu|$ is finite. Let $\epsilon > 0$ be arbitrary. For each i choose a measurable dissection $\{E_{i,1}\ E_{i,2} \ldots E_{i,n_i}\}$ of A_i such that

$$|\mu|(A_i) - \frac{\epsilon}{2^i} < \sum_{k=1}^{n_i} |\mu(E_{i,k})|.$$

So, for any fixed positive integer m

$$\sum_{i=1}^{m} |\mu|(A_i) < \sum_{i=1}^{m} \left[\frac{\epsilon}{2^i} + \sum_{k=1}^{n_i} |\mu(E_{i,k})| \right] < \epsilon + \sum_{i=1}^{m} \sum_{k=1}^{n_i} |\mu(E_{i,k})|$$

$$\leq \epsilon + \left| \mu\left(\bigcup_{i=m+1}^{\infty} A_i \right) \right| + \sum_{i=1}^{m} \sum_{k=1}^{n_i} |\mu(E_{i,k})| \quad (10.14)$$

Since $\{E_{i,1}, E_{i,2}, \ldots, E_{i,n_i}\}$ is a measurable dissection of A_i, the collection $\{E_{i,k}\}$, $k = 1, 2, \ldots n_i$, $i = 1, 2, \ldots m$, is a measurable dissection of $\bigcup_{i=1}^{m} A_i$. So, this $\{E_{i,k}\}$ together with $\bigcup_{i=m+1}^{\infty} A_i$ is a measurable dissection of $A = \bigcup_{i=1}^{\infty} A_i$. Hence

$$\sum_{i=1}^{m} \sum_{k=1}^{n_i} |\mu(E_{i,k})| + |\mu(\bigcup_{i=m+1}^{\infty} A_i)| \leq |\mu|(A)$$

So, from (10.14)

$$\sum_{i=1}^{m} |\mu|(A_i) \leq \epsilon + |\mu|(A)$$

Since m is arbitrary letting first $m \to \infty$ and then letting $\epsilon \to 0$,

$$\sum_{i=1}^{\infty} |\mu|(A_i) \leq |\mu|(A). \quad (10.15)$$

From (10.13) and (10.15) $|\mu|(A) = \sum_{i=1}^{\infty} |\mu|(A)$, completing the proof. □

Definition 10.11

(i) A complex-valued function f defined on a measurable space $(X, \mathcal{A})$ is said to be measurable if its real part Ref and imaginary part Imf are both measurable.

(ii) A complex-valued measurable function f defined on a measure space $(X, \mathcal{A}, \mu)$ is said to be integrable if both Ref and Imf are integrable and the integral of f is defined by

$$\int_E f d\mu = \int_E Re f d\mu + i \int_E Im f d\mu \quad \text{for } E \in \mathcal{A}.$$

Theorem 10.30 *Let f be a complex-valued measurable function defined on a measure space $(X, \mathcal{A}, \mu)$. Then $|f|$ is measurable and f is integrable if and only if $|f|$ is integrable.*

Proof Since f is measurable, $Re f$ and $Im f$ are measurable. Therefore since $|f| = \sqrt{(Re f)^2 + (Im f)^2}$, $|f|$ is measurable. Let f be integrable. Then $Re f$ and $Im f$ are integrable and so $|Re f|$ and $|Im f|$ are integrable and therefore $|Re f| + |Im f|$ is integrable. Since $|f| \leq |Re f| + |Im f|$, $|f|$ is integrable. Conversely, let $|f|$ be integrable. Then since $|Re f| \leq |f|$ and $|Im f| \leq |f|$, $|Re f|$ and $|Im f|$ are integrable and since $Re f$ and $Im f$ are measurable $Re f$ and $Im f$ are integrable and so f is integrable. $\square$

Definition 10.12 Let μ be a complex measure on a measurable space $(X, \mathcal{A})$ and let $\mu = \mu_1^+ - \mu_1^- + i(\mu_2^+ - \mu_2^-)$ be the Jordan decomposition of μ. A complex-valued measurable function f is said to be integrable with respect to μ if f is integrable with respect to each of $\mu_1^+, \mu_1^-, \mu_2^+, \mu_2^-$ and the integral of f is defined by

$$\int_E f d\mu = \int_E f d\mu_1^+ - \int_E f d\mu_1^- + i\left(\int_E f d\mu_2^+ - \int_E f d\mu_2^-\right) \quad \text{for } E \in \mathcal{A}.$$

Theorem 10.31 *Let μ be a complex measure on a measurable space of $(X, \mathcal{A})$ and let $\mu = \mu_1^+ - \mu_1^- + i(\mu_2^+ - \mu_2^-)$ be the Jordan decomposition of μ. Then a complex-valued function f on X is integrable with respect to $|\mu|$ if and only if f is integrable with respect to each of $\mu_1^+, \mu_1^-, \mu_2^+, \mu_2^-$.*

Proof Let f be integrable with respect to $|\mu|$. Then by Theorem 10.30 $|f|$ is integrable with respect to $|\mu|$. Since by Theorem 10.28 (iv) each of $\mu_1^+, \mu_1^-, \mu_2^+$, and μ_2^- is less than or equal to $|\mu|$ by Theorem 10.12, $|f|$ is integrable with respect to each of $\mu_1^+, \mu_1^-, \mu_2^+$, and μ_2^-.

Conversely, let f be integrable with respect to each of $\mu_1^+, \mu_1^-, \mu_2^+$, and μ_2^-. Then $|f|$ is integrable with respect to each of $\mu_1^+, \mu_1^-, \mu_2^+$, and μ_2^- and so by Theorem 10.10 $|f|$ is integrable with respect to $\mu_1^+ + \mu_1^- + \mu_2^+ + \mu_2^-$. Since by Theorem 10.28 (iii) $|\mu| \leq \mu_1^+ + \mu_1^- + \mu_2^+ + \mu_2^-$, by Theorem 10.12 $|f|$ is integrable with respect to $|\mu|$. Hence f is integrable with respect to $|\mu|$. $\square$

Theorem 10.32 (Radon-Nikodym theorem) *Let $(X, \mathcal{A}, \mu)$ be a finite measure space and let ν be a complex measure on $(X, \mathcal{A})$ which is absolutely continuous with respect to μ (i.e if $\mu(E) = 0$ for $E \in \mathcal{A}$ then $\nu(E) = 0$). Then there exists a μ-integrable complex function f on X such that*

$$\nu(E) = \int_E f d\mu \text{ for } E \in \mathcal{A}.$$

10.8 Complex Measure and Integration

Proof Let $\nu = \nu_1^+ - \nu_1^- + i(\nu_2^+ - \nu_2^-)$ be the Jordan decomposition of ν. Then $\nu_1^+, \nu_1^-, \nu_2^+, \nu_2^-$ are finite measures on $(X, \mathcal{A})$. Applying analog of Theorem 10.14 (*iii*). $\nu_i^+ \ll \mu$ and $\nu_i^- \ll \mu$ for $i = 1, 2$. So, by Theorem 10.18 there exist non-negative μ-integrable functions f_1, f_2, f_3 and f_4 such that for all $E \in \mathcal{A}$

$$\nu_1^+(E) = \int_E f_1 d\mu, \; \nu_1^-(E) = \int_E f_2 d\mu, \; \nu_2^+(E) = \int_E f_3 d\mu \text{ and } \nu_2^-(E) = \int_E f_4 d\mu$$

and so

$$\nu(E) = \nu_1^+(E) - \nu_1^-(E) + i(\nu_2^+(E) - \nu_2^-(E))$$
$$= \int_E ((f_1 - f_2) + i(f_3 - f_4)) d\mu = \int_E f d\mu$$

where

$$f = f_1 - f_2 + i(f_3 - f_4).$$

Clearly f is μ-integrable. This complete the proof. □

Example 10.6 Let $(X, \mathcal{A}, \mu)$ be a finite measure space and let f be a complex-valued μ-integrable function on X. Then the set function ϕ defined on $\mathcal{A}$ by

$$\phi(E) = \int_E f d\mu, \; E \in \mathcal{A}$$

is a complex measure on $(X, \mathcal{A})$ and ϕ is absolutely continuous with respect to μ.

Solution Since f is integrable its real part Ref and imaginary part Imf are integrable on $(X, \mathcal{A})$. Writing

$$\phi_1(E) = \int_E Ref d\mu \text{ and } \phi_2(E) = \int_E Imf d\mu$$

we have

$$\phi(E) = \int_E f d\mu = \int_E Ref d\mu + i \int_E Imf d\mu = \phi_1(E) + i\phi_2(E)$$

By Theorem 10.16, ϕ_1 and ϕ_2 are signed measures which are absolutely continuous with respect to μ. So, if $\mu(E) = 0$ then $\phi_1(E) = 0$ and $\phi_2(E) = 0$ and so $\phi(E) = 0$ showing that ϕ is absolutely continuous with respect to μ.

10.9 Point Wise Differentiation of Measures

If μ and ν are signed measures on a measurable space such that ν is absolutely continuous with respect to μ then by Radon-Nikodym theorem ν has a derivative with respect to μ which is called Radon-Nikodym derivative and is denoted by $\frac{d\nu}{d\mu}$, Radon-Nikodym derivative is not a pointwise derivative. We consider pointwise differentiation of measures with respect to the Lebesgue measure.

Theorem 10.33 *Let $\mathcal{A}$ be the σ - algebra of all Borel sets in (a, b) and let μ be the Lebesgue measures on $\mathcal{A}$. Let ν be a complex measure on $\mathcal{A}$ and let $f(x) = \nu((a, x))$ for $x \in (a, b)$. Then the following statements are equivalent :*

(i) *f is differentiable at $\xi \in (a, b)$ and $f'(\xi) = A$*
(ii) *for every $\epsilon > 0$ there is $\delta > 0$ such that $\left|\frac{\nu(I)}{\mu(I)} - A\right| < \epsilon$ for every interval $I \subset (a, b)$ such that $\xi \in I$ and $\mu(I) < \delta$.*

Proof Clearly f is a complex-valued function of real variable. We first prove the theorem for the special case when $A = 0$.

(i) $\Rightarrow$ (ii) Let ϵ be arbitrary. Since f is differentiable at ξ and $f'(\xi) = 0$, there is $\delta > 0$ such that

$$|f(t) - f(\xi)| \leq \frac{1}{2}\epsilon|t - \xi| \text{ for } |t - \xi| < \delta. \tag{10.16}$$

Let $I \subset (a, b)$ be any interval such that $\xi \in I$ and $\mu(I) < \delta$. If I is left closed and ξ is the left endpoint of I then $I = [\xi, t)$ where $\xi < t < b$ and $|t - \xi| < \delta$ and so by (10.16)

$$|\nu(I)| = |\nu([\xi, t))| = |f(t) - f(\xi)| \leq \frac{1}{2}\epsilon|t - \xi| < \epsilon\mu(I)$$

and hence $\left|\frac{\nu(I)}{\mu(I)}\right| < \epsilon$. So, suppose that ξ is not the left endpoint of I and let $I = (s, t)$. Then $|t - s| < \delta$ and $s < \xi$. Let $\{s_n\}$ be a decreasing sequence such that $s < s_n < \xi$ for all n and $s_n \to s$ as $n \to \infty$. Then by (10.16)

$$|\nu([s_n, t))| = |f(t) - f(s_n)| \leq |f(t) - f(\xi)| + |f(\xi) - f(s_n)|$$
$$\leq \frac{\epsilon}{2}|t - \xi| + \frac{\epsilon}{2}|\xi - s_n| = \frac{\epsilon}{2}|t - s_n|. \tag{10.17}$$

Since $\{[s_n, t)\}$ is an increasing sequence of intervals and $\bigcup_{n=1}^{\infty}[s_n, t) = (s, t)$, taking limit as $n \to \infty$ we get from (10.17) $|\nu((s, t))| \leq \frac{\epsilon}{2}|t - s| < \epsilon\mu(I)$ and hence $\left|\frac{\nu(I)}{\mu(I)}\right| < \epsilon$. The case when I is right closed and ξ is the right endpoint of I can be similarly treated. So in any case (ii) holds.

(ii) $\Rightarrow$ (i) Let $\epsilon > 0$ be arbitrary. Then there is $\delta > 0$ such that $\left|\frac{\nu(I)}{\mu(I)}\right| < \epsilon$ for every interval $I \subset (a, b)$ such that $\xi \in I$ and $\mu(I) < \delta$. Take $I = [\xi, t)$ such that

10.9 Point Wise Differentiation of Measures

$I \subset (a, b)$ and $0 < |t - \xi| < \delta$. Then $\left|\frac{v(I)}{\mu(I)}\right| < \epsilon$ and so $|v(I)| < \epsilon\mu(I)$. Hence $|f(t) - f(\xi)| < \epsilon|t - \xi|$. Since this is true for all t such that $\xi < t < b$ and $t - \xi < \delta$, the right-hand derivative $f'_+(\xi)$ exists and $f'_+(\xi) = 0$. Similarly take $I = (t, \xi]$ such that $I \subset (a, b)$ and $0 < |\xi - t| < \delta$. Then $\left|\frac{v(I)}{\mu(I)}\right| < \epsilon$ and so $|v(I)| < \epsilon\mu(I)$. Hence $|f(\xi) - f(t)| < \epsilon|t - \xi|$. Since this is true for all t such that $a < t < \xi$ and $\xi - t < \delta$, the left-hand derivative $f'_-(\xi)$ exists and $f'_-(\xi) = 0$. Hence $f'(\xi)$ exists and $f'(\xi) = 0$ and so (i) holds. This completes the proof when $A = 0$. If $A \neq 0$, let $\lambda = v - A\mu$ and $g(x) = \lambda((a, x))$. Then λ is a complex measure on $\mathcal{A}$ and $\lambda((a, x)) = v((a, x)) - A\mu((a, x))$ and $g(x) = v((a, x)) - A\mu((a, x)) = f(x) - A(x - a)$. So, if $f'(\xi) = A$ then $g'(\xi) = 0$ and so by the above proof $\left|\frac{\lambda(I)}{\mu(I)}\right| < \epsilon$ and hence $\left|\frac{v(I)}{\mu(I)} - A\right| < \epsilon$ and conversely, if $\left|\frac{v(I)}{\mu(I)} - A\right| < \epsilon$ then $\left|\frac{\lambda(I)}{\mu(I)}\right| < \epsilon$ and so by the above proof $g'(\xi) = 0$ and hence $f'(\xi) = A$. This completes the proof. □

It appears from the above theorem that the differentiation of measure is analogous to that of real function.

Definition 10.13 A sequence of intervals $\{I_n\}$ in $\mathbb{R}$ is said to converge to $x \in \mathbb{R}$ if $x \in I_n$ for all n and $\lim_{n \to \infty} |I_n| = 0$ where $|I_n|$ denotes the length of I_n.

Definition 10.14 Let ψ be a real-valued set function defined on a σ-algebra which contains the class of all intervals in $\mathbb{R}$. The upper derivative of ψ at $x \in \mathbb{R}$ is defined by

$$\overline{\psi}'(x) = \sup \limsup_{n \to \infty} \frac{\psi(I_n)}{|I_n|}$$

where '$\sup$' is taken over all sequences of intervals $\{I_n\}$ converging to x. Similarly the lower derivate of ψ at $x \in \mathbb{R}$ is defined by

$$\underline{\psi}'(x) = \inf \liminf_{n \to \infty} \frac{\psi(I_n)}{|I_n|}$$

when '$\inf$' is taken over all sequences of intervals $\{I_n\}$ converging to x. If $\overline{\psi}'(x) = \underline{\psi}'(x)$ then the common value denoted by $\psi'(x)$, is called the derivative of ψ at x. If $\psi'(x)$ is finite then ψ is called differentiable at x.

Theorem 10.34 *The upper derivate $\overline{\psi}'$ and the lower derivate $\underline{\psi}'$ of ψ are Borel measurable function on $\mathbb{R}$.*

Proof Let a be an arbitrary real number and let $E = \{x \in \mathbb{R} : \overline{\psi}'(x) > a\}$. For each pair of positive integers m, n, let E_{mn} be the union of all intervals $I \subset \mathbb{R}$ such that $|I| < \frac{1}{n}$ and $\frac{\psi(I)}{|I|} > a + \frac{1}{m}$. Then E_{mn} is Borel measurable. To see this, let I^0 be the interior of the constituent interval I of E_{mn} which satisfies $|I| < \frac{1}{n}$ and $\frac{\psi(I)}{|I|} > a + \frac{1}{m}$ and let E^0_{mn} be the union of all such open intervals I^0. Then E^0_{mn} is an open set and so E^0_{mn} is a countable union of open intervals. Now adjoining the left out endpoints of the constituent intervals I of E_{mn} to E_{mn}^0 we get E_{mn} and so E_{mn} is a countable union of intervals in $\mathbb{R}$. Since intervals are Borel sets, E_{mn} is a Borel set. Also

$$E = \bigcup_{m=1}^{\infty} \bigcap_{n=1}^{\infty} E_{mn} \qquad (10.18)$$

In fact, if $\xi \in E$ then $\overline{\psi}'(\xi) > a$ and so there is $m_\circ$ such that $\overline{\psi}'(\xi) > a + \frac{1}{m_0}$. So there is a sequence $\{I_n\}$ of intervals converging to ξ such that $\limsup_{n \to \infty} \frac{\psi(I_n)}{|I_n|} > a + \frac{1}{m_0}$. Hence for every n there is an interval I such that $\xi \in I$, $|I| < \frac{1}{n}$ and $\frac{\psi(I)}{|I|} > a + \frac{1}{m_0}$. Hence $I \subset E_{m_0 n}$ and so $\xi \in E_{m_0 n}$. Since this is true for every n, $\xi \in \bigcap_{n=1}^{\infty} E_{m_0 n}$ and so $\xi \in \bigcup_{m=1}^{\infty} \bigcap_{n=1}^{\infty} E_{mn}$. Conversely let $\xi \in \bigcup_{m=1}^{\infty} \bigcap_{n=1}^{\infty} E_{m,n}$. Then there is m_0 such that $\xi \in E_{m_0 n}$ for all n. So for each n there is an interval $I_n \subset E_{m_0 n}$ such that $\xi \in I_n$, $|I_n| < \frac{1}{n}$ and $\frac{\psi(I_n)}{|I_n|} > a + \frac{1}{m_0}$ which gives $\overline{\psi}'(\xi) \geq a + \frac{1}{m_0} > a$ and hence $\xi \in E$. So, (10.18) holds. Since E_{mn} is Borel measurable for all m and n. E is Borel measurable by (10.18). Since a is arbitrary, $\overline{\psi}'$ is Borel measurable. Similarly $\underline{\psi}'$ is Borel measurable. $\square$

Theorem 10.35 *Let $\mathcal{B}$ be the σ algebra of all Lebesgue measurable sets in a finite interval $[a, b]$ and let μ be the Lebesgue measure on $\mathcal{B}$. Let ν be a finite (signed) measure on $\mathcal{B}$ which is absolutely continuous with respect to μ. Then the derivative ν' of ν exists finitely a.e.$[\mu]$ and is Lebesgue integrable on $[a, b]$ and*

$$\nu(E) = \int_E \nu' d\mu \quad \text{for every } E \in \mathcal{B}.$$

Proof Since μ is Lebesgue measure and ν is finite by Theorem 10.19 there is a Lebesgue integrable function f on $[a, b]$ such that

$$\nu(E) = \int_E f d\mu \quad \text{for } E \in \mathcal{B}. \qquad (10.19)$$

Let $F(x) = \int_a^x f d\mu$. Then $F' = f$ a.e. in $[a, b]$. Let $A = \{x : x \in [a, b]; F'(x) = f(x)\}$. Then $\mu(A) = b - a$. Let $x \in A$. Let $\{I_n\}$ be any sequence of intervals converging to x. Then since $I_n \in \mathcal{B}$, by (10.19)

$$\frac{\nu(I_n)}{|I_n|} = \frac{1}{|I_n|} \int_{I_n} f d\mu \to f(x) \text{ as } n \to \infty. \qquad (10.20)$$

For if x is one of endpoints of I_n, this is obvious. Otherwise let $I_n = (a_n, b_n)$. Then $a_n, b_n \to x$ as $n \to \infty$ and so $\frac{1}{x - a_n} \int_{a_n}^x f d\mu \to f(x)$ and $\frac{1}{b_n - x} \int_x^{b_n} f d\mu \to f(x)$ as $n \to \infty$. So writing $\frac{1}{x - a_n} \int_{a_n}^x f d\mu = f(x) + \epsilon_n$ and $\frac{1}{b_n - x} \int_x^{b_n} f d\mu = f(x) + \eta_n$ where $\epsilon_n, \eta_n \to 0$ as $n \to \infty$ we have

10.9 Point Wise Differentiation of Measures

$$\frac{1}{|I_n|}\int_{I_n} f d\mu = \frac{1}{|I_n|}\int_{a_n}^{x} f d\mu + \frac{1}{|I_n|}\int_{x}^{b_n} f d\mu = \frac{x-a_n}{|I_n|}\frac{1}{x-a_n}\int_{a_n}^{x} f d\mu +$$

$$\frac{b_n - x}{|I_n|}\frac{1}{b_n - x}\int_{x}^{b_n} f d\mu = \frac{x-a_n}{|I_n|}(f(x)+\epsilon_n) + \frac{b_n-x}{|I_n|}(f(x)+\eta_n)$$

$$= f(x) + \frac{x-a_n}{|I_n|}\epsilon_n + \frac{b_n-x}{|I_n|}\eta_n \to f(x) \text{ as } n \to \infty.$$

So, by (10.20) $\frac{\nu(I_n)}{|I_n|} \to f(x)$ as $n \to \infty$. Since ν is a finite signed measure on $\mathcal{B}$ and $\mathcal{B}$ contains all intervals, ν is a set function and so applying Definition (10.14) $\nu'(x)$ exists and $\nu'(x) = f(x)$. Since x is arbitrary and $\mu(A) = b - a$, $\nu' = f$ a.e. in $[a, b]$. Since f is Lebesgue integrable, ν' is Lebesgue integrable and $\int_E \nu' d\mu = \int_E f d\mu$ for all $E \in \mathcal{B}$. So, by (10.19) $\nu(E) = \int_E \nu' d\mu$ for $E \in \mathcal{B}$. Since f is Lebesgue integrable, f is finite a.e. and so ν' is finite a.e. completing the proof. □

Note that as in Definition (10.6) a complex measure ν is said to be absolutely continuous with respect to μ if $\nu(E) = 0$ whenever $|\mu|(E) = 0$. So, if ν is a complex measure and if ν_1 and ν_2 are the real and imaginary parts of ν then ν_1 and ν_2 are absolutely continuous with respect to μ if ν is absolutely continuous with respect to μ. So, if ν is absolutely continuous with respect to μ then by above theorem

$$\nu_1(E) = \int_E \nu_1' d\mu \text{ and } \nu_2(E) = \int_E \nu_2' d\mu \text{ for } E \in \mathcal{B}.$$

and hence

$$\nu(E) = \nu_1(E) + i\nu_2(E) = \int_E \nu_1' d\mu + i\int_E \nu_2' d\mu$$

$$= \int_E (\nu_1 + i\nu_2)' d\mu = \int_E \nu' d\mu.$$

So, we have

Corollary 10.36 *Under the hypotheses of Theorem 10.35 if ν is a complex measure which is absolutely continuous with respect to μ then the derivative ν' of ν exists a.e. $[\mu]$ and $\nu(E) = \int_E \nu' d\mu$ for any $E \in \mathcal{B}$.*

10.10 Exercises

1. If μ and ν are measures on a measurable space $(X, \mathcal{A})$ then show that $\mu + \nu$ is a measure on $(X, \mathcal{A})$ but $\mu - \nu$ may not be a measure but $\mu - \nu$ is a signed measure when at least one of μ and ν is finite.
2. Show that there exist complex measures μ and ν such that (i) $\mu + \nu$ is a measure (ii) $\mu + \nu$ is a signed measure.
3. If ν_1, ν_2 and μ are measures on the measurable space $(X, \mathcal{A})$ and if $\nu_1 \perp \mu$ and $\nu_2 \perp \mu$ then show that $\nu_1 + \nu_2 \perp \mu$.
4. If μ and ν are measures on $(X, \mathcal{A})$ such that $\nu << \mu$ and $\nu \perp \mu$ then show that ν is identically 0.
5. If $\lambda_1, \lambda_2,$ and μ are signed measures on $(X, \mathcal{A})$ and if $\lambda_1 << \mu$ and $\lambda_2 << \mu$ then show that $\lambda_1 + \lambda_2 << \mu$.
6. If λ_1, λ_2 and μ are measures on $(X, \mathcal{A})$ and if $\lambda_1 << \mu$ and $\lambda_2 \perp \mu$ then show that $\lambda_1 \perp \lambda_2$.
7. Let $f(x) = \sin x$ for $x \in (0, 1]$ and $g(x) = \sin \frac{1}{x}$ for $x \in (0, 1]$. Let $\mathcal{A}$ be the σ- algebra of all Lebesgue measurable sets in $(0, 1]$ and let $\mu(E) = \int_E f dx$ and $\nu(E) = \int_E g dx$ for $E \in \mathcal{A}$. Show that μ is a measure on $\mathcal{A}$ and ν is a signed measure on $\mathcal{A}$. Find the Hahn decomposition of ν.
8. Let ν be a signed measure on a measurable space $(X, \mathcal{A})$. If (A_1, B_1) and (A_2, B_2) are Hahn decomposition of X with respect to ν then show that $A_1 \triangle A_2$ is a set of ν - measure zero.
9. Let $(X, \mathcal{A})$ be a measurable space and let ν be a signed measure on $\mathcal{A}$. If $\{E_n\}$ is a sequence of disjoint sets in $\mathcal{A}$ such that $|\nu(\bigcup_{n=1}^{\infty} E_n)| < \infty$ then show that $\sum_{n=1}^{\infty} \nu(E_n)$ is absolutely convergent.
10. Let $(X, \mathcal{A})$ be a measurable space and let ν be a signed measure on $\mathcal{A}$. If $E \in \mathcal{A}, |\nu(E)| < \infty$ and $F \subset E$ for $E, F \in \mathcal{A}$ then prove that $|\mu(F)| < \infty$.
11. Consider the measurable space $(X, \mathcal{A})$ where $X = [-\pi, \pi]$ and $\mathcal{A}$ is the σ- algebra of all Lebesgue measurable sets in X. Let $\nu(E) = \int_E e^{ix} dx, E \in \mathcal{A}$. Then show that ν is a complex measure on $\mathcal{A}$ and verify that $|\nu| \leq \nu_1^+ + \nu_1^- + \nu_2^+ + \nu_2^-$ where $\nu_1^+, \nu_1^-, \nu_2^+, \nu_2^-$ are as in Theorem 10.28.
12. Show, by suitable examples that sum of two complex measures may be (i) a measure and (ii) a signed measure.

References

1. Halmos, P.R.: Measure Theory, Van Nostrand (1950)
2. Halmos, P.R.: Naive Set Theory, Van Nostrand (1960)
3. McShane, E.J.: Integration. Princeton University Press (1944)
4. Saks, S.: Theory of the Integral, Worszawa (1937)
5. Titchmarsh, E.C.: Theory of Functions. Oxford University Press (1939)
6. McLeod, R.M.: The Generalized Riemann Integral, Carus Monograph No. 20, M A A (1980)
7. Natanson, I.P.: Theory of Function of a Real Variable, vols. I and II. Frederick Ungar, New York (1964)
8. Munroe, M.E.: Measure and Integration. Addison-Wesley (1971)
9. Kamke, E.: Theory of Sets. Dover (1950)
10. Pfeffer, W.F.: The Riemann Approach to Integration. Cambridge University Press (1993)
11. Henstock, R.: Theory of Integration. Butterworths, London (1963)
12. Caratheodory, C.: Algebraic Theory of Measure and Integration, Chelsea (1963)
13. Hobson, E.W.: The Theory of Functions of a Real Variable and the Theory of Fourier Series, vols. Cambridge University Press, I and II (1927)
14. Rogers, C.A.: Hausdorff Measure. Cambridge University Press (1970)
15. Gordon, R.A.: The Integrals of Lebesgue, Denjoy, Perron and Henstick, Graduate Studies in Mathematics, vol. 4. AMS (1994)
16. Lee, P.Y.: Lanzhou Lectures on Henstock Integration. Word Scientific, Singapore (1989)
17. Bruckner, A.M.: Differentiation of Real Functions. CRC Monograph Series, vol 5. AMS (1994)
18. Jeffery, R.L: Theory of functions of real variable, Second edition, University of Toronto Press (1953)